Sexual Interactions in
Eukaryotic Microbes

This is a volume in
CELL BIOLOGY
A series of monographs

A complete list of the books in this series appears at the end of the volume.

O'DAY, DANTON H
SEXUAL INTERACTIONS IN EUKARY
000410058

HCL   QR86.O21

WITHDRAWN

Sexual Interactions in Eukaryotic Microbes

Edited by

Danton H. O'Day
Department of Zoology at Erindale College
University of Toronto in Mississauga
Mississauga, Ontario, Canada

Paul A. Horgen
Department of Botany at Erindale College
University of Toronto in Mississauga
Mississauga, Ontario, Canada

1981

ACADEMIC PRESS
A Subsidiary of Harcourt Brace Jovanovich, Publishers
New York   London   Toronto   Sydney   San Francisco

COPYRIGHT © 1981, BY ACADEMIC PRESS, INC.
ALL RIGHTS RESERVED.
NO PART OF THIS PUBLICATION MAY BE REPRODUCED OR
TRANSMITTED IN ANY FORM OR BY ANY MEANS, ELECTRONIC
OR MECHANICAL, INCLUDING PHOTOCOPY, RECORDING, OR ANY
INFORMATION STORAGE AND RETRIEVAL SYSTEM, WITHOUT
PERMISSION IN WRITING FROM THE PUBLISHER.

ACADEMIC PRESS, INC.
111 Fifth Avenue, New York, New York 10003

United Kingdom Edition published by
ACADEMIC PRESS, INC. (LONDON) LTD.
24/28 Oval Road, London NW1 7DX

Library of Congress Cataloging in Publication Data
Main entry under title:

Sexual interactions in eukaryotic microbes.

(Cell biology)
Includes bibliographies and index.
1. Protozoa--Reproduction. 2. Fungi--
Reproduction. 3. Algae--Reproduction.
4. Cell interaction. I. O'Day, Danton H.
II. Horgen, Paul A. [DNLM: 1. Algae--
Physiology. 2. Fungi--Physiology. 3.
Protozoa--Physiology. 4. Cell membrane--
Physiology. 5. Sex attractants. 6. Cells.
QW 180 S518]
QH481.S49   576'.116   80-39593
ISBN 0-12-524160-7

PRINTED IN THE UNITED STATES OF AMERICA

81 82 83 84   9 8 7 6 5 4 3 2 1

Contents

List of Contributors xi
Preface xiii

PART I. INTRODUCTION

1. Modes of Cellular Communication and Sexual Interactions in Eukaryotic Microbes

DANTON H. O'DAY

    I. Introduction 3
    II. Modes of Cellular Communication 5
    III. The Value of Using Eukaryotic Microbes for the Study of Cell Communication 14
    References 16

PART II. PHEROMONAL INTERACTIONS

2. The Isolation, Characterization, and Physiological Effects of the *Saccharomyces cerevisiae* Sex Pheromones

T. R. MANNEY, W. DUNTZE, AND RICHARD BETZ

    I. Introduction 21

II. Isolation and Analysis of Mating Pheromones 27
III. Response of Cells to Mating Pheromones 33
IV. Perspectives 47
References 49

3. The Role of Sexual Pheromones in *Allomyces*
JEFFREY POMMERVILLE

I. Introduction 53
II. Survey of the Literature 56
III. Current Research 59
IV. Perspectives 69
References 70

4. Sexual Pheromones in *Volvox* Development
GARY KOCHERT

I. Introduction and Survey of the Literature 73
II. Current Research 81
III. Perspectives 88
References 92

5. Cell Interactions by Gamones in *Blepharisma*
AKIO MIYAKE

I. Introduction 95
II. Survey of the Literature 100
III. Current Research 101
IV. Perspectives 123
References 126

6. Sex Pheromones in *Neurospora crassa*
M. S. ISLAM

I. Introduction 131
II. Survey of the Literature 134
III. Current Research 138
IV. Perspectives 150
References 151

Contents

7. The Role of the Steroid Sex Pheromone Antheridiol in Controlling the Development of Male Sex Organs in the Water Mold, *Achlya*

PAUL A. HORGEN

    I. Introduction   155
    II. Survey of the Literature   159
    III. Current Research   161
    IV. Perspectives   172
    References   174

8. Sex Pheromones in *Mucor*

B. E. JONES, I. P. WILLIAMSON, AND G. W. GOODAY

    I. Introduction   179
    II. Survey of the Literature and Current Research   188
    III. Conclusion   195
    References   195

9. Pheromonal Interactions during Mating in *Dictyostelium*

DANTON H. O'DAY AND KEITH E. LEWIS

    I. Introduction   199
    II. Survey of the Literature   204
    III. Current Research   206
    IV. Perspectives   218
    References   220

PART III. CELL SURFACE INTERACTIONS

10. The Cell Wall as Sex Organelle in Fission Yeast

G. B. CALLEJA, BYRON F. JOHNSON, AND B. Y. YOO

    I. Introduction   225
    II. Survey of the Literature   230
    III. Current Research   242
    IV. Perspectives   253
    References   255

11. **Sexual Interactions in *Saccharomyces cerevisiae* with Special Reference to the Regulation of Sexual Agglutinability**

NAOHIKO YANAGISHIMA AND KAZUO YOSHIDA

    I. Introduction 261
   II. Survey of the Literature 263
  III. Current Research 266
  IV. Perspectives 289
      References 291

12. **Sexual Interactions in the Green Alga *Chlamydomonas eugametos***

H. VAN DEN ENDE

    I. Introduction 297
   II. Description of the Sexual Process 299
  III. Cultivation and Gametogenesis 301
  IV. The Flagellar Surface 302
   V. Nature of the Agglutination Process 311
  VI. Prospects 315
      References 316

13. **Preconjugant Cell Interactions in *Oxytricha bifaria* (Ciliata, Hypotrichida): A Two-Step Recognition Process Leading to Cell Fusion and the Induction of Meiosis**

NICOLA RICCI

    I. Introduction 319
   II. Literature Review and Current Research 325
  III. Perspectives 343
      References 347

14. **Sexual Interactions of the Cell Surface in *Paramecium***

KOICHI HIWATASHI

    I. Introduction 351
   II. Processes of Sexual Interactions 353
  III. Mating Reaction and Mating Substances 358
  IV. Output of the Mating Substance Interactions and Activation-Initiating Mechanisms 364
   V. Control of Mating Type and Mating Activity 370

VI. Perspectives 373
   References 375

15. The Genetics and Cellular Biology of Sexual Development in *Ustilago violacea*

ALAN W. DAY AND JOSEPH E. CUMMINS

   I. Introduction 379
   II. Summary of the Literature 382
   III. Current Research 388
   IV. Perspectives 398
      References 400

      Index 403

List of Contributors

Numbers in parentheses indicate the pages on which the authors' contributions begin.

RICHARD BETZ (21), Institut für Physiologische Chemie, Ruhr-Universität Bochum, D-4630 Bochum-Querenburg, West Germany

G. B. CALLEJA (225), Division of Biological Sciences, National Research Council of Canada, Ottawa, Ontario, Canada K1A 0R6

JOSEPH E. CUMMINS (379), Department of Plant Sciences, University of Western Ontario, London, Ontario, Canada, N6A 5B7

ALAN W. DAY (379), Department of Plant Sciences, University of Western Ontario, London, Ontario, Canada N6A 5B7

W. DUNTZE (21), Institut für Physiologische Chemie, Ruhr-Universität Bochum, D-4360 Bochum-Querenburg, West Germany

G. W. GOODAY (179), Department of Microbiology, University of Aberdeen, Marischal College, Aberdeen AB9 1AS, Scotland

KOICHI HIWATASHI (351), Biological Institute, Tohoku University, Aoba-Yama, Sendai 980, Japan

PAUL A. HORGEN (155), Department of Botany at Erindale College, University of Toronto in Mississauga, Mississauga, Ontario, Canada L5L 1C6

M. S. ISLAM (131), Division of Genetics, Irradiation and Pest Control Research Institute, Bangladesh Atomic Energy Commission, Dacca, Bangladesh, India

BYRON F. JOHNSON (225), Division of Biological Sciences, National Research Council of Canada, Ottawa, Ontario, Canada K1A 0R6

B. E. JONES (179), Department of Microbiology, University of Aberdeen, Marischal College, Aberdeen AB9 1AS, Scotland

GARY KOCHERT (73), Botany Department, University of Georgia, Athens, Georgia 30602

KEITH E. LEWIS (199), Department of Zoology at Erindale College, University of Toronto in Mississauga, Mississauga, Ontario, Canada L5L 1C6

T. R. MANNEY (21), Department of Physics, Kansas State University, Manhattan, Kansas 66502

AKIO MIYAKE (95), Zoologisches Institut, Universität Münster, D-4400 Münster, West Germany

DANTON H. O'DAY (3, 199), Department of Zoology at Erindale College, University of Toronto in Mississauga, Mississauga, Ontario, Canada L5L 1C6

JEFFREY POMMERVILLE (53), Botany Department, University of Georgia, Athens, Georgia 30602

NICOLA RICCI (319), Istituto di Zoologica e Anatomia Comparata, Università di Pisa, 56100 Pisa, Italy

H. VAN DEN ENDE (297), Department of Plant Physiology, University of Amsterdam, Amsterdam, The Netherlands

I. P. WILLIAMSON (179), Department of Biochemistry, University of Aberdeen, Marischal College, Aberdeen AB9 1AS, Scotland

NAOHIKO YANAGISHIMA (261), Biological Institute, Faculty of Science, Nagoya University, Chikusa-ku, Nagoya 464, Japan

B. Y. YOO (225), Department of Biology, University of New Brunswick, Fredericton, New Brunswick, Canada E3B 5A3

KAZUO YOSHIDA (261), Biological Institute, Faculty of Science, Nagoya University, Chikusa-ku, Nagoya 464, Japan

Preface

We decided to compile a book dedicated to the sexual processes of eukaryotic microorganisms because previously no complete, up-to-date volume existed on this subject. Although there have been many individual recent reviews on various aspects of microbial mating, both specific and general, these works do little to provide the reader with a detailed understanding of the individual organisms and their real and potential value as research tools.

Essentially all of the microbes whose sexual cycles have been detailed to a significant extent in the past are examined in this book. In keeping with our intent to impart some special knowledge to our readers, each chapter has been written according to a more-or-less specific organizational plan. Each begins with a summary of aspects of the lifestyles, life cycles, and availability of the organism under analysis. After a subsequent brief review of previous work done on the sexual processes of the specific organisms and its relatives, the authors describe in detail the current research of their laboratories. Finally, they discuss what they consider to be important areas for future work.

Because of its design, this book should be of value on a multitude of levels: from a general reference text to a source of research ideas. We feel it should appeal to a wide spectrum of readers in a large number of disciplines, but will be particularly valuable to cell biologists, microbiologists, protozoologists, and mycologists interested in the study of cellular communication.

We thank Elinor Foden for her exceptional secretarial assistance and editorial comments. We also thank the staff of Academic Press for their assistance in the compilation of this work. We sincerely hope that it will fill the existing void.

Danton H. O'Day
Paul A. Horgen

Part I
INTRODUCTION

1

Modes of Cellular Communication and Sexual Interactions in Eukaryotic Microbes

DANTON H. O'DAY

I.	Introduction	3
II.	Modes of Cellular Communication	5
	A. Communication via Diffusible Molecules	5
	B. Communication via Cellular Continuities	7
	C. Cell Contact-Mediated Communication	10
	D. Extracellular Matrix-Mediated Communication	12
III.	The Value of Using Eukaryotic Microbes for the Study of Cell Communication	14
	References	16

I. INTRODUCTION

Although the topic under consideration in this book is sexual interactions in eukaryotic microbes, the essential problem under analysis is the ways in which eukaryotic cells communicate with one another. The survival of cells as individuals and as integral parts of multicellular organisms requires that they be able to communicate with other cells. Cell communication is important in the regulation of cell movement, morphogenesis, cellular differentiation, cell division, and cellular adhesivensss. Whether cells interact over long distances or short, two essential components must comprise their communications systems: (1) a signaler, which produces and transmits the message, and (2) a responder, which receives the message and translates it into an action.

There appear to be a limited number of ways in which eukaryotic cells communicate. These are diagrammatically represented in Fig. 1 and each system is outlined in the following section. Rather than provide an exhaustive review, the aim of this chapter is to set the stage for the chapters that follow by giving an overview of what we understand about intercellular communication. One should keep in mind that a cell may, at a point in time or throughout its life, utilize several and possibly all methods of communication. The use of one communication system does not exclude another.

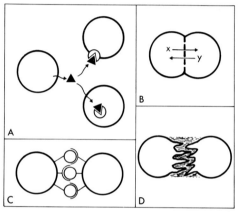

Fig. 1. A diagrammatic representation of the modes of communication used by eukaryotic cells. (A) Communication via diffusible molecules. One cell or a group of cells synthesizes and secretes a message molecule which travels in the extracellular environment. The message is received by a target cell which has surface receptors (upper cell) and/or intracellular receptors (below) that are specific for the molecule. (B) Communication via cellular continuities. In this system direct cytoplasmic coupling exists between cells. These vary in size and may be large enough to permit penetration of entire organelles or small enough to restrict the flow of all but specific ions. They may or may not reveal special structural differentiations. Gap junctions are special structures which allow cells to be metabolically and/or ionically coupled. (C) Cell contact-mediated communication. Cell contact regulates many biological phenomena. The example shown represents the most widely accepted concept of cell-cell contact which involves complementary molecules bound to the surface of the adjacent cells. Generally, the molecular interaction is believed to be a molecule-receptor or an enzyme-substrate (glycosyltransferase) interaction. Other types of contact-mediated communication may also exist. (D) Extracellular matrix-mediated communication. This is similar in some ways to the situation discussed in (C) except that extracellular components are interposed between and essential for the communication between adjacent cells. The similarity lies in models that propose adhesive molecular interactions between the cells and the extracellular matrix. The difference lies in models that propose that the matrix provides other functions such as collecting and concentrating extracellular factors of communication. In all of these models the message that is received by the target cell(s) must be interpreted and transduced into a subsequent biological response.

II. MODES OF CELLULAR COMMUNICATION

A. Communication via Diffusible Molecules

1. *Hormones versus Pheromones*

Distance alone should not be a priori a means of inferring that a specific method of communication is employed. However, if cells are to communicate over long distances they must employ chemicals which diffuse in the extracellular environment. The messages that are sent can be received either by the cell surface or by an internal receptor of the target or responder cell (Fig. 1A). Subsequently the message-receptor complex will dictate a specific cellular response.

Specific molecules of communication that are produced and secreted within one part of an organism which influence other parts of that same organism are termed hormones. Although the microbial literature is full of references to hormones, it is clear that the term is not being correctly used since these molecules are released from one organism to influence others. Karlson and Luscher (1959) proposed the term pheromone for such interorganismic molecules of communication and this is the term which is employed throughout this volume regardless of historical usage. Part II of this book focuses on pheromonal communication during sexual development in eukaryotic microbes. While most hormones that serve the same function in different multicellular organisms are very similar in structure and often cross-immunoreactive, pheromones do not necessarily show such molecular relatedness even in closely related species. In his chapter, Kochert expresses the view that this reflects the role of pheromones "which serve as a vital communication mechanism through an external medium between individuals of the same species." It would be disadvantageous to an organism's survival to respond to the pheromones of other species. Thus, although all *Volvox* sex pheromones appear to be glycoproteins they are species specific (see Chapter 4 by Kochert). Similarly, species-specific sex pheromones are produced by *Dictyostelium* (see Chapter 9 by O'Day and Lewis). In *Blepharisma,* Miyake shows that the glycoprotein gamones (gamone 1) which control the mating interactions are also species specific. In contrast, several species of *Achlya* all seem to use identical female (antheridiol) and male (oogoniol) sterol sex pheromones (see Chapter 7 by Horgen). The most unusual situation occurs in the Mucorales where all genera appear to use trisporic acid as their sex pheromone. However, as is detailed by Jones *et al.,* the control in this sexual system resides at the level of the collaborative biosynthesis of the specific propheromones which ultimately lead to the final end product, trisporic acid.

2. Surface Reception

A multitude of peptide hormones have been well studied whose function is mediated by surface reception. All vertebrate polypeptide hormones and growth factors apparently interact with their target cells through cell membrane bound receptors (Bradshaw and Frazier, 1977). Examples of such molecules range from the tripeptide thyrotropin-releasing hormone to the macromolecular (191 amino acids) placental lactogen. Peptide hormones bind to surface macromolecular receptors (usually glycoprotein) to produce a cellular response often through the mediation of membrane bound adenylate cyclase which generates intracellular cyclic AMP as a second messenger (Robinson et al., 1971). The means by which hormone reception is transduced to generate these intracellular signals is still poorly understood (Helmreich et al., 1976). In Chapter 4, Kochert relates the sequence of events that results in the intracellular establishment of increased cyclic AMP (cAMP) levels after surface-mediated hormone reception in vertebrate cells. Using this as a model he is pursuing the mode of action of the glycoprotein sex pheromones of the green alga *Volvox*.

In contrast, some peptide hormones (e.g., nerve growth factor and insulin) do not seem to use cAMP as the second messenger and both their surface reception as well as their internalization may be critical for their function (Bradshaw and Frazier, 1977). In keeping with this, receptors for nerve growth factor have been detected on the nuclear membrane. Brown and Goldstein (1979) have proposed that the internalization and subsequent degradation that occur after the surface-receptor-mediated binding, may also be critical in the action of some hormones. Using the low density lipoprotein (LDL) receptor system, which is involved in the regulation of cholesterol metabolism, they show that the regulation of metabolic events depends on the degradation products that result after the internalized LDL-receptor complexes interact with lysosomes. Although no experimental evidence exists to support this model it is interesting to speculate that, in systems where a hormone (or pheromone) has a multitude of different effects, specific steps in the sequence from hormone (pheromone) binding to the final events of its degradation may each play a role in mediating the pleiotropic effects of that hormone (or pheromone).

3. Intracellular Reception

The action of lipid soluble hormones is mediated by intracellular receptors (Fig. 1A). Of these the vertebrate sexual hormones, estrogen and progesterone, have been well studied. In these cases, the steroid sex hormone couples with a cytoplasmic receptor in the target cell and the resultant steroid–receptor complex migrates into the nucleus ultimately regulating gene activity through its binding to chromatin acceptor sites (Jensen and DeSombre, 1973). Recently, Toyoda and Spelsberg (1979) have shown that (A-T)-rich sequences in the chromatin of hen oviduct may be involved in binding the progesterone-receptor nonhistone acceptor proteins. The steroid sexual hormones of *Achlya,* antheridiol and oogoniol, are very similar in structure to the mammalian sexual hormones and their mode of action also appears to be regulated by intracellular recognitive entities. This is discussed in detail in Chapter 7 by Horgen.

4. Microbial Sex Pheromones

The eukaryotic microbes discussed in this volume use a variety of different molecules in their sexual communications. These include small molecules, such as steroids, other lipids, peptides, and derivatives of organic acids, as well as large molecules, such as glycoproteins. These pheromones also have a multitude of diverse roles ranging from being chemotactic attractants to acting as inducers of cell fusion. The pheromones, their elucidated or putative chemical nature, their mode of production and reception, and their functions are listed in Table I and are detailed in Part II of this volume.

As a final note to this discussion of communication via diffusible molecules we should not forget to note that interspecific communication also occurs in eukaryotic microbes. Of particular interest, is the interaction that occurs in the plant pathogen *Ustilago violacea* whereby a phenolic product (silenin) of the host species triggers the dikaryotic mycelial stage of the fungus to develop into its parasitic form. The induction of the parasitic phase is regulated by the mating-type locus. This subject is discussed in Chapter 15 by Day and Cummins.

B. Communication via Cellular Continuities

For years, cytoplasmic continuities between cells in the tissues of both plants and animals have been known to exist (Fig. 1B). These might be extremely large, as in the plasmodesmata of plants which can even allow the complete penetration of organelles from one cell to another. Usually they are small, like the ion gates of membranes, which lie below the resolution of our present day visual tools. Although intercellular junctions are widespread during embryogenesis and their existence might imply that they serve as pathways for intercellular communication, there is no experimental evidence to prove that they do (deLaat *et al.,* 1976; Sheridan, 1977).

TABLE I

The Pheromonal Systems and Their Functions in the Eukaryotic Microorganisms Discussed in Part II

Organism	Common name	Chemical structure	Source (mating type)	Pheromone Production[a]	Reception	Functions
Achlya ambisexualis	Antheridiol	Sterol	♀	Constitutive	Cytoplasmic[b]	Induces secretion of oogoniol and formation and chemoattraction of antheridial branches in ♂
	Oogoniol	Sterol	♂	Inducible	Cytoplasmic[c]	Induces oogoniol initial in ♀
Allomyces macrogynus	Sirenin	Oxygenated sesquiterpene		Constitutive	Unknown	Directs chemotaxis of ♂ gametes
Blepharisma japonicum	Gamone 1 (blepharmone)	Glycoprotein	I	Constitutive	Surface[b]	Induces gamone 2 and pairing by type II cells
	Gamone 2 (blepharismone)	3-(2'-formyl-amino-5-hy-droxybenzoyl) lactate	II	Inducible	Unknown	Attracts and induces cell union in type I cells; enhances gamone I secretion by type I cells
Dictyostelium discoideum	None	Unidentified volatile molecule	NC4	Constitutive	Unknown	Probably directs cell fusion[b]

Mucor mucedo	None	Trisporic acid (18-carbon terpenoid acid)	+ and −	Inducible ("collaborative biosynthesis by mating type pairs")	Unknown	Induces zygophore formation, and carotene biosynthesis
Neurospora crassa	None	Long-chain, unsaturated hydrocarbon	**A, a** **A** × **a**	Constitutive	Unknown	Induces protoperethecia development in **a** mating type
Saccharomyces cerevisiae	**a** factor	Oligopeptide (?) (11 amino acids)	**a**	Constitutive	Surface	Induces G_1 arrest, "shmoo" morphology, surface agglutinins change in cell wall structure, and inhibits membrane-bound adenylcyclase in opposite mating type
	α factor	Oligopeptide (2 forms; 12–13 L-amino acids)	α	Constitutive	Surface	
Volvox carteri	None	Glycoprotein	♂	Constitutive	Surface	Induce gonidia to form sexual spheroids

[a] In this context, constitutive and inducible refer only to the production of the pheromone in the presence or absence of the opposite mating type; thus, if a single mating type produces pheromone its production is constitutive, but if one mating type requires the opposite mating type to produce pheromone it is inducible.

[b] Data suggest this but conclusive evidence lacking.

Of current appeal to the cell biologist interested in intercellular communication is a specially differentiated structure that occurs widely in animal cells. The structure is called a gap junction or nexus. The gap junction mainly serves as the pathway for intercellular transmission of small molecules and inorganic ions, resulting in the metabolic or ionic coupling of the cells (Gilula, 1977). Whatever their role in intercellular communication, Pitts (1977) argues that the signals carried by gap junctions must only be "trigger molecules" which invoke or inhibit a preexisting response because "informational" macromolecules are much too large to traverse these intercellular pathways. So far, gap junctions have not been found in any eukaryotic microbe.

Since eukaryotic microbes generally exist as individuals, communication via direct cellular continuities would appear to be of little significance. However, some species of the colonial alga, *Volvox,* possess fine, intercellular bridges which allow cytoplasmic continuity and thus communications between the cells of that community. These cytoplasmic strands appear during cleavage and may play a morphogenetic role during inversion (Viamontes *et al.*, 1979). Whether or not cytoplasmic communication occurs via these connections during sexual interactions has not been investigated.

Other examples of cytoplasmic continuities in eukaryotic microbes represent short-lived communicative events such as the transient intercellular bridges that occur for micronuclear exchange in ciliates (see Chapter 14 by Hiwatashi on *Paramecium,* Chapter 5 by Miyake on *Blepharisma,* and Chapter 13 by Ricci on *Oxytrichia*). These represent cases of intercellular communication where cells exchange both nuclear and cytoplasmic information rather than a specific molecule. However, during mating in eukaryotic microbes such sexually elicited cytoplasmic continuities usually result in the total mixing of the cytoplasmic contents of the consenting pair. Cell fusion could thus be considered the ultimate in cell communication since the coalescing cells exchange every bit of information they possess.

C. Cell Contact-Mediated Communication

Normally, contact between cells results in behavioral change in the contacting cells. The classic example for this is "contact inhibition of locomotion" (Abercombie, 1967; Heaysman, 1978): when two cells contact, cell movement is inhibited in the region of contact. Although there has been much work on this phenomenon the mechanism by which the locomotory function of the contacting cells is inhibited is still not known (Heaysman, 1978). Cell contact also plays an important role in the induction of cellular differentiation and morphogenesis, and in regulating cell division (Moscona, 1974; Lehtonen, 1976). Grunz and Staubach (1979) have recently proposed that cell to cell contact between chordamesoderm and neural ectoderm may play a critical role in primary embryonic induction by restricting the intercellular movement of the diffusible neuralizing substances. Cell contact also plays a key part in the mating process in many of the motile microbes discussed in this book.

Cell contact is a general term which by its inexplicitness must be examined. The essence of the term is the word "contact" which does not define an all-or-none proposition. There are many levels of contact. Chapter 13 by Ricci clearly delineates the problem. During mating in *Oxytrichia bifaria* increasing levels of contact-mediated behavior are reached. Cells first make orienting, transient contacts during what can anthropomorphically be termed as a precise mating dance. Subsequently, ciliary contact is established which becomes progressively more strong. Finally, localized membrane fusion occurs at the tightly contacting regions. The behavioral complexity of this unicellular organism is truly astounding and may serve as a good system to study the molecular differences between weak and strong cell contacts.

We know little about the diverse ways the surface properties of cells are defined or how they mediate intracellular events. How is the apparently tactile sensation of the transient filopodial contact of nerve cells so rapidly transduced to the immediate avoidance reaction of filopodial withdrawal (Dunn, 1971)? Are altered electrical surface signals important in cellular communication? What role does surface charge play in the contact process itself? Although Gingell and Fornés (1975) have suggested that van der Waals forces may play a role in cell contact there is little research being done to answer such basic questions. Today, our understanding of cellular contact is based on analyses of adhesivity which can be defined in the more approachable terms of molecular biology. The currently accepted view is that cell adhesivity operates through specific ligand molecules. Since cell adhesion occurs in a series of steps (Grinnell, 1978) it is clear that a multitude of "forces" work together in the adhesion process.

Adhesive cells have been shown to have specific recognition sites (Hynes, 1976; Loomis, 1979; Grady and McGuire, 1976). A symbolic representation of a molecular cell–cell contact is shown in Fig. 1C. Cells may contact through specific surface molecules and their receptors. Many cell adhesion factors have been purified and have been shown to be glycoproteins. Recently, Kinders and Johnson (1979) have purified a glycoprotein from the surface of mouse brain cells that inhibits cell growth and protein synthesis. Another means by which adhesivity might be mediated is through surface glycosyltransferases. Roseman (1970) proposed that cell adhesion could result from groups of these enzymes binding to their substrates on the surfaces of adjacent cells. Surface glycosyltransferases have been studied in a number of cell systems but, in particular, the association of sialyl-, gluconyl-, and galactosyltransferase activity with the aggregation factor of sponges provides the best support so far for Roseman's (1970) hypothesis (Muller *et al.,* 1977).

Table II, part A summarizes the cell–surface interactions during mating in three eukaryotic microbes that are discussed in detail in Part III of this volume. It is interesting that in all cases a protein or glycoprotein molecule plays a role in the adhesion of the mating type cells.

Intracellular adhesion also varies during the cell cycle. Hellerqvist (1979) has shown that Chinese hamster ovary cells are most adhesive during the G_1 phase. It is interesting that most eukaryotic microbes must be arrested in the G_1 phase before mating occurs (Crandall, 1977). Thus, it is possible that cell surface phenomena essential for cell to cell contacts are directly coupled to the cell cycle such that all cells must be the G_1 phase to be contact competent. In Chapter 2, Manney *et al.* show that one role of the sex pheromones is to induce G_1 arrest in *Saccharomyces cerevisiae* and several other chapters in this book discuss the role of G_1 arrest in the mating process.

D. Extracellular Matrix-Mediated Communication

Work on the problem of embryonic induction indicated that extracellular matrix was important during certain communicative events. Grobstein (1955) considered the matrix interaction as "something intermediate between full cellular contact and free diffusion." Two of the most important macromolecules of the extracellular matrix of animal tissues are the glycosaminoglycans (GAG) and collagen (Wessells, 1977). Hay and Meir (1974) have shown that the induction of somites by notochord and spinal cord tissues is mediated by GAG. However, very little has been done on the communicative function of extracellular matrix in nonembryonic systems except in eukaryotic microbes.

In many ways the cell walls of plants and eukaryotic microbes must be considered to be the equivalent of the extracellular matrix of higher animal systems. During sexual interactions in eukaryotic microbes that possess a cell wall, the initial site of cell contact is the cell wall and the role of this entity as a sex organelle during mating in *Schizo-*

TABLE II
The Surface-Mediated Interactions and Their Functions during Mating in the Eukaryotic Microbes Discussed in Part III

Organism	Sexual structure	Surface interaction			
		Recognitive unit	Chemical nature	Source (mating type)	Functions
A. Direct surface contact					
Chlamydomonas eugametos	Flagella	Membrane vesicles (isoagglutinins)	Glycoprotein	Both[a]	Induce flagellar agglutination
Oxytrichia bifaria	Cilia (peristomial ciliary organelles)	Unknown	Protein	All[a]	Directs cell contact and cellular orientation prior to cell adhesion
Paramecium caudatum	Cilia (ventral cilia)	Membrane vesicles	Protein	Both[a]	Causes agglutination of cells and inactivation of ciliary movement
B. Contact via extracellular matrix					
Saccharomyces cerevisae	Cell wall	Agglutination substances	Glycoprotein	Both[a]	Sexual agglutination between opposite mating type
	Cell wall	α pheromone binding substance	Protein	a cells	Binds α pheromone
Schizosaccharomyces pombe	Cell wall	Sex hairs	Protein	Both[a]	Cell–cell adhesion between mating types
Ustilago violacea	Cell wall	Fimbriae	Protein	Both[a]	Regulates cell communication during mating

[a] Each mating type produces its own factor.

saccharomyces pombe is discussed in detail in Chapter 10 by Calleja *et al.* Their work clearly shows the importance of the cell wall in mediating cell–cell adhesion (agglutination) and suggests that certain cell wall proteins are involved in the process. A similar situation exists in *Saccharomyces cerevisiae* and is detailed in Chapter 11 by Yanagishima and Yoshida. During mating in the smut fungus *Ustilago violacea* special differentiated structures called fimbriae project from the cell wall to mediate communication between pairs of mating-type cells. As Day and Cummins elaborate in Chapter 15, continuous fimbrial contact is essential during the early phases of mating for the reciprocal exchange of biochemical information for the regulation of the synthesis of the conjugation tube components. Subsequently, the fimbriae act as a scaffolding for the posttranslational formation of the conjugation tube itself.

When considering extracellular matrix one is immediately confronted by the enormous amount of gelatinous matrix in which the colonial residents of a *Volvox* spherical aggregate are embedded. The matrix contains a protein component plus some sugars and sulfated polysaccharides in an, as yet, undefined organization. In Chapter 4, Kochert suggests that the matrix may function in "collecting" pheromone from the environment where it would be in a dilute concentration. In this way it might intervene in the mating event by acting as a signal amplifier for the pheromonal communication system. A similar suggestion was earlier applied to embryonic induction systems wherein the extracellular matrix was suggested to act as an agent for trapping factors which were essential for cellular differentiation (Konigsberg and Hauscka, 1965). Possibly the knowledge we gain about the roles of extracellular matrix in microbial development will shed more light on its embryonic significance. A summary of the extracellular matrix-mediated mating processes discussed in Part III of this volume is presented in Table II (part B). In all of these cases a protein component associated with the cell wall or with cell wall modifications is involved in the interaction process and its effects.

III. THE VALUE OF USING EUKARYOTIC MICROBES FOR THE STUDY OF CELL COMMUNICATION

Eukaryotic microbes display an individuality not possible for multicellular organisms. In a multicellular organism it is essential for survival that cell populations coordinate their cellular activities. As a result at any one time the cell of a multicellular organism is being besieged by a multitude of messages of different types which must be interpreted and summarized into a response that benefits the whole cellular community. On the other hand, in unicellular organisms such complex interactions are usually not essential and as a result it is possible to essentially deal with a single event of communication at a point in time. This is not possible with multicellular systems, the physiological interdependence of the cells precludes it. Thus, the natural reductionism provided by eukaryotic microbes can assist the cellular communications scientist in deciphering the critical events that are associated with a specific mode or event of communication.

The argument that multicellular tissues can also be rendered to individual cells for study is not a particularly good one since the mechanical and chemical dissociation of tissues must be traumatic for the dispersed cells. Furthermore, the new artificial environment that the cells are subjected to cannot truly mimic the normal *in vivo* environs the cells typically live in. Although such dispersed cells may act in a way which implies normality, the investigator cannot be sure that certain critical physiological events are suppressed or enhanced due to the loss of the normal cell to cell associations or to the lack of certain molecules which are not provided by the culture medium. These problems are minimized with eukaryotic microbial systems. There is also a multitude of additional benefits that are provided by the use of nucleated microorganisms for the study of cellular communication. Rather than deal with these exhaustively, they will simply be mentioned.

In these austere times, one must economize. In this regard, eukaryotic microbes, the non-brand-name organisms, are cheap to culture generally yielding impressive amounts of physiologically equivalent cells in a short period of time. The ability to easily clone them allows the researcher to produce, routinely, large volumes of cells of identical genotype. Similarly mutants are easily generated and selected. In most microbial systems, genetic analysis is easy but this is not always so. The ability to induce stable dormant phases allows the researcher to store the organism, thus permitting him to study the same cells over a long period of time or to easily ship the beast to enlightened colleagues.

One of the most exciting aspects of working with microbes is the ability to manipulate the environment to induce the cells to undergo alternative pathways of development. Thus, cells of identical genotype might be induced to divide, to orient chemotactically toward a chemical, to encyst, to undergo sexual development, to sporulate, etc. Generally each cellular event can often be separated in time from other cellular events, thus providing the worker with a "simplified" problem. As a result, the analysis of the events involved in different types of communicative events can be analyzed with cells of the same genotype.

Generally, the synchrony and timing of specific cellular events rivals that or surpasses that of any nonmicrobial eukaryotic system.

We know very little about cells and the ways they communicate. There is no such thing as a simple biological system. However, eukaryotic microbes are the closest to simplicity that nucleated organisms get. Thus, it seems sensible that they will serve as valuable instruments in our attempts to decipher the details of cellular communication and other aspects of the biology of cells.

ACKNOWLEDGMENTS

I would like to thank Abdul Chagla, Paul Horgen, Keith Lewis, and Gary Paterno for their constructive criticism of this chapter.

REFERENCES

Abercrombie, M. (1967). Contact inhibition: The phenomenon and its biological implications. *Natl. Cancer Inst., Monogr.* 26, 249–277.

Bradshaw, R. A., and Frazier, W. A. (1977). Hormone receptors as regulators of hormone action. *Curr. Top. Cell. Regul.* 12, 1–37.

Brown, M. S., and Goldstein, J. L. (1979). Receptor-mediated endocytosis: Insights from the lipoprotein receptor system. *Proc. Natl. Acad. Sci. U.S.A.* 76, 3330–3337.

Crandall, M. (1977). Mating-type interactions in micro-organisms. In "Receptors and Recognition" (P. Cautrecasas and M. F. Greaves, eds.), Ser. A., Vol. 3, pp. 45–100. Chapman & Hall, London.

deLaat, S. W., Barts, P. W. J. A., and Bakker, M. I. (1976). New membrane formation and intercellular communication in the early *Xenopus* embryo. *J. Membr. Biol.* 27, 109–129.

Dunn, G. A. (1971). Mutual contact inhibition of chick sensory nerve fibres *in vitro*. *J. Comp. Neurol.* 14, 491–500.

Gilula, N. B. (1977). Gap junctions and cellular communication. In "International Cell Biology, 1976–1977" (B. R. Brinkley and K. R. Porter, eds.), pp. 61–69. Rockefeller Univ. Press, New York.

Gingell, D., and Fornés, J. A. (1975). Interaction of red blood cells with a polarized electrode: evidence for long-range intermolecular forces. *Biophys. J.* 16, 1131–1153.

Grady, S. R., and McGuire, J. (1976). Species selectivity of embryonic liver intercellular adhesion. *J. Cell Biol.* 71, 96–106.

Grinnell, F. (1978). Cellular adhesiveness and extracellular substrata. *Int. Rev. Cytol.* 53, 65–144.

Grobstein, C. (1955). Tissue interactions in the morphogenesis of mouse embryonic rudiments *in vitro*. In "Aspects of Synthesis and Order in Growth" (D. Rudnick, ed.), pp. 233–256. Princeton Univ. Press, Princeton, New Jersey.

Grunz, H., and Staubach, J. (1979). Cell contacts between chorda-mesoderm and the overlaying neuroectoderm (presumptive central nervous system) during the period of primary embryonic induction in amphibians. *Differentiation* 14, 59–65.

Hay, E. D., and Meier, S. (1974). GAG synthesis in notochord, spinal cord, and lens, all of which evoke similar synthesis in responding tissue. *J. Cell Biol.* 62, 889–898.

Heaysman, J. E. M. (1978). Contact inhibition of locomotion: A reappraisal. *Int. Rev. Cytol.* 55, 49–66.

Hellerqvist, C. G. (1979). Intercellular adhesion as a function of the cell cycle traverse. *J. Cell Biol.* 82, 682–687.

Helmreich, E. J., Zenner, H. P., Pfevffer, T., and Cori, C. F. (1976). Signal transfer from hormone receptor to adenylate cyclase. *Curr. Top. Cell. Regul.* 10, 41–87.

Hynes, R. O. (1976). Cell surface proteins and malignant transformation. *Biochim. Biophys. Acta* 458, 73–107.

Jensen, E. V., and DeSombre, E. R. (1973). Estrogen-receptor interaction. *Science* 182, 126–134.

Karlson, P., and Luscher, M. (1959). 'Pheromones': A new term for a class of biologically active substances. *Nature (London)* 183, 55–56.

Kinders, R. J., and Johnson, T. C. (1979). An inhibitor of cell growth and protein synthesis from mouse brain cerebral cortex. *J. Cell Biol.* 82, 68a.

Konigsberg, I. R., and Hauscka, S. D. (1965). Cell and tissue interactions in the reproduction of cell type. *In* "Reproduction: Molecular, Subcellular and Cellular" (M. Locke, ed.), pp. 243–290. Academic Press, New York.

Lehtonen, E. (1976). Transmission of signals in embryonic induction. *Med. Biol.* 54, 108–158.

Loomis, W. F. (1979). Biochemistry of aggregation in *Dictyostelium. Dev. Biol.* 70, 1–12.

Moscona, A. A. (1974). Surface specification on embryonic cells: Lectin receptors, cell recognition, and specific cell ligands. *In* "The Cell Surface in Development" (A. A. Moscona, ed.), pp. 67–99. Wiley, New York.

Muller, W. E. G., Arendes, J., Kurelec, B., Zahn, R., and Muller, I. (1977). Species-specific aggregation factors in sponges. Sialyltransferase associated with aggregation factor. *J. Biol. Chem.* 252, 3836–3842.

Pitts, J. D. (1977). Direct communication between animal cells. *In* "International Cell Biology, 1976–1977" (B. R. Brinkly and K. R. Porter, eds.), pp. 43–49. Rockefeller Univ. Press, New York.

Robinson, A. G., Butcher, W., and Sutherland, E. W. (1971). "Cyclic AMP." Academic Press, New York.

Roseman, S. (1970). The synthesis of complex carbohydrates by multi-glycosyltransferase systems and their potential function in intercellular adhesion. *Chem. Phys. Lipids* 5, 270–297.

Sheridan, J. D. (1977). Cell coupling and cell communication during embryogenesis. *In* "The Cell Surface in Animal Embryogenesis and Development" (G. Poste and G. L. Nicolson, eds.), pp. 80–92. Elsevier/North-Holland, Amsterdam.

Toyoda, H., and Spelsberg, T. C. (1979). Evidence for DNA sequence specificity of the acceptor protein which is bound by the progesterone receptor. *J. Cell Biol.* 83, 239a.

Viamontes, G. I., Fochtmann, L. J., and Kirk, D. L. (1979). Morphogenesis in *Volvox:* Analysis of critical variables. *Cell* 17, 537–550.

Wessells, N. K. (1977). "Tissue Interactions and Development." Benjamin, New York.

Part II
PHEROMONAL INTERACTIONS

ns, and
The Isolation, Characterization, and Physiological Effects of the *Saccharomyces cerevisiae* Sex Pheromones

T. R. MANNEY, W. DUNTZE, AND RICHARD BETZ

I.   Introduction 21
     A. The Organism and Its Life Cycle 21
     B. Culture Media 26
II. Isolation and Analysis of Mating Pheromones 27
     A. Strains and Growth Conditions 27
     B. Bioassays 28
     C. General Problems of Mating Pheromone Purification 29
     D. Structure of α Factor Peptides 30
     E. Analysis of α Factor Peptides 31
     F. Purification of a Factor 32
III. Response of Cells to Mating Pheromones 33
     A. Current State of Knowledge 33
     B. A Transient Differentiation Pathway 35
     C. Cell Volume Changes 43
IV. Perspectives 47
     References 49

I. INTRODUCTION

A. The Organism and Its Life Cycle

Among the unicellular fungi commonly designated as "yeasts," the species *Saccharomyces cerevisiae* is most widely used in industrial microbiology, especially in the preparation of alcoholic beverages and in baking. Moreover, *S. cerevisiae* has several properties that make it a particularly suitable object for biochemical, cytological, and genetic analysis of basic processes of cell physiology. It is because of these properties that during the past two or three decades *S. cerevisiae* has become widely accepted as an important eukaryotic model organism in cell biology. In particular, progress over the past decade toward understanding the events of the cell cycle, mating, and sporulation has made this organism an especially attractive model system for the study of cellular differentiation. Recent development of the techniques for transforming this yeast with a variety of plasmids containing DNA from *Escherichia coli* as well as from *S. cerevisiae*, together with molecular cloning techniques have brought the full power of contemporary molecular biology to bear on this system.

According to the taxonomic criteria applied by Kreger-van Rij (1969), *Saccharomyces cerevisiae* is defined as an ascosporogenous yeast that is fermentation positive and unable to assimilate nitrate as its sole source of nitrogen. The round or oval cells which contain a nucleus and other organelles characteristic of eukaryotic cells reproduce vegetatively by budding.

Originally, all *S. cerevisiae* strains used in the laboratory were obtained as single-cell clones from commercial bakers' or brewer's yeasts. Some of these isolates have been propagated as pure clones for many years under laboratory conditions. Although generally designated as "wild-type" these strains bear considerable genetic polymorphism. This has been a source of considerable confusion as conflicting results have been reported from laboratories using different "wild-type" strains. Fortunately, this source of confusion has been greatly reduced by the wide dissemination of a set of well-behaved isogenic strains which were developed by R. K. Mortimer. These strains have been informally accepted by many geneticists, biochemists, and molecular biologists as "standard types." The diploid strain, X2180, was isolated from the mating type α haploid strain S288C. Diploids occur spontaneously at a relatively high frequency in cultures of this strain. They apparently result from cells that switch mating type from α to **a** and mate with one of the original α cells in the culture. Two haploid strains X2180-1A (mating type **a**) and X2180-1B (mating type α) were isolated from the diploid by ascus dissection following sporulation. These "standard type" strains, as well as an extensive collection of genetically marked strains, are available from the Yeast Genetic Stock Center (Donner Laboratory, University of California, Berkeley, California 94720).

Figure 1 illustrates the life cycle of *Saccharomyces cerevisiae* with emphasis on the role of the peptide mating pheromones, **a** factor and α

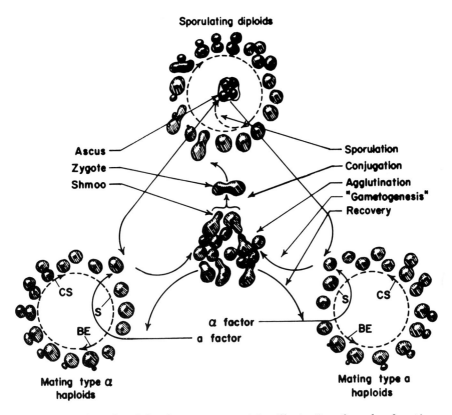

Fig. 1. Life cycle of *Saccharomyces cerevisiae* illustrating the role of mating pheromones. Vegetatively growing cultures of haploids of the two mating types are shown at the bottom. The sketches illustrate the morphologies of the cells throughout the cell division cycle, and the visible landmarks as described by Hartwell (1974). The events defined as "cell separation" (CS), "start" (S), and "bud emergence" (BE) are marked. Also illustrated are the progressive increase in the size of the bud throughout the cycle, and the division of the nucleus. The arrows labeled "a factor" and "α factor" illustrate the point in the cell division cycle at which the pheromones act. The group in the center represents the cell types, found in a mating mixture, which result from the action of the pheromones. At the top is illustrated the cell division cycle for a diploid and the entry of a newly formed zygote into the cycle just beyond "start." Finally, the sporulation step, which produces haploid ascospores which germinate and enter the haploid cell division cycle just before "start," is illustrated.

factor. Although the word hormone has been used extensively in the yeast literature, for uniformity, we will employ the more correct term pheromone. The groups at the bottom of Fig. 1 represent asynchronously growing cultures of mating type a and α haploids. Mating type is determined by two alternative alleles at the mating type locus, designated MATa and MATα. Each of these cell types produces, constitutively, a mating pheromone which is an oligopeptide. These peptides are designated after the mating type of the cells producing them, as a factor and α factor. Cells of each mating type respond specifically to the pheromone produced by the opposite mating type, and exhibit a variety of changes. This response is executed at a point in the G_1 phase of the cell division cycle that is termed "start" (Hartwell, 1974), where the cells are unbudded and have not initiated DNA synthesis. Responding cells are arrested at that point in the cycle and undergo a transient differentiation to a cell type possessing the characteristic features of a gamete; in addition to having unreplicated haploid genomes, they have the capacity for specifically recognizing and fusing with appropriate mates.

The most impressive aspect of this response is an alteration of the normally spherical cell shape to an elongated pear-shaped form which is generally referred to as "shmoo" morphology. The morphological response, "shmooing," provides the most generally useful basis for detecting and estimating the concentration of a factor and α factor (Duntze et al., 1970; MacKay and Manney, 1974a; Betz et al., 1977). This characteristic morphology, as revealed by scanning electron microscopy, is illustrated in Fig. 2A. The polar elongation of the cells which is characteristic of the shmoo morphology is clearly visible. In freeze-etched electron micrographs of shmoos (Fig. 2B and C) it can be seen that in the elongated tips of the shmoos there is an aggregation of small, spherical vesicles, which appear to fuse with the cytoplasmic membrane (Fig. 2C). Similar vesicles have been observed to accumulate under the cell wall in budding cells in the region where the bud is emerging (Matile et al., 1969). From these observations it may be inferred that shmooing results from a disturbance of the normal budding process.

In mating mixtures the first visible event that can be attributed to the action of mating pheromones is sexual agglutination, in which cells of the two mating types stick together in large aggregates (this aspect of the mating process is detailed in Chapter 11 by Yanagishima and Yoshida). In spite of the apparently disorganized character of these aggregates, the cells become organized into specific mating pairs, which fuse into true diploid zygotes. These enter the diploid cell division cycle just after the "start" point and continue to form a diploid clone.

Diploid cells formed by this process of sexual conjugation are heterozygous for the two alternative alleles at the *MAT* locus. They are relatively stable and can be propagated indefinitely under proper

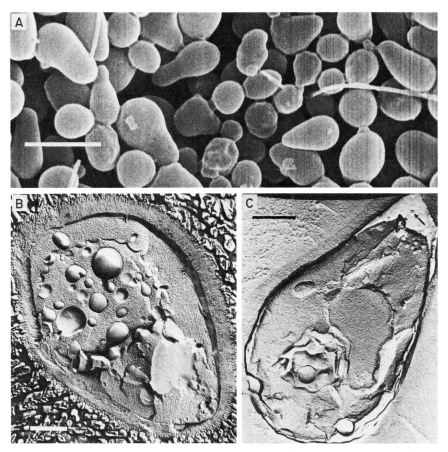

Fig. 2. Electron micrographs of mating type a cells exhibiting the shmoo morphology. (A) Scanning electron micrograph of mating type a cells which were treated with 10 units/ml of α factor for 5 hr. Scale line corresponds to 10 μm. (Photograph courtesy of Prof. W. Breipohl, Bochum.) (B) and (C) Electron micrograph of a freeze-etched mating type a cell treated with 5 units/ml of α factor for 4 hr. Scale line corresponds to 1 μm. (Photographs courtesy of Dr. V. Speth, Freiburg.) The arrow in (C) indicates a vesicle fusing with the cytoplasmic membrane.

laboratory conditions. However, they also have the ability to sporulate (reviewed in Fowell, 1969), and will do so if placed on a suitable medium. During sporulation the diploid nucleus undergoes meiosis, producing four haploid nuclei, which reside in four ascospores. These in turn are enclosed in a membrane, forming an ascus. When placed on adequate growth medium the ascospores will germinate, either within the ascus or having been freed from it, and enter the cell division cycle just before "start." In the laboratory the ascus wall may be removed with one of several enzyme preparations and the spores separated by micromanipulation. The ease with which this can be done has facilitated extensive genetic mapping of the chromosomes by tetrad analysis (Mortimer and Schild, 1980).

The peptide mating pheromones appear to play a central role in mediating the transient differentiation of vegetative haploid cells into gametes. All the changes that are observed in mating mixtures can also be seen in pure cultures of cells of either mating type that have been treated with low levels of the purified pheromones which have been isolated from cells of the opposite mating type. The cellular changes produced by these pheromones include arrest in the G_1 phase of the cell-division cycle, development of surface agglutinins, specific and localized changes in cell-wall structure, changes in cell morphology, and inhibition of membrane-bound adenylcyclase.

B. Culture Media

The following media and appropriate variations of them are used for most genetic and physiological studies with this organism. YEPD (Difco Yeast Extract, 10 gm/liter; Difco Bacto-Peptone, 20 gm/liter; dextrose, 20 gm/liter) is a rich complex medium generally used for routine culture and for stock maintenance. Other carbon sources may replace the glucose. Supplementary adenine (80 mg/liter) is often added for strains that carry red adenine mutations (*ade1* and *ade2*) to prevent selection for white mutants. MV (Difco Yeast Nitrogen Base without Amino Acids, 6.7 gm/liter; dextrose, 20 gm/liter; after Wickerham, 1946) is a defined minimal salts medium supplemented with vitamins most generally used for culturing prototrophic strains used in the studies of mating and mating pheromones. It is a universal minimal medium for genetic studies, and with appropriate supplements, is used to score most auxotrophic genetic markers. Suitable concentrations of amino acid, purine, pyrimidine and other supplements for many strains are the following (all quantities are given in mg/liter): adenine, 20; anthranilic acid, 10; arginine, 30; aspartic acid, 100; choline, 50; histidine, 10; homoserine, 200; indole, 10; isoleucine, 20; leucine, 60; lysine, 40; methionine, 20; ornithine, 200; phenylalanine, 50; serine, 375; threonine, 350; tryptophan, 30; tyrosine, 20; uracil, 20; and valine, 65. Many and varied sporulation media have been described (Fowell, 1969; McClary *et al.*, 1959), but a 1% solution of potassium acetate supplemented with 0.25% Difco Yeast Extract is quite effective for many strains. Any of these media may be prepared in either liquid or solid form. In the latter case 2% Difco Bacto-agar is satisfactory. In general, autoclaving for 15 min at 15 psi is sufficient, although YEPD media require 25 min. Most of the supplements can be autoclaved with the media, but the tryptophan and threonine should be filter-sterilized and added after autoclaving.

II. ISOLATION AND ANALYSIS OF MATING PHEROMONES

A. Strains and Growth Conditions

Both a factor and α factor have been detected in the growth media of all fertile haploid *S. cerevisiae* strains which have been examined for mating factor production. An α factor has also been isolated from *Saccharomyces kluyveri* (McCullough and Herskowitz, 1979). In *S. cerevisiae* the pheromones have not been detected in cell extracts, probably because of rapid degradation by liberated cellular proteases. Unfortunately, the concentrations of mating factors in the culture media are rather low (5–30 μg/liter). The lack of a simple, precise, and sensitive assay for these pheromones has hampered the isolation of strains producing increased amounts and the definition of optimum conditions for higher yields.

In our experience, the highest yields are obtained with the isogenic standard strains, described above (X2180-1B for α factor and X2180-1A for a factor; Betz *et al.*, 1977). Both strains behave as weak homothallics. They diploidize spontaneously at low frequencies, forming sterile diploids which, upon prolonged propagation, outgrow the haploid cells. Since the diploids produce neither mating factor, one must be careful to use defined haploids for mating factor preparations. To assure this, we routinely select single-cell isolates and test for mating factor production.

Pheromone purification is easiest with cultures grown on a defined glucose salt medium such as MV. Separation of the mating factors from the organic constituents of complex media is difficult, so they should be avoided. A critical condition for optimal mating factor production is intensive aeration of the culture. However, it is not clear whether this stimulation is due to a specific oxygen requirement for mating-factor production, or whether it reflects the general stimulation of cell metabolism.

There are few and somewhat contradictory studies on the kinetics of mating factor production. In the case of α factor, Scherer *et al.* (1974) observed that the pheromone concentration in the culture medium continuously increased during exponential growth and even during stationary phase. In contrast, Tanaka and Kita (1977) reported that the pheromone is produced predominantly during exponential growth and rapidly degraded when the cells enter stationary phase. Consequently, they have observed a maximum of α factor concentration at the transition from exponential to stationary phase. This apparent discrepancy may be explained by differences in the culture conditions and in methods used for α factor determination. Apparently the optimal conditions for mating factor production must be verified, and perhaps adjusted, for each strain. However, the originally published conditions for the X2180 strains (Duntze et al., 1973) give reasonably reproducible yields of mating factors and have been used successfully in a number of laboratories (Lipke et al., 1976; Chan, 1977; Ciejek et al., 1977; Maness and Edelman, 1978; Udden and Finkelstein, 1978; McCullough and Herskowitz, 1979).

B. Bioassays

The morphological response—shmooing—provides the most generally useful basis for detecting and estimating the concentration of a factor and α factor (Duntze et al., 1973; MacKay and Manney, 1974a; Betz et al., 1977). In the case of α factor, the test is readily carried out on agar medium where the peptide, either from a streak of growing cells or from a solution placed in a well in the agar, can diffuse into the surrounding agar and affect a cells which have been placed there. It can also be carried out in liquid cultures. However, a factor in crude preparations will not readily diffuse through agar because the peptide is associated with a high molecular weight, carbohydrate-containing component. As a result, crude preparations must be tested in liquid cultures.

Quantitation, within about a factor of two, can easily be achieved by serial dilution (usually in twofold steps), using as a dilution endpoint the failure to observe shmoos after a fixed incubation time (e.g., 3 hr). The sensitivity depends on the exact period of time selected. More precise quantitation can be achieved only to a limited extent using smaller dilution because of ambiguity in determining the endpoint. The most commonly used unit of activity for these pheromones is the reciprocal of the endpoint dilution for shmooing described above. However, because of the variety of culture conditions and endpoint criteria used, there appears to be considerable discrepancy in the absolute value of the unit defined. Thus, use of the dilution endpoint to quantitate these activities has the advantage of being specific, simple, and relatively direct, but lacks sensitivity and precision. Several other approaches have been proposed that take advantage of the apparent dose dependence of various response parameters.

Aspects of the response that have been shown to depend on the initial concentration of peptide include the length of the transient inhibition time as judged by cell number increase (Betz et al., 1977) or unbudded cells (Chan, 1977; Wilkinson and Pringle, 1974), the sensitivity to cell lysis by glusulase (Lipke et al., 1976), and induction of agglutinability (Hartwell, 1980). A sensitive radioimmunoassay has been developed by Jones-Brown, Thorner, and Ciejek (personal communication). In a later section we will explore in more detail the kinetics of the response of cells treated with **a** factor and α factor and their implications for bioassays.

C. General Problems of Mating Pheromone Purification

Because of the small amounts of mating pheromones accumulated in the medium, it is uneconomical to use culture volumes smaller than 100 liters for routine purifications. In our laboratories we routinely grow approximately 100 or 200 liters of culture in 6 or 12 carboys, each containing 17 liters of MV medium. The cultures are inoculated with approximately 2×10^5 cells/ml and vigorously aerated by bubbling filter-sterilized air (12.5 l/min/carboy) through the medium. The cultures are grown at 30°C for 48 hr (for α factor) and 36 hr (for **a** factor). Removal of the yeast cells is rapidly achieved by filtration of the cultures through porcelain filter candles or by centrifugation. With 6 filter candles connected in parallel to the vacuum of a water pump, a cell-free filtrate is obtained from 200 liters of culture within 3 to 5 hr.

Basically, the purification procedures for α factor and **a** factor from culture filtrates are rather similar. The activities are first adsorbed to ion-exchange resin and eluted in a smaller, more easily handled volume of a volatile solvent. α factor is quantitatively adsorbed to Amberlite CG50, and completely eluted with 0.01 N HCl in 80% ethanol, and **a** factor is adsorbed to SP-Sephadex and eluted with 2 M pyridine. The eluates are further concentrated by rotatory evaporation and/or lyophilization. It is thus possible to concentrate the mating pheromone contained in several hundred liters of culture filtrate to about 10 ml within 2 days. The next step is to extract the mating pheromones into methanol. This yields highly active preparations of partially purified peptides that are suitable for many biological purposes, and may be stored for weeks without loss of activity. Further purification is achieved by gel filtration on Sephadex LH-20 in organic solvents, yielding highly purified active preparations from which homogeneous mating factor can be obtained by a final partition chromatography on silica gel plates or a suitable resin. On the basis of biological activity, the recovery of α factor is usually very high, often nearly 100%. The recovery of a factor activity, however, is only about 5%. Owing to the sensitivity of the biological assays to factors that cannot be controlled in impure preparations, these values may not reflect the actual recovery of the peptides.

The details of the purification schemes have been published for both α factor (Duntze et al., 1973) and a factor (Betz and Duntze, 1979). Variations on the procedure for purifying α factor have been published in the references cited above. The basic difference between the mating pheromones that influences their purification appears to lie in their hydrophobicity, and in their tendency to aggregate and to adsorb to other material. Whereas α factor behaves as a freely diffusible, low molecular weight substance in culture filtrates and partially purified preparations, a factor shows a strong tendency to associate with high molecular weight carbohydrate material present in the culture medium. Even when purified to homogeneity both factors tend to form insoluble aggregates in aqueous systems. It is therefore useful to keep purified preparations in methanol.

D. Structure of α Factor Peptides

Isolation of α factor from X2180-1B yields a group of four closely related trideca- and dodecapeptides, designated as $\alpha 1$, $\alpha 2$, $\alpha 3$, and $\alpha 4$, which differ only slightly in their specific biological activities as determined by the standard morphogenetic assay (i.e., shmoo formation; Stötzler and Duntze, 1976). These peptides can be separated and purified to homogeneity in small amounts by thin layer chromatography on silica gel plates (Duntze et al., 1973) or on a preparative scale by chromatography on Biorex 70 as described by Stötzler and Duntze (1976). In some α factor preparations which were obtained by thin layer chromatography we found that the most active compound exhibited a distinct blue color due to the formation of a complex with cupric ions (Duntze et al., 1973). However, in later preparations which were purified by chromatography on Biorex 70 the active α factor peptides were isolated as colorless white compounds, which did not turn blue upon addition of small amounts of cupric ions. Nevertheless, since these peptides possessed high specific α factor activities they were assumed to represent active species of α factor and their primary structure was determined (Stötzler et al., 1976). It was found that $\alpha 2$ and $\alpha 4$

2. Saccharomyces cerevisiae Sex Pheromones

α1:   NH$_2$-Trp-His-Trp-Leu-Gln-Leu-Lys-Pro-Gly-Gln-Pro-Met-Tyr-COOH

α2:   NH$_2$-His-Trp-Leu-Gln-Leu-Lys-Pro-Gly-Gln-Pro-Met-Tyr-COOH

α3:   NH$_2$-Trp-His-Trp-Leu-Gln-Leu-Lys-Pro-Gly-Gln-Pro-Met(SO)-Tyr-COOH

α4:   NH$_2$-His-Trp-Leu-Gln-Leu-Lys-Pro-Gly-Gln-Pro-Met(SO)-Tyr-COOH

Fig. 3. Primary structures of the four α factor peptides.

are dodecapeptides which differ from the tridecapeptides α1 and α3 solely by the lack of an N-terminal tryptophan residue. Furthermore, it was found that α3 and α4 are oxidation products of α1 and α2, respectively, containing a methionine sulfoxide residue instead of methionine in the penultimate position. Figure 3 reveals the amino acid sequences of the α factor peptides. Identical structures were determined by other workers (Ciejek et al., 1977; Tanaka et al., 1977), although Tanaka et al. (1977) were unable to detect the α2 peptide in their preparations. It is likely that α2 is produced in the cultures by the action of an aminopeptidase on α1. Similarly, α3 and α4 probably arise by spontaneous oxidation of α1 and α2, respectively. Definite proof for the structures of the α factor peptides was provided by chemical synthesis of biologically active peptides of the proposed structures by several groups (Ciejek et al., 1977; Masui et al., 1977; Samokhin et al., 1979). The activity of these completely synthetic peptides demonstrates that these compounds indeed represent the sole primary signals for eliciting the biological effects of α factor.

E. Analysis of α Factor Peptides

Characteristics of a particular α factor peptide can be achieved by several means. Thin layer chromatography on silica gel plates in n-butanol, propionic acid, and water (50:25:35, by volume) resolves all four peptides. This method is, therefore, most suitable to assess the homogeneity of a preparation. Conversion of the oxidized peptides to their reduced forms is readily achieved by treatment with 1% thioglycolic acid. Likewise, it is possible to oxidize the reduced peptides by treatment with 1% hydrogen peroxide. The composition of a specific α factor peptide may be determined by standard amino acid analysis. However, the standard method of hydrolysis in 6% HCl, which leads to a rapid destruction of tryptophan, usually does not allow distinction between the trideca- and the dodecapeptides, which differ by a single tryptophan residue. More reliable results are obtained by hydrolysis of the peptides in 3 M mercaptoethane sulfonic acid, as described by Stötzler and Duntze (1976) and Ciejek et al. (1977). Best results are obtained by digestion of an α factor peptide with aminopeptidase M and determination of the Trp:His ratio (Stötzler et al., 1976). This gentle procedure allows the distinction of $\alpha 1$ and $\alpha 3$ (Trp:His = 2) from $\alpha 2$ and $\alpha 4$ (Trp:His = 1). Enzymatic cleavage with carboxypeptidase is the most convenient way to distinguish the reduced α factor peptides from the oxidized ones. As shown by Stötzler and Duntze (1976), methionine sulfoxide is completely preserved by digestion of the peptides with carboxypeptidase C. In contrast, during hydrolysis in HCl or in mercaptoethane sulfonic acid methionine sulfoxide is totally reduced to methionine. Carboxypeptidase A is apparently unable to liberate methionine sulfoxide. Upon incubation with this enzyme only tyrosine is obtained from $\alpha 3$ and $\alpha 4$, whereas from $\alpha 1$ and $\alpha 2$ tyrosine and methionine are liberated in a ratio of 1 (Stötzler and Duntze, 1976). Likewise, only the reduced peptides $\alpha 1$ and $\alpha 2$ are cleaved by BrCN, giving rise to free tyrosine, while the oxidized peptides are not affected (Stötzler and Duntze, 1976).

F. Purification of a Factor

a Factor is isolated from culture filtrates of X2180-1A cells which are grown on MV medium as described above for α cells. The pheromone level in crude culture filtrates (approximately 0.5 to 2 units/ml) is often too small to be detected without concentration. Fortunately a factor is readily adsorbed to a variety of ion exchangers, of which SP-Sephadex and phosphocellulose have been found most useful (Betz et al., 1977; Betz and Duntze, 1979). For routine preparations we use a slightly modified version of the originally published purification procedure (Betz and Duntze, 1979).

The untreated cell-free culture filtrate is applied directly to a column of SP-Sephadex C-25, which has been equilibrated with 20 mM citrate buffer, pH 3.0. For each liter of filtrate a gel bed of 25 ml is used. After washing the column with 2 column volumes of water, a factor is quantitatively eluted with one volume of 2 M pyridine. This eluate is concentrated to about 300 ml by rotatory evaporation and then lyophilized. The dried powder is readily dissolved in a small volume of dimethyl sulfoxide. Water is added to this solution so that the final ratio of DMSO:H_2O is 1:10. To this solution 3 volumes of methanol are slowly added, with stirring, in an ice bath. During the addition of the methanol a white precipitate is formed which contains only traces of a factor activity. The precipitate is removed by centrifugation. Usually the total a factor activity recovered in the methanol supernate is about 30%. However, full activity is recovered when the inactive precipitate and the methanol supernate are recombined. The nature of the material responsible for this stimulation of a factor activity is not known. However, the precipitate contains large amounts of mannose-containing carbohydrate. As a working hypothesis we assume that this material serves as a nonspecific carrier for the very hydrophobic a factor peptide. The methanol supernate may be conveniently used as a concentrated source of partially purified a factor for most biological experiments, and can be stored in the freezer for several weeks without loss of activity. From the methanol extract a factor is further purified by chromatography on Sephadex LH-20 in 70% propan-1-ol. Three different peptides are routinely found in the fractions containing the a factor activity. These peptides, referred to as band 1, band 2, and band 3, are separated by thin-layer chromatography on silica gel in butan-1-ol, acetic acid, ethyl acetate, and water (80:10:10:20 by volume). All three exhibit a factor activity, the most active being band 2, with a specific activity of 1.2×10^5 units/mg in the standard morphogenetic assay. Furthermore, the band 2 peptide exhibits all biological activities that have been attributed to a factor. It is, therefore, assumed to represent the native hormone. aFactor is a very hydrophobic peptide having the structure H_2N-Tyr (Asx_1, Gly_1, Ala_1, Val_1, Ile_2, Phe_1, Lys_1, Trp_1, Pro_1). The exact compositions of band 1 and band 3 have not yet been established. However, they are similar to that of band 2 and may, therefore, represent degradation products of a factor similar to those described for α factor.

III. RESPONSE OF CELLS TO MATING PHEROMONES

A. Current State of Knowledge

The exact point in the cell division cycle at which cells respond to α factor (or a factor) has been investigated by a variety of means. Bücking-Throm *et al.* (1973) used time-lapse photography and measurement of DNA synthesis to show that arrested cells were blocked at a specific site in G_1, and that the arrested cells accumulated at that site. When the peptide was removed from the medium the cells resumed progress through the division cycle in synchrony. This synchrony is reflected both in the appearance of buds and in the onset of DNA synthesis. These observations led to the hypothesis that the biological role of these pheromones is to synchronize the two mating haploid cells prior to conjugation to assure their orderly entrance into the new diploid cell division cycle. Hartwell (1973) presented supporting evidence for this hypothesis by demonstrating the necessity for a period of courtship in mating mixtures before conjugation could occur.

Subsequent work by Hartwell (1974) and collaborators has led to the identification of a point in the cell cycle, which they term "start," which was defined by use of a temperature-sensitive cell division cycle mutant *cdc28*. Cells that have passed this point in the cycle are committed to another round of division, but cells approaching this point in the presence of the pheromone isolated from the opposite mating type will respond and enter the pathway that leads to conjugation (Reid and Hartwell, 1977; Hartwell, 1978). Byers and Goetsch (1975) have obtained electron microscopic evidence with conjugating cells that supports the synchronization hypothesis. They documented that there is a continuity of the development of the spindle plaque from haploid to diploid during conjugation. Cells arrested at "start" have a satellite-bearing, single plaque which persists through conjugation to become a double plaque in the diploid zygote. Accordingly, the arrest of the conjugating haploid cells at this exact point in the cell division cycle provides for a continuous transition into the diploid cell division cycle at the point where the spindle plaque would normally double.

Several other features of the response of haploid cells to the mating pheromones can be understood as direct consequences of this cell division cycle arrest. The inhibition of cell cycle events that are normally initiated after "start," such as DNA synthesis and bud initiation, would seem to fall into this category. The specificity of this action of the pheromones as a cell cycle regulation was demonstrated by Throm and Duntze (1970), who showed that only DNA synthesis, and not net RNA or protein synthesis, is arrested by α factor. As will be detailed later, the increase in cell volume observed in α factor-arrested cultures (Throm and Duntze, 1970; Johnston *et al.*, 1977) may also be accounted for by the specific arrest of progress through the cell division cycle without arrest of cell growth.

Other features of the response to these pheromones, however, do not appear to be consequences of cell cycle arrest. Instead, the changes in the cell wall, including shmooing, development of surface agglutinins as discussed in Chapter 11 by Yanagishima and Yoshida, changes in the mannan/glucan composition (Tkacz and MacKay, 1979; Lipke *et al.*, 1976) and inhibition of membrane-bound adenylate cyclase (Liao and Thorner, 1980) would seem to be more properly viewed as parallel responses, or perhaps as steps in the transient "gametogenesis" pathway. A consistent feature of the response is its transient nature (Throm and Duntze, 1970; Wilkinson and Pringle, 1974). In the case of **a** cells, this recovery is, at least in part, the result of destruction of the α factor by the cells (Hicks and Herskowitz, 1976; Chan, 1977; Finkelstein and Strausberg, 1979; Ciejek and Thorner, 1979). A similar destruction of **a** factor by α cells may be inferred.

B. A Transient Differentiation Pathway

The response of cells to a factor and α factor is of considerable interest for a variety of reasons. However, the idea that it represents a model system of some generality for the study of hormone-controlled eukaryotic differentiation has special appeal. We know of no comparable system in any organism that is as approachable by the combined tools of biochemistry, genetics, and molecular biology. Accordingly, we will explore the evidence and arguments that support the idea that cells responding to these peptides are undergoing a specific differentiation, a transient and reversible gametogenesis.

If this view is correct, then it should be possible to arrange the events associated with the response in a reasonable temporal sequence leading from a vegetative cell to a specialized gamete, capable of conjugating with an appropriate partner to form a zygote. Although there have been a number of reports describing various aspects of the response in morphological, physiological, and biochemical terms, few of these reports have been concerned with their temporal relationships. Furthermore, nearly all the published studies of biological effects of these pheromones have employed extremely high concentrations of the peptides in comparison with the concentrations found in cultures under conditions that favor efficient mating. These abnormal conditions very likely distort the picture of the actual biological roles of these changes. Indeed several authors have observed that addition of such levels of α factor to a mating mixture is actually inhibitory (Sena *et al.*, 1975; Udden and Finkelstein, 1978).

The events of most interest are the biochemical and molecular changes that must begin with the initial recognition of the specific peptide and ultimately lead either to the fusion of cells into zygotes or to recovery. Most of these events have been elusive so far; little can be said with certainty about what occurs, let alone when it occurs. But in an attempt to provide a framework in which to look at the temporal order of events in this process, we have undertaken a detailed study of the kinetics of the "visible" changes—these are the changes in cell number, in the fraction of budded cells, in cell volume, and in cell morphology.

We know, from previous published work, when the process begins. If a cell has passed the point in the division cycle called "start" it completes that cycle in the presence of a high concentration of pheromone. On the other hand, if a cell has not passed "start" when pheromone is added, it responds, and is arrested with respect to cell cycle events, at that point. This suggests that a cell will respond if it has not executed the decisive events at that point, but it does not tell us when the cell and pheromone molecule actually interact physically, or when the earliest changes occur in the cell as a result of that interaction. On one extreme, it is possible that there are pheromone receptors or binding sites present on the cell throughout the cycle but that only events associated with "start" are affected, or that these events must be affected first. At the other extreme, it is possible that the specific association with and response to pheromone occurs only during a short period of time just before "start."

In spite of these uncertainties, we can infer that an event occurs at or near "start." The first visible manifestation of that event is an increase in the fraction of unbudded cells, which becomes apparent at the time when the affected cells would otherwise have budded. It is possible to study the kinetics of response by counting budded and unbudded cells, but extreme perseverance is required to get precise data. Large numbers of cells must be scored in short periods of time. But such tedious direct methods are necessary to define the events. Hopefully, once this is done, it will be possible to obtain more detailed information by indirect chemical and physical methods.

1. Temporal Sequence of Events

Although the various visible manifestations of response to a factor and α factor have been described by a number of authors, the variety of conditions and strains used makes it difficult to determine the timing of these events from published data. Consequently, we have determined the time of occurrence of the major visible changes associated with the response under uniform culture conditions for the standard strains X2180-1A and X2180-1B. We have measured events both in mating mixtures of these strains and in pure cultures which have been treated with purified preparations of the pheromone produced by the opposite mating type. The results are summarized in Fig. 4.

a. Arrest at "Start." From evidence reviewed above it is well established that cells that have not executed the events at "start" respond at that point in their cell division cycle and enter the "gametogenesis pathway." Cells that have executed those events continue through the division cycle. The time of "start" relative to the time of bud emergence

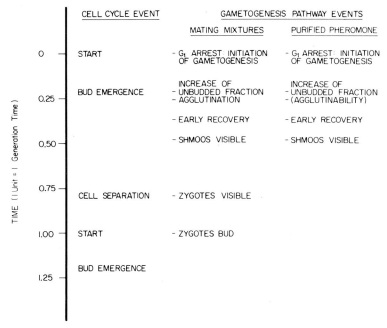

Fig. 4. Sequence of visible events induced by mating pheromones. Events observed in mating mixtures and in pure haploid cultures exposed to purified mating pheromones are indicated at the times at which the first cells are observed executing the event. The cell cycle event "start" is taken as the position in the cycle of the cells that are immediately inhibited when pheromone is added. The time between "start" and "bud" emergence is determined from the time after pheromone addition when the fraction of unbudded cells begins to increase. The time between "cell separation" and "bud emergence" is calculated from the fraction of unbudded cells (50%) in the untreated exponentially growing cultures from the equation

$$f_u = 2(1 - 2^{-t_{BE}\tau})$$

where f_u is the fraction of unbudded cells, t_{BE} is the time of bud emergence, τ is the generation time of the culture (1.9 hr).

is determined from the time at which the fraction of unbudded cells begins to increase in a culture after the addition of pheromone. Figure 5 shows the percent unbudded cells as a function of time after addition of α factor or **a** factor to cultures growing at 2.2×10^6 cells/ml. The time at which the proportion of unbudded cells begins to increase, 22 min after addition of the pheromone, is independent of the pheromone concentration, and is the same for both **a** factor and α factor. The excess unbudded cells that appear at this time correspond to the cells that were arrested immediately after the pheromone was added to the culture. A similar pattern of arrest is observed in a mating mixture at the same cell density, with the proportion of unbudded cells increasing after the cells have been mixed for 22 min.

b. Agglutination. Sexual agglutinability in these strains must be induced by exposure to the pheromones (Betz et al., 1978). The time at which this induced agglutinability occurs in mating mixtures can be inferred from the lag in appearance of aggregates. Under the conditions used in this study this lag has been found to be 30 min for a variety of strains, including X2180-1A and X2180-1B (Manney and Meade, 1977). We have not measured the time required for induction of agglutinability by purified pheromones under these conditions; however, Fehrenbacher et al. (1978) have detected increased agglutinability within 30 to 60 min after addition of α factor to **a** cells. Other, possibly related, cell wall changes also occur. Lipke et al. (1976) have reported α factor-induced changes in the carbohydrate composition of the cell wall and an increased susceptibility to lysis by glucanases (e.g., glusulase). Since they used comparable conditions, we can estimate the time of this change from their data. A plot of glusulase sensitivity (reciprocal of half-time for lysis) as a function of time of treatment with 1 unit of α factor/ml, made from the data in Table 4 of their paper, shows an initial increase in glusulase sensitivity at about 3/4 hr.

c. Early Recovery. When cultures are treated with low levels of α factor or **a** factor (Fig. 5), the frequency of unbudded cells drops below the frequency expected for total inhibition. This early recovery, which in some cases only involves part of the cells in the culture, appears to result from a different mechanism than the later recovery, observed at higher concentrations, which is generally attributed to destruction of the pheromone (see discussion of recovery below). The time of this early recovery is only slightly dependent on α factor concentration; doubling the concentration increases the time by only about 8 min. A similar recovery of budding is observed in mating mixtures.

d. Morphological Change. The characteristic morphological change (shmooing) associated with response to these phermones varies with the pheromone concentration, both in the fraction of cells responding and in the severity of the morphological change (Lipke et al., 1976). However, Fig. 6A shows that the time at which shmoos begin to appear in a treated culture is essentially independent of the pheromone concentration. For initial α factor concentrations between 0.5 and 5.0 units/ml, shmoos begin to accumulate in the culture at approximately 1 hr.

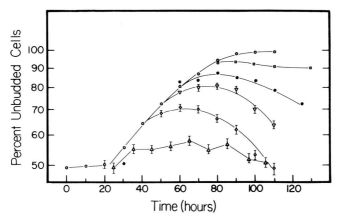

Fig. 5. Percent unbudded cells plotted on a logarithmic scale against time after addition of various concentrations of pheromone. Closed circles (●): **a** factor, 1.0 unit/ml. Open symbols: α factor, (△) 0.03, (◇) 0.04, (▽) 0.22, (□) 1.0, and (○) 3.4 units/ml. Error bars are standard deviation, σ_f, calculated from the equation

$$\sigma_f = \frac{f(1-f)}{n-1}$$

where f is the fraction of unbudded cells, and n is the total number of budded and unbudded cells scored. Strain X2180-1A (**a**) or X2180-1B (α) was grown in SC medium overnight to a density of 2.2×10^6 cells/ml on a rotary shaker at 200 rpm and 30°C. Mating pheromone from the opposite mating type, purified as described in the text, was added to the growing culture at the final concentration indicated. 1-ml samples were sonicated for 12 sec with a Biosonic IV equipped with needle probe operated on low power range. Samples were immediately scored microscopically without fixation. 1000 to 2000 cells were scored in a period of 5 min. Single cells were scored as unbudded and pairs which were resistant to sonication were scored as budded.

e. Zygote Formation. In mating mixtures (Fig. 6B) zygotes become distinguishable by morphological criteria by 1.5 hr after mixing, and budded zygotes occur at about 2 hr. The lag of approximately 30 min between the appearance of the first zygote and the first budded zygote indicates that the diploid fusion nucleus enters the cell division cycle at approximately the same point at which the haploids were arrested. The number of zygotes formed is limited by the number of arrested cells availale for mating. The plateau in the frequency of zygotes reflects the recovery of cells. This is a direct consequence of previous observations that only cells that are unbudded (specifically those arrested at "start") are able to mate (Hartwell, 1974). Indeed, Sena *et al.* (1975) have shown that by selecting unbudded cells from a growing culture, using zonal centrifugation, it is possible to achieve much higher zygote frequencies.

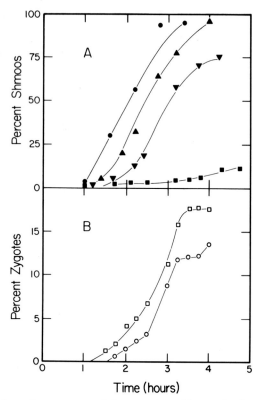

Fig. 6. (A) Time of appearance of shmoos after addition of α factor to an exponentially growing culture of X2180-1A (a); (■) 0.5, (▼) 1.0, (▲) 2.0, and (●) 5.0 units of α factor/ml. (B) Formation of zygotes in a mating mixture of X2180-1A (a) and X2180-1B (α); (□) total zygotes scored on basis of characteristic morphology, and (○) budded zygotes. Experimental procedures are as described in the legend to Fig. 5.

The above observations were summarized in Fig. 4. In this figure the events are ordered along a time axis which is labeled in units of the mean generation time of these strains under the conditions used, 1.9 hr. The times of several related cell cycle events are also shown. Each pheromone-induced event is placed at the time of its earliest observation. This corresponds to the time the event would be executed by the first cells that enter the gametogenesis pathway. The parallel between the events and their timing in pure cultures treated with purified peptide preparations and those in the mating mixture is striking. This correspondence strongly supports the widely held contention that these peptides are the sole chemical mediators of the events that lead to zygote formation.

2. Recovery from Arrest by Mating Pheromone

The transient nature of the arrest of the cell division cycle by these pheromones, the synchronous recovery observed when the pheromone is removed by changing the culture medium, and the observed enzymatic destruction of α factor by mating type **a** cells have led to the view that recovery is largely due to destruction of pheromone. But while a relatively strong case has been made for this view at high initial concentrations of α factor (Ciejek and Thorner, 1979; Finkelstein and Strausburg, 1979), it is not clear that the same mechanism is responsible for the recovery observed at lower concentrations. The time constant of the pheromone-concentration dependence for this early recovery differs by two orders of magnitude from that observed for the later recovery. Furthermore, the observation of two times of recovery in the same culture (Fig. 7) indicates that the sensitivity of the cells is not constant and homogeneous. The problem is to distinguish between changes in the concentration of the pheromone and changes in the sensitivity of the cells used to measure it.

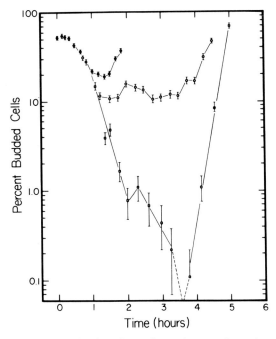

Fig. 7. Percent budded cells plotted on a logarithmic scale against time after addition of various concentrations of α factor. (●) 0.22, (○) 0.94, and (□) 3.38 units/ml. The experimental procedures are as described in the legend to Fig. 5. The data for 0.22 units/ml are the same as those plotted in Fig. 5.

Let us first consider the influence of the initial α factor concentration (at constant cell density for the moment) on the general pattern of the transient inhibition of budding. In Fig. 7 we have plotted percent budded cells, on a logarithmic axis, against time after addition of α factor, on a linear axis, for three different initial concentrations. We observe three different patterns of response. At the lowest concentration shown (0.2 units/ml), only a fraction of the cells are arrested, and these recover within one generation time. At the highest concentration shown (3.4 units/ml), all the cells are eventually inhibited and subsequently recover with a high degree of synchrony after about two generation times. At the intermediate concentration (0.9 units/ml), a mixed response occurs. Some of the cells are inhibited and recover synchronously after more than one generation time, but a smaller fraction of the cells either are never inhibited at all, or recover, again with some synchrony, within the first generation time.

The behavior shown in Fig. 7 is highly reproducible and raises many questions, but we will focus for now on the question of the destruction of the α factor. The response to the highest concentration, with its sharply synchronous recovery, looks very much like the recovery observed when arrested cells are washed free of α factor (Bücking-Throm et al., 1973), and we know from previous work (Chan, 1977; Stötzler et al., 1977) that the inhibition time for this range of concentration is related to the initial concentration of the pheromone and to the initial cell density. The nature of these relationships is illustrated in Fig. 8. At constant cell density the inhibition time is proportional to the log of the initial pheromone concentration, and at constant initial pheromone concentration it is inversely proportional to the log of the initial cell density. For cell densities between 2.2 and 2.6×10^6 cells/ml the recovery time increases by approximately 1.4 hr for each doubling of the α factor concentration. A plausible interpretation of these data is that the pheromone is degraded by the culture with a half-life of 1.4 hr, and that recovery begins when the pheromone has dropped below some critical level. The observation by Ciejek and Thorner (1979) that certain protease inhibitors prolong the inhibition strongly supports this view.

However, at lower initial pheromone concentrations some cells were more sensitive than others (Fig. 7); part of the culture either recovered earlier or was not inhibited at all. This can be seen qualitatively when scoring the budding fraction in treated cultures. At initial α factor concentrations between about 0.7 and 2 units/ml (at approximately 2.3×10^6 cells/ml) there are no longer any cells with extremely small (newly emerged) buds. Then at about 1.25 hr cells with newly emerged buds reappear.

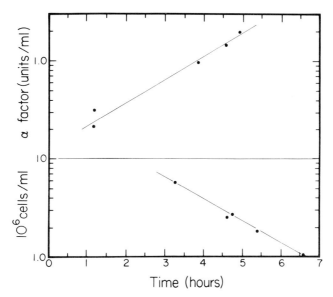

Fig. 8. Recovery times of α factor-inhibited cultures plotted against the logarithm of the initial α factor concentration at constant cell density (2.4×10^6 cells/ml) in the top panel, and against the logarithm of the initial cell density at constant initial α factor concentration (3.38 units/ml) in the lower panel. The recovery time was taken as the time when the percent budded cells had returned to 10% in experiments as described and illustrated in Fig. 7.

Destruction of the pheromone with a 1.4-hr half-life does not account for this early recovery. On the contrary, it appears that at the lower concentrations characteristic of levels found in growing cultures and mating mixtures, the time of recovery is not determined exclusively by the pheromone concentration. At these lower concentrations some cells recover in the presence of pheromone levels that are sufficient to produce cell cycle arrest in others.

C. Cell Volume Changes

Cells that are arrested by these pheromones become conspicuously larger; the difference is easily observed under the microscope. This volume increase results from arresting cell division without blocking the net synthesis of RNA and protein (Throm and Duntze, 1970). Indeed, Fig. 9 shows that the total volume of cells/ml of culture (number of cells ml × mean cell volume) is the same in an α factor-arrested culture as in the untreated exponentially growing culture for up to two generation times. Consequently, measurement of cell volume, which can be done quickly and with great precision, provides a useful tool for

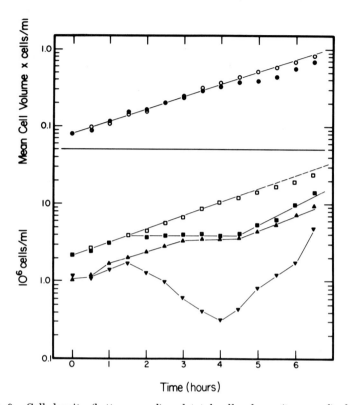

Fig. 9. Cell density (bottom panel) and total cell volume (top panel) plotted on logarithmic scales against time after addition of α factor to an exponentially growing culture of X2180-1A (a). Bottom panel: (□) untreated control, total cells/ml; closed symbols, 2.0 units/ml. (■) total cells, (▲) cells with volumes greater than median cell volume of untreated culture, and (▼) cells with volumes less than median. Top panel: (○) untreated control, total cells/ml × mean cell volume, and (●) treated cells, total cells/ml × mean cell volume. The experimental procedures are as described in the legend to Fig. 5. Mean cell volumes were calculated from cell volume distributions (see Fig. 10) measured with a Particle Data Celloscope with a 60-μm aperture, equipped for hydrodynamic focusing. Pulse height distributions were accumulated with a Canberra 8100e pulse-height analyzer and computer analyzed using an Itel AS5 computer. Volume calibration was achieved using standard polystyrene latex spheres (9.87 μm diameter lot no. 2130) obtained from Coulter Electronics, Incorporated. Cell counts were determined with the same instrument without hydrodynamic focusing.

quantitative studies of the arrest of cell division and subsequent recovery.

The instrument we have chosen for cell volume analysis is a Particle Data Celloscope electric zone-sensing device (which operates on the same principle as a Coulter Counter), which we have modified to provide hydrodynamic focusing of the sample stream. This modification causes all the sample particles to pass through the center of the sensitive zone of the detector (orifice) parallel to its axis. As a result, pulse shape variations, normally associated with particles passing through different regions of the inhomogeneous sensing zone at different angles, are minimized and the resolution is dramatically increased. This refinement makes it possible to detect and measure changes in the volume distribution within 30 min of the addition of pheromone to a growing culture.

Figure 10 illustrates the dynamics of the cell volume distribution of an exponentially growing culture of X2180-1A after addition of 2 units/ml (2×10^6 cells/ml) of α factor. The mean cell volume is plotted against time in Fig. 11A. Although the mean cell volume remains constant for the first hour, significant changes in the distribution are apparent at the earliest times. Initially, there is a decrease in the number of cells in the middle of the distribution, followed for a time by a progressive decrease in the number of larger cells. But, after 1.5 hr there is an increase, both in the mean volume and in the maximum volume; all the cells become larger. By 4.5 hr the mean cell volume reaches a maximum. The distribution is then very broad and flat, with fewer than 10% of the cells having volumes less than the median of the

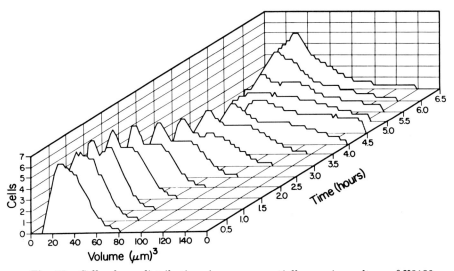

Fig. 10. Cell volume distributions in an exponentially growing culture of X2180-1A plotted against time after addition of 2.0 units/ml of α factor. Measurement of volume distributions is described in the legend to Fig. 9. All the distributions have been normalized to the same total cell number. The small-particle noise peak has been removed and the 0 hr distribution extrapolated to 0 counts.

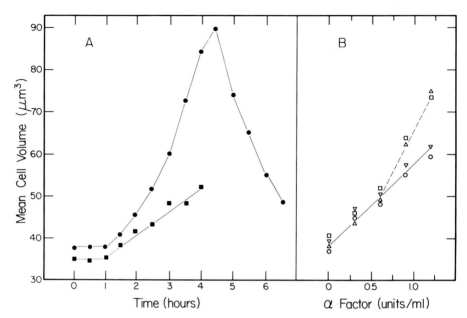

Fig. 11. Mean volumes of α factor treated X2180-1A (a) cells and a mating mixture of X2180-1A (a) and X2180-1B (α) plotted against time (A) and against α factor concentration (B). (●) Mean volumes from cell-volume distributions illustrated in Fig. 10. (■) Mean volumes of distributions obtained as described in legend to Fig. 9 on mating mixtures described in legend to Fig. 6. Mean volumes for different α factor concentrations were measured after 3 hr (▽), 4 hr (□), and 5 hr (△). The median volume was also measured after 3 hr (○). Experimental procedures are as described in the legends to Figs. 5 and 9.

original culture. After 4.5 hr the recovery of the culture is reflected in the shift of the distribution to smaller volumes, with an especially pronounced increase in the number of normal-size cells. Figure 11A also shows the mean cell volume for a mating mixture as a function of time. As with the α factor-treated pure culture, the mean volume remains constant for the first hour and then increases. In both cases the initial increase appears to be exponential.

Complete volume distributions require the availability of a pulse-height analyzer. Their analysis, except at a superficial level, requires the use of a computer. However, much information about inhibition and recovery can be obtained with a simple cell counter equipped with one or two volume threshold discriminators (e.g., a Particle Data Celloscope or a Coulter Counter). If dual counts are taken with one threshold setting that counts the total population and one that counts only those cells larger than the median for the untreated culture, results such as those illustrated in Fig. 9 are obtained. The correspondence between the changes in these counts and the changes in the volume distribution is clear. Cell volume changes can also be used as a quantitative assay for the pheromone activities. Figure 11B shows the mean cell volume as a function of α factor added to a culture for various times. By selecting an appropriate time for counting one can obtain a linear relationship between any suitable volume parameter and the pheromone concentration.

IV. PERSPECTIVES

The study of the mating pheromones of *Saccharomyces cerevisiae* appears to be at a commonly encountered awkward stage. Most of the easy experiments have been done, the results are exciting, and there is promise of even more interesting discoveries yet to be made. But the studies that lie ahead promise to be more difficult, for many of them have already been tried by simple approaches and have yielded frustration.

The initial isolation and purification of α factor was laborious, and often exasperating, but it turned out to be basically simple. The material is relatively stable and the bioassay, while not precise, is simple and reliable. The puzzle of its primary structure offered a satisfying combination of challenging pitfalls and good fortune. The chemistry of a factor is more elusive, but now appears to be nearly within reach. The similarities between the two pheromones provide a pleasing basis for the symmetry of their actions, while the differences in their particular compositions pose an interesting puzzle regarding their specific modes of action and parallel evolution.

The unraveling of the general features of their actions has similarly provided pleasurable satisfaction to a number of workers in a variety of disciplines. The initial appreciation of the fact that α factor specifically blocks the progress of cells in the cell division cycle was indeed fortuitously timed. The techniques and perspectives that had been developed by Hartwell and collaborators, brought to bear on this question, quickly added considerably to understanding both of the action and biological role of mating pheromones, and of the events associated with the "start" of the cell division cycle. Indeed, α factor has become a useful reagent for a variety of cellular and molecular biological studies. Ironically, even the yet unsuccessful attempts to decipher the genetic control of these pheromones (MacKay and Manney, 1974a,b) helped sketch out the general outlines of the mating type control and provided some of the raw material, in the form of the first extensively characterized collection of nonmating mutants, for the detailed understanding of the genetic organization of the mating type locus (Herskowitz et al., 1980).

Indeed, during the past decade the study of the diverse aspects of mating type and related processes in this organism has become a popular field that has attracted a number of enthusiastic and talented workers from a variety of backgrounds, who unfortunately, seem to share a modest reluctance to express themselves in print. The task ahead is formidable. Only a single paper has been published on the entire subject of the synthesis of these pheromones (Scherer et al., 1974). This paper establishes the likelihood that α factor is synthesized on ribosomes, but nothing further is known about it. It is often speculated that such a small peptide must be formed as part of a larger precursor protein, but we know of no evidence bearing on the existence of such a precursor. However, there is a growing body of indirect evidence that points to one or more processing functions being involved in the release of α factor into the medium. This comes from the studies of pleiotropic mutants.

Although hundreds of mutants that affect mating functions have been isolated (MacKay and Manney, 1974a,b; Manney and Woods, 1976; Manney and Meade, 1977; I. Herskowitz, personal communication; L. H. Hartwell, 1980), the great majority of such mutants have proved to be pleiotropic and to affect functions that are not specific for one or the other mating type. In particular, there is a common association of deficiency in the ability to release α factor into the medium and other phenotypic defects that are associated with changes in the cell envelope (Manney and Meade, 1977; Lemontt et al., 1980). These findings suggest that there are processing steps involved in the release of the peptides that involve other aspects of cell wall or membrane development and metabolism.

Finally, the major unanswered questions are, "In molecular terms, what exactly do these peptides do to sensitive cells, and how do they do it? Where are the specific receptors, and how are the multiple aspects of the response mediated within the cell?" The techniques are available to search for answers to these questions; hopefully, it should be only a matter of time.

So, despite the amount of information that has been generated about the mating pheromones in this yeast and the mating process they seem to mediate, some of the most fundamental biological questions are still to be addressed; in fact, we do not even know if they enjoy it.

ACKNOWLEDGMENTS

The original research reported in this paper was supported by a grant of the Deutsche Forschungsgemeinschaft, and by a grant GM-19175 (to T.R.M.) awarded by the U.S. Public Health Service. We are also grateful to Patricia Jackson, who carried out the experiments on response to the pheromones, to Natalya Hall, who made the drawing in Fig. 1, and to Dr. Kenneth Conrow, who guided the computer analysis of the cell volume data.

REFERENCES

Betz, R., and Duntze, W. (1979). Purification and partial characterization of a-factor, a mating hormone produced by mating-type-a cells from *Saccharomyces cerevisiae*. *Eur. J. Biochem.* 95, 469–475.

Betz, R., MacKay, V. L., and Duntze, W. (1977). a-Factor from *Saccharomyces cerevisiae:* Partial characterization of a mating hormone produced by cells of mating type a. *J. Bacteriol.* 132, 462–472.

Betz, R., Duntze, W., and Manney, T. R. (1978). Mating-factor-mediated sexual agglutination in *Saccharomyces cerevisiae*. *FEMS Microbiol. Lett.* 4, 107–110.

Byers, B., and Goetsch, L. (1975). Behavior of spindles and spindle plaques in the cell cycle and conjugation of *Saccharomyces cerevisiae*. *J. Bacteriol.* 124, 511–523.

Bücking-Throm, E., Duntze, W., Hartwell, L. H., and Manney, T. R. (1973). Reversible arrest of haploid yeast cells at the initiation of DNA synthesis by a diffusible sex factor. *Exp. Cell Res.* 76, 99–110.

Chan, R. K. (1977). Recovery of *Saccharomyces cerevisiae* mating-type a cells from G_1 arrest by α-factor. *J. Bacteriol.* 130, 766–774.

Ciejek, E., and Thorner, J. (1979). Recovery of *Saccharomyces cerevisiae* a cells from G_1 arrest by α-factor pheromone requires endopeptidase action. *Cell* 18, 623–635.

Ciejek, E., Thorner, J., and Geier, M. (1977). Solid phase peptide synthesis of α-factor, a yeast mating pheromone. *Biochem. Biophys. Res. Commun.* 78, 952–961.

Duntze, W., MacKay, V., and Manney, T. R. (1970). *Saccharomyces cerevisiae:* A diffusible sex factor. *Science* 168, 1472–1473.

Duntze, W., Stötzler, D., Bücking-Throm, E., and Kalbitzer, S. (1973). Purification and partial characterization of α-factor, a mating-type specific inhibitor of cell reproduction from *Saccharomyces cerevisiae*. *Eur. J. Biochem.* 35, 357–365.

Feherenbacher, G., Perry, K., and Thorner, J. (1978). Cell–cell recognition in *Saccharomyces cerevisiae:* Regulation of mating-specific adhesion. *J. Bacteriol.* 134, 893–901.

Finkelstein, D., and Strausberg, S. (1979). Metabolism of α-factor by **a** mating type cells of *Saccharomyces cerevisiae*. *J. Biol. Chem.* 254, 796–803.

Fowell, R. R. (1969). Sporulation and hybridization of yeast. In "The Yeasts" (A. Rose and J. S. Harrison, eds.), pp. 303–385. Academic Press, New York.

Hartwell, L. H. (1973). Synchronization of haploid yeast cell cycles, a prelude to conjugation. *Exp. Cell Res.* 76, 111–117.

Hartwell, L. H. (1974). *Saccharomyces cerevisiae* cell cycle. *Bacteriol. Rev.* 38, 164–198.

Hartwell, L. H. (1978). Cell division from a genetic perspective. *J. Cell Biol.* 77, 627–637.

Hartwell, L. H. (1980). Mutants of *Saccharomyces cerevisiae* unresponsive to cell division control by polypeptide mating hormone. *J. Cell Biol.* **85,** 811-822.

Herskowitz, I., Blair, L., Forbes, D., Hicks, J., Kassir, Y., Rine, J., Sprague, G., Jr., and Strathern, J. (1980). Control of cell type in the yeast *Saccharomyces cerevisiae* and a hypothesis for development in higher eukaryotes. *In* "The Molecular Genetics of Development" (T. Leighton and W. Loomis, eds.), pp. 80-118. Academic Press, New York.

Hicks, J. B., and Herskowitz, I. (1976). Interconversion of yeast mating types III. Action of the homothallism (HO) gene in cells homozygous for the mating-type locus. *Genetics* **85,** 395-405.

Johnston, G. C., Pringle, J. R., and Hartwell, L. H. (1977). Coordination of growth with cell division in the yeast *Saccharomyces cerevisiae*. *Exp. Cell Res.* **105,** 79-98.

Kreger-van Rij, N. J. W. (1969). Taxonomy and systematics of yeast. *In* "The Yeasts" (A. Rose and J. S. Harrison, eds.), pp. 5-78. Academic Press, New York.

Lemontt, J. F., Fugit, D. R., and MacKay, V. L. (1980). Pleiotropic mutations at the *tup1* locus that affect the expression of mating-type-1 dependent functions in *Saccharomyces cerevisiae*. *Genetics* **94,** 899-920.

Liao, H., and Thorner, J. (1980). Yeast mating pheromone alpha-factor inhibits adenylate cyclase. *Proc. Natl. Acad. Sci. U.S.A.* **77,** 1898-1902.

Lipke, P. N., Taylor, A., and Ballou, C. E. (1976). Morphogenic effects of α-factor on *Saccharomyces cerevisiae*. *J. Bacteriol.* **127,** 610-618.

McClary, D. O., Nulty, W. L., and Miller, G. R. (1959). Effect of potassium versus sodium in the sporulation of *Saccharomyces*. *J. Bacteriol.* **78,** 362-368.

McCullough, J., and Herskowitz, I. (1979). Mating Phermones of *Saccharomyces kluyveri* and *Saccharomyces cerevisiae*. *J. Bacteriol.* **138,** 146-154.

MacKay, V. L., and Manney, T. R. (1974a). Mutations affecting sexual conjugation and related processes in *Saccharomyces cerevisiae*. I. Isolation and phenotypic characterization of nonmating mutants. *Genetics* **76,** 255-271.

MacKay, V. L., and Manney, T. R. (1974b). Mutations affecting sexual conjugation and related processes in *Saccharomyces cerevisiae*. II. Genetic analysis of nonmating mutants. *Genetics* **76,** 273-288.

Maness, P. F., and Edelman, G. M. (1978). Inactivation and chemical alteration of mating factor α by cells and spheroplasts of yeast. *Proc. Natl. Acad. Sci. U.S.A.* **75,** 1304-1308.

Manney, T. R., and Meade, J. H. (1977). Cell-cell interactions during mating in *Saccharomyces cerevisiae*. *In* "Microbial Interactions" (J. L. Reissig, ed.), pp. 283-321. Chapman & Hall, London.

Manney, T. R., and Woods, V. (1976). Mutants of *Saccharomyces cerevisiae* resistant to the α mating-type factor. *Genetics* **82,** 639-644.

Masui, Y., Chino, N., Sakakibara, S., Tanaka, T., Murakami, T., and Kita, H. (1977). Synthesis of the mating factor of *Saccharomyces cerevisiae* and its truncated peptides: The structure-activity relationship. *Biochem. Biophys. Res. Commun.* **78,** 534-538.

Matile, P., Moor, H., and Robinow, C. F. (1969). Yeast cytology. *In* "The Yeasts" (A. Rose and J. S. Harrison, eds.), pp. 219-297. Academic Press, New York.

Mortimer, R. K., and Schild, D. (1980). The genetic map of *Saccharomyces cerevisiae*. *Microbiol. Rev.* **44** (in press).

Reid, B. J., and Hartwell, L. H. (1977). Regulation of mating in the cell cycle of *Saccharomyces cerevisiae*. *J. Cell Biol.* **75,** 355-365.

Samokhin, G. P., Lizlova, L. V., Bespalova, J. D., Titov, M. I., and Smirnov, V. N. (1979).

Substitution of Lys^7 by Arg does not affect biological activity of α-factor, a yeast mating pheromone. *FEMS Microbiol. Lett.* 5, 435–438.

Scherer, G., Haag, G., and Duntze, W. (1974). Mechanism of α factor biosynthesis in *Saccharomyces cerevisiae*. *J. Bacteriol.* 119, 386–393.

Sena, E. P., Radin, D. N., Welch, J., and Fogel, S. (1975). Synchronous mating in yeast. *Methods Cell Biol.* 11, 71–88.

Stötzler, D., and Duntze, W. (1976). Isolation and characterization of four related peptides exhibiting α-factor activity from *Saccharomyces cerevisiae*. *Eur. J. Biochem.* 65, 257–262.

Stötzler, D., Kiltz, H., and Duntze, W. (1976). Primary structure of α-factor peptides from *Saccharomyces cerevisiae*. *Eur. J. Biochem.* 69, 397–400.

Stötzler, D., Betz, R., and Duntze, W. (1977). Stimulation of yeast mating hormone activity by synthetic oligopeptides. *J. Bacteriol.* 132, 28–35.

Tanaka, T., and Kita, H. (1977). Degradation of mating factor by a-mating type cells of *Saccharomyces cerevisiae*. *J. Biochem. (Tokyo)* 82, 1689–1693.

Tanaka, T., Kita, H., Murakami, T., and Narita, K. (1977). Purification and amino acid sequence of mating factor from *Saccharomyces cerevisiae*. *J. Biochem. (Tokyo)* 82, 1681–1687.

Throm, E., and Duntze, W. (1970). Mating-type-dependent inhibition of deoxyribonucleic acid synthesis in *Saccharomyces cerevisiae*. *J. Bacteriol.* 104, 1388–1390.

Tkacz, J. S., and MacKay, V. L. (1979). Sexual conjugation in yeast. Cell surface changes in response to the action of mating hormones. *J. Cell Biol.* 80, 326–333.

Udden, M. M., and Finkelstein, D. B. (1978). Reaction order of *Saccharomyces cerevisiae* alpha-factor-mediated cell cycle arrest and mating inhibition. *J. Bacteriol.* 133, 1501–1507.

Wickerham, L. J. (1946). A critical evaluation of the nitrogen assimilation tests commonly used in classification of yeasts. *J. Bacteriol.* 52, 293–301.

Wilkinson, L. E., and Pringle, J. R. (1974). Transient G_1 arrest of *S. cerevisiae* cells of mating type α by a factor produced by cells of mating type a. *Exp. Cell Res.* 89, 175–187.

3

The Role of Sexual Pheromones in *Allomyces*

JEFFREY POMMERVILLE

I. Introduction	53
II. Survey of the Literature	56
III. Current Research	59
A. Timing of Pheromone Secretion	59
B. Sirenin Production and Macromolecular Synthesis	60
C. Aspects of Pheromone Reception	61
D. Signal Transduction and Cell Behavior	62
E. Conclusion	68
IV. Perspectives	69
References	70

I. INTRODUCTION

Chemical communication between cells requires a sensory system having four major steps: (1) synthesis and secretion of a communication signal in the form of a chemical substance; (2) transmission of the signal through a medium that is in contact with the cells; (3) reception of the signal by the target cell; and (4) transduction or processing mechanisms necessary to convert the signal into a change in cell behavior. In order to understand more clearly these events in higher eukaryotes, simpler organisms having primitive sensing systems have been used as models. The study of chemotaxis in the prokaryotes *Escherichia coli* and *Salmonella typhimurium* is one example which has provided clues to better understand more complex chemosensory mechanisms (Koshland, 1974, 1977; Macnab, 1978). Among eukaryotic microbes, the slime mold *Dictyostelium discoideum* is one of the best studied chemotactic systems (Loomis, 1979). However, it represents an attraction system based on amebal movement. On the other hand, the potential of flagellated eukaryotic microbes as model systems for the study of cell communication has not been utilized to its fullest. This is unfortunate because many unicellular, eukaryotic microbes use sensory systems for their daily existence, are more easily manipulated for experimentation than higher organisms, and represent a group of organisms intermediate in complexity between the prokaryotes and the higher eukaryotes. If we are to understand cell communication thoroughly, a comprehensive knowledge of the sensory systems in eukaryotic microbes will prove extremely valuable.

One of the best systems for studying cell communication in eukaryotic microbial cells is through the interaction of sexual pheromones which, being produced by one gamete (or mating type), act as locational signals that allows the other gamete (or mating type) to find the cell producing the pheromone. Pheromones thus provide an efficient mechanism to facilitate the events of the sexual process. Although several sexual pheromones involved in cell communication and recognition have been isolated and characterized in the fungi and algae (see Kochert, 1978), a complete and detailed understanding of the effect and role of these substances on the target cells (male gametes) has hardly begun.

This chapter will focus on the mode of action and role of sexual pheromones in communication between gametes of the fungus *Allomyces* (order Blastocladiales). The following introduction will serve to describe the organism and set the stage for the discussion of chemosensory activities and cell communication.

Allomyces is an aquatic fungus that grows as a saprophyte on animal or plant debris (Emerson, 1941, 1954). The subgenus *Euallomyces* can alternate between an asexual, diploid generation and a sexual, haploid generation (Fig. 1). The sexual phase, which is of concern here, consists of a branching, coenocytic vegetative mycelium. At the tips of many of the hyphae sexual reproductive structures (gametangia) are differentiated. These gametangia are multinucleate and are cut off from the rest of the vegetative mycelium by complete cross walls (septa). When the species *A. macrogynus* is cultured on agar it will form a terminal pair of gametangia; the orange-colored male gametangium being epigynous to the colorless female gametangium (Fig. 1). Often a chain of reproductive structures will form below the terminal pair and these may be male, female, or most commonly, some combination of the

3. Pheromones in *Allomyces*

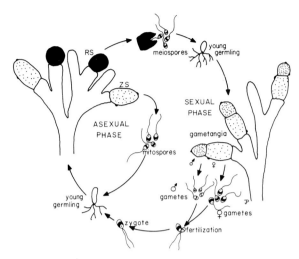

Fig. 1. Life cycle of *Allomyces macrogynus*. The resistant sporangia (RS) produce haploid meiospores that give rise to a mature sexual mycelium (gametothallus) that bears male and female gametangia. Male and female gametes emerge from the respective gametangia and undergo fertilization to form a diploid zygote. The zygote develops into a mature asexual mycelium (sporothallus) bearing thin-walled zoosporangia (ZS) and thick-walled RS. Diploid mitospores emerge from the ZS and develop into another sporothallus. Reproductive structures are not to scale.

two types. In agar culture once the gametangia have reached this mature state, further development and gametogenesis will not occur until they are placed in or flooded with a nutrient-deficient medium. When flooded with a dilute salts (DS) medium (Machlis, 1958a) the gametangial cytoplasm cleaves into uninucleate segments and gamete differentiation occurs synchronously. After the gametes are formed they are released from the gametangia as a result of the dissolution of exit pores (papillae) located in the gametangial wall. The motile, posteriorly flagellate male and female gametes are easily identified: the male gametes are orange in color (due to the presence of γ-carotene; Emerson and Fox, 1940), highly motile, and relatively small in size (5–6 μm long), while the female cells are colorless, sluggishly motile, and about twice the size of the male cells. Cell communication in this fungus occurs through the attraction of male gametes to the female cells brought about by a sexual pheromone secreted by the female cells. [Although the word hormone has been used to describe the attractant molecule in *Allomyces,* the term pheromone (Karlson and Luscher, 1959) is more appropriate for the chemical communication between individual cells described in this chapter.] As a result contact is made between the sex cells and fertilization can occur in a highly efficient manner to produce a population of biflagellate, motile zygote cells (Fig. 1).

Allomyces is ideal for studies on cell communication and chemosensory activities because (1) knowledge of the sexual phase is understood fairly well at the morphological level (Turian, 1969; Pommerville and Fuller, 1976; Morrison, 1977); (2) the gametes do not have external coverings, thick glycoprotein coats, or cell walls (Pommerville and Fuller, 1976) which would hinder physiological examination of the cell communication system; (3) the structure and properties of the pheromone are known and the attractant can be synthesized in the laboratory; and (4) the fungus is easily grown in culture. For cytological studies, where large numbers of gametes are not required, male and female gametangia from the wild-type strain have been physically separated to obtain small populations of male or female gametes (Pommerville, 1977, 1978a,b). Superior methods also have been perfected for synchronously obtaining large populations of gametes. This involves using strains of *A. macrogynus* that produce greater than 90% pure male or female gametangia (Rønne and Olson, 1976). With these strains large populations of male or female gametes can be obtained for physiological and biochemical analyses.

II. SURVEY OF THE LITERATURE

The presence of a pheromone in the sexual phase of *Allomyces* first was demonstrated by Machlis (1958a,b) and called sirenin. He demonstrated that the male gametes would cluster around female gametangia which had not yet released female gametes (Machlis, 1958a) and, by constructing a multicell bioassay apparatus, he showed that male gametes preferentially were attracted to a dialysis membrane through which the pheromone was diffusing (Machlis, 1958a).

Procedures for the isolation, purification, and characterization of sirenin required large numbers of female gametes. Since all natural species of the fungus are monoecious, wild-type strains which produce both male and female gametes could not be used. However, by using predominantly female hybrids obtained from interspecies crosses between *A. macrogynus* and *A. arbuscula* (Emerson and Wilson, 1954), this obstacle was overcome. Through careful and rigorous procedures sirenin has been completely characterized (Machlis *et al.*, 1966, 1968; Nutting *et al.*, 1968). The pheromone is an oxygenated sesquiterpene with a molecular weight of 236 and consists of a cyclopropyl ring attached to an isohexenyl side chain (Fig. 2). The pheromone can be synthesized in the laboratory by a number of routes (Plattner et al., 1969; Bhalerao et al., 1970; Plattner and Rapoport, 1971), and in doing so a number of pheromone isomers and analogs have been isolated and purified (Fig. 2). Machlis (1973b) tested the biological activity of these forms for their ability to attract male gametes. The results show that only natural sirenin and the chemically synthesized *l*-enantiomer are active and the addition of *d*-sirenin to the *l* form does not inhibit male attraction. All other derivatives are inactive demonstrating the importance of the allylic alcohol and the stereochemistry of the side chain. The role of the cyclopropyl ring remains unknown since the bicyclic keto form and the β-ester form used by Machlis (1973b) have a methyl group substituted for the alcohol function on the side chain (Fig. 2). Methylation leads to a loss of biological activity as shown with monodeoxysirenin.

The biosynthetic route for sirenin synthesis has not been established, although a pathway from farnesyl pyrophosphate has been suggested (Nutting et al., 1968; Jaenicke, 1972) and a hypothetical pathway from acetyl-CoA has been proposed (Pommerville, 1977).

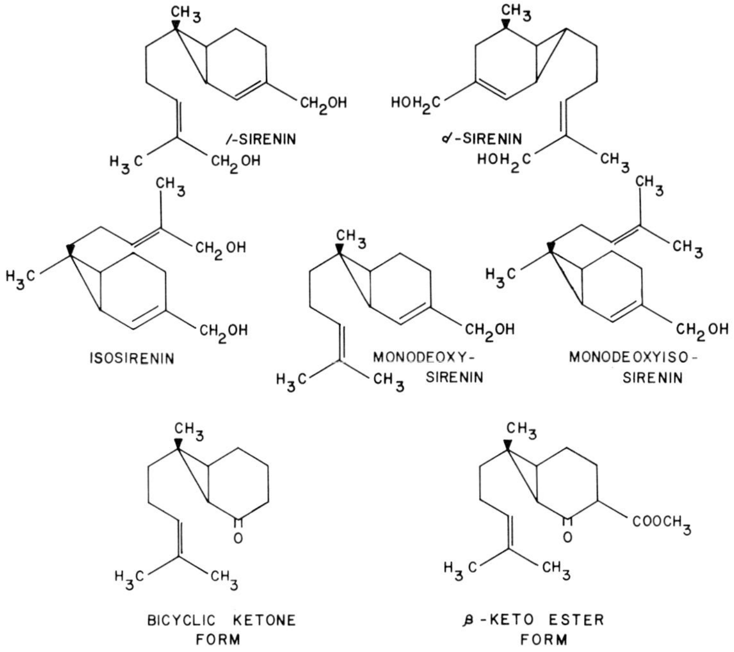

Fig. 2. The structure of sirenin and the analogs and isomers tested for attractive activity.

Carlile and Machlis (1965a) examined the attractive properties of the isolated sirenin. Replotting these data as a concentration–response curve (Fig. 3), the number of male gametes attracted increases with the size of the pheromone stimulus over a broad range. The threshold concentration is about 10^{-10} M while the peak concentration (saturation) for attraction is 10^{-6} M with half-saturation at 2×10^{-7} M. It is interesting that the peak concentration corresponds to an estimated secretion of the pheromone from the female gametangia of $1-2 \times 10^{-6}$ M (Carlile and Machlis, 1965a). In order to observe optimal chemotaxis 3×10^{-3} M Ca^{2+} and other chelated elements are required (Machlis, 1973a).

Exactly what happens to the pheromone once it is received by the male gamete has only been partially investigated. Only the male cells of *Allomyces* are attracted to sirenin (Carlile and Machlis, 1965a,b). Pheromone interaction with the male gametes involves first-order kinetics (Machlis, 1973b) and the concentration of sirenin in a suspension of male gametes declines the longer the male gametes are left in the solution (Carlile and Machlis, 1965a,b), suggesting that the male gametes in some way inactivate or metabolize the pheromone. The

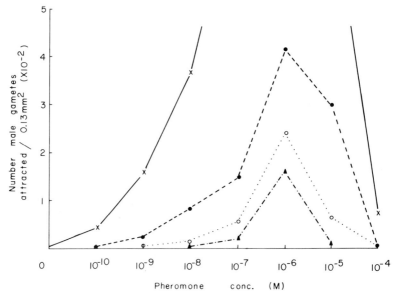

Fig. 3. Concentration–response curve for the attraction of male gametes to sirenin. The gametes counted represent the number settled on 0.13 mm² of membrane after a 1-hr incubation. ×, 10^6 cells; ●, 2×10^5 cells; ○, 1×10^5; ▲, 5×10^4 cells. (Adapted from Carlile and Machlis, 1965a, with permission from *Am. J. Bot.* and the senior author.)

other motile cells of *Allomyces* are not able to inactivate the attractant (Carlile and Machlis, 1965a,b). Sirenin cannot be extracted or recovered in a biologically active form from the male cells and the existence of an extracellular enzyme (sireninase?) secreted by the male gametes to inactivate the pheromone has not been found (Carlile and Machlis, 1965a; Machlis, 1973b). Interestingly, Klapper and Klapper (1977) have partially purified an inhibitor of male attraction and termed the low molecular weight substance keerosin.

Machlis (1958a, 1973b) has shown using the attractant bioassay that male gametes are attracted to the dialysis membrane for about 90 min, after which time they start to swim back into the chamber. This demonstrates that once the pheromone concentration equilibrates across the bioassay, attraction ceases, suggesting that attraction and the oriented movement of the male gametes are governed by a pheromone gradient and that without such a gradient male attraction cannot occur. If male gametes are placed in high concentrations of the pheromone, a refractory period ensues that lasts for as long as 45 min (Machlis, 1973b). At these high pheromone concentrations the male gametes cannot orient to a concentration gradient but do continue to inactivate the pheromone (Machlis, 1973b). Therefore, this refractory period, as pointed out by Bean (1979), is caused by a saturation of specific recognition or receptor sites on the male gamete's plasma membrane and not by the inability of the gamete to inactivate the pheromone.

III. CURRENT RESEARCH

This section discusses the problem of communication between the gametes of *Allomyces macrogynus*. Current work concerning the timing of the initial synthesis and secretion of the sirenin molecule will be presented as well as some evidence for elucidating the mechanism of pheromone reception and transduction at the cell surface of the male gamete.

A. Timing of Pheromone Secretion

One of the interesting problems not yet addressed in regard to cell communication in *Allomyces* is the time when the pheromone is first secreted. Machlis (1958a) described the attraction of the male gametes to unreleased female gametangia but no mention was made as to the developmental stage of these gametangia at the time of male attraction. This has been analyzed by incubating mature female gametangia with male gametes in the DS medium. Light microscope observations of these cells show that male attraction does not occur until 2–3 min before the female gametes are released from the gametangia. Therefore at the first sign of attraction, the female gametangia were removed from DS, placed in 1% glutaraldehyde, and prepared for transmission electron microscopy. Thin sections of these gametangia show the gametes to be fully differentiated, demonstrating that it is the female gametes themselves that produce the attractant prior to gamete release. Once released the female gametes continue to secrete sirenin for more than 6 hr (Pommerville, 1977). Since generation of the pheromone does not occur until the very end of gametogenesis, some late developmental event may be responsible for "activating" pheromone secretion. Unfortunately, a complete and detailed ultrastructural study of gametogenesis in *Allomyces* has not been done.

B. Sirenin Production and Macromolecular Synthesis

Experiments are being conducted to determine the sensitivity of sirenin secretion to RNA and protein synthesis during gametogenesis. When mature female gametangia are incubated with a suspension of male gametes in a solution of actinomycin D (AD), the completion of gametogenesis and the number of gametangia able to attract male gametes are reduced (Table I). If gametangia are induced in DS such that the first 20 min of gametogenesis occurs in the absence of AD, there is a dramatic increase in male attraction to these gametangia. Secretion of the pheromone is extremely sensitive to AD during the first 10 min of gametogenesis.

These experiments have been repeated with cycloheximide (CHX) and show that male attraction to female gametangia is reduced when mature female gametangia are incubated in CHX (Table I). A dramatic change in attraction occurs when the gametangia are induced in DS for at least 30 min prior to being put in CHX. These observations are in agreement with Fähnrich (1974) who has shown that protein synthesis in the gametangia of *A. arbuscula* ceases about 20–30 min after the induction of gametogenesis.

Although more work is required to fully understand the morphological and macromolecular events necessary for the initiation of sirenin secretion from the female gametes, the data presented here show that early events in gametogenesis are sensitive to RNA and protein synthesis inhibitors, resulting in a reduction of pheromone secretion. The enzymes of the sirenin biosynthetic pathway must be

TABLE I

The Effect of Actinomycin D (AD) and Cycloheximide (CHX) on the Process of Gametogenesis and Sirenin Secretion by the Female Gametangia of *A. macrogynus*

Time inhibitor added	Release (%)[a]	Female gametangia attracting male gametes (%)
Control[b]	90 ± 4[d]	94 ± 5
0 min[c] + AD	0	20 ± 11
10	0	13 ± 9
20	0	54 ± 5
30	3 ± 2	60 ± 5
40	34 ± 9	85 ± 8
50	50 ± 9	90 ± 6
0 min + CHX	0	0
10	0	0
20	0	0
30	6 ± 3	50 ± 10
40	52 ± 8	95 ± 5
50	83 ± 5	100

[a] Percentage of female gametangia releasing female gametes.
[b] Represents female gametangia incubated in DS for 60 min.
[c] Inhibitor was added by flooding the culture of *A. macrogynus* with 20 μg ml^{-1} AD or CHX at 0 min. For other times a culture was flooded at 0 min with DS and scraped with a microscope slide to dislodge gametangia (Pommerville, 1978a). At the appropriate times 100 gametangia were placed in the inhibitor. All observations were made with a dissecting microscope.
[d] ± = Standard error.

synthesized prior to or during early gametogenesis while actual secretion does not begin for at least another 30 min, at the end of gametogenesis.

C. Aspects of Pheromone Reception

Once secretion of the pheromone occurs and the female gametes are released from the gametangia, sirenin will diffuse through the liquid medium and make contact with the male gametes. Exactly how the male gametes recognize and bind the pheromone is not known. In an effort to better understand these events experiments are being carried out to block pheromone reception. Male gametes have been incubated in solutions of proteases (trypsin, Pronase), phospholipases (A_2, C, D), or substances collectively referred to as "membrane stabilizing agents" (hydrocortisone, diphenhydramine, chloroquine; Greenham and Poste, 1971) after which male gamete motility and attraction to female cells were observed by light microscopy. None of the enzymes or stabilizers tried affects pheromone reception or male attraction even though they do prevent, with the exception of the phospholipases, gamete fusion and fertilization (Pommerville, in preparation). These experiments show that either the pheromone receptors are insensitive to these chemicals, are inaccessible to enzymatic action (e.g., buried deep within the plasma membrane or are in the cytoplasm), or that receptors for the pheromone do not exist.

D. Signal Transduction and Cell Behavior

The final phase of cell communication deals with transduction phenomena whereby plasma membrane and intracellular events govern the changes in cell behavior. Before we can hope to understand these events at the physiological and molecular levels, we must understand exactly how the male gametes are attracted to the female cells. This has been accomplished recently by using dark field microscopy to analyze the motility paths of male gametes (Pommerville, 1978a). Male gamete motility in the absence of pheromone shows an organized swimming pattern. This pattern is characterized by short, smooth swimming paths (runs) that are interrupted by brief periods (less that 1 sec) when the gamete undergoes a behavior change involving a temporary cessation of motility and a turning of the cell body so that the next motility run is a new direction (Figs. 4 and 5). All the turns in a motility path are right- or left-handed in pitch, but not both (Figs. 4 and 5) and thus the cell body acts as a rudder to steer the cell into a new direction. Motility runs do not usually consist of straight paths but rather arcs that bend opposite to the turning direction causing the swimming pattern to take on a loop or spiral form.

When female gametes are added to a population of male cells and photographed with dark field optics, changes in the male motility runs are evident. These changes result in attraction of the male gametes to the pheromone source—the female gametes (Figs. 6 and 7). Although a spiral pattern is retained in the motility runs, fewer directional changes occur and the length of the runs both in time and distance traveled toward the female gametes is much longer than the runs where no female cells are present (compare Figs. 4 and 6; Table II). Since there is a natural bend to the motility runs, eventually a male gamete will begin moving away from the female cell (down pheromone gradient). At this time the male cell will stop swimming, undergo one or two changes in direction that reorient the cell into the gradient (Fig. 6), and swim again on another long, smooth run toward the female cell.

3. Pheromones in *Allomyces* 63

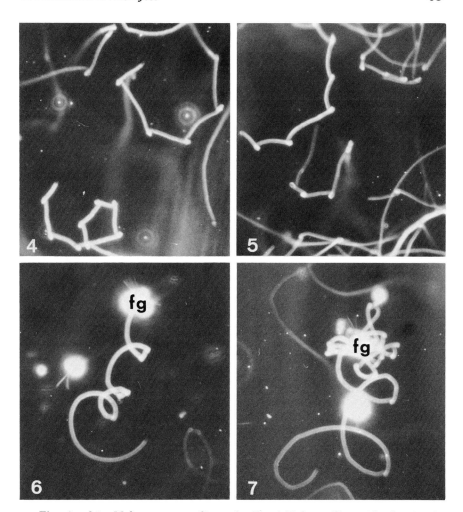

Figs. 4 and 5. Male gamete motility paths. Fig. 4: Male motility tracks showing the smooth nature of the short runs and the interruptions of these runs by directional changes. ×200. Fig. 5: Male motility runs showing bends or arcs in the swimming path. ×210. (From Pommerville, 1978a, with permission from Academic Press.)

Figs. 6 and 7. Male attraction to female gametes. The spiral motility pattern of the male gametes can be seen as they approach the female gametes (fg). See text for full explanation. Both ×210. (From Pommerville, 1978a, with permission from Academic Press.)

TABLE II

Response Times and Length of Motility Runs in the Absence and Presence of Sirenin

Treatment	Mean length of run (μm)	Mean run time (sec)
No pheromone	47.6 ± 7.8[a]	0.9 ± 0.1[a]
Addition of pheromone	219.4 ± 21.6	2.1 ± 0.4

[a] Means and standard errors ($\pm$) based on the measurement of at least five motility runs from ten separate male gamete motility paths.

The pheromone has directly affected male swimming by causing an inhibition or reduction in directional changes when approaching female gametes.

Direct contact with a female gamete can be made by a male from a long run (Fig. 6). More often these long runs approach close to but pass by the female cell. At this close range the male gamete undergoes a number of directional changes and through a random hit process, contact is made (Fig. 7). The pheromone effectively acts to increase the chances that a hit will result in contact with a female gamete.

Experiments involving the rapid mixing of male gametes with only the pheromone have added to the understanding of male atraction (Pommerville, 1978a). When the pheromone is added, male gametes immediately change their swimming response to one where little net movement occurs (Fig. 8). Within 15–30 sec after addition of the pheromone, male gametes start moving in large circular patterns (Fig. 9), similar to that described when male gametes are approaching female gametes from a distance (Fig. 6). This situation leads to long, smooth swimming patterns with few directional changes. Full circles would be seen in the male attraction photographs (Figs. 6 and 7) if the sirenin gradient were absent. Within 60–120 sec after pheromone addition, male gametes begin to return to their nonstimulated motility pattern (Fig. 10). As the pheromone is inactivated normal motility returns and directional changes reappear.

The data presented demonstrate how the male gametes respond to and monitor the pheromone gradient. Once attraction begins directional changes are inhibited and long, smooth runs occur due to the continued increase in the pheromone gradient (Fig. 11). While the cells are moving up the gradient smooth swimming will occur, but as the gamete swims parallel to and then down the sirenin gradient into lower concentrations of the pheromone normal turns and directional

3. Pheromones in *Allomyces*

changes appear. These continue until the cell is again moving up the gradient (Fig. 11). When the male gamete approaches very close to the female cell (which represents the point where the pheromone concentration gradient is greatest) the male gamete does not have to move a great distance in order to be moving down the gradient and therefore directional changes will occur. This is why there are many directional changes when the male gamete is very near the female cell. The runs are shorter and the number of directional changes have increased. The

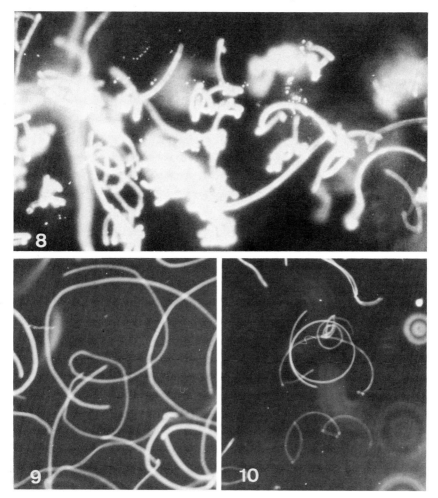

Figs. 8–10. Male gamete responses with time to rapid mixing with sirenin. Note the gradual reappearance of normal swimming (Figs. 9 and 10). Fig. 8: ×235; Fig. 9: ×210; Fig. 10: ×135. (From Pommerville, 1978a, with permission from Academic Press.)

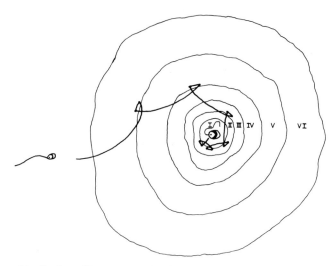

Fig. 11. Idealized motility pattern of male gamete attraction. Although a pheromone gradient would be continuous, in this figure circles have been drawn to identify areas of the gradient (I through VI). See text for details.

important point is that no matter where the gamete is in the pheromone gradient, the responses are the same, and a reduction in the concentration of pheromone will cause directional changes and reorientation of the male cell. The motility patterns show that the male gamete does not change its rate of locomotion during attraction (orthokinesis). Rather sirenin influences the frequency of directional changes, so that the oriented responses of the male gamete are more accurately defined by the term klinokinesis than the general term chemotaxis.

The male gametes of *A. macrogynus* could monitor the pheromone gradient by either a spatial or temporal sensing system. In a spatial system, receptors at the head and tail of the cell would simultaneously compare the pheromone concentration and determine whether the cell was moving up or down the gradient. In a temporal sensing system, the gamete would compare sirenin concentrations at various times to determine the direction the cell is traveling. The observations described here argue for a temporal sensing system. If the male gametes used a spatial sensing system they should swim normally if placed in a uniform concentration of sirenin because the head and tail receptors would not register any change in pheromone concentration. But in such experiments there is an altered motility pattern when pheromone is added (Figs. 8–10) and this pattern is not identical to the random pattern (Figs. 4 and 5). Male gametes subjected to an increase in pheromone concentration swim in longer runs with fewer directional changes for an extended time but gradually return to the normal swimming pattern described for nonstimulated male gametes. Therefore, the male cells must make temporal comparisons of the pheromone concentration at various times as they travel through space with the result that if there is an increase in pheromone concentration the directional changes will decrease and the gametes will move toward higher concentrations of pheromone. The interval between comparisons cannot be accurately established from the available data. From the dark field micrographs it is obvious that almost immediately on moving down a gradient, directional changes are initiated. Therefore, the temporal sensing interval must be fairly short since longer intervals would allow enough time for the gamete to move farther down the gradient.

A sensory signal must be induced and analyzed by the male gamete during the transduction phase. In *Allomyces,* transduction events are being studied by attempting to interfere with the normal functioning of the plasma membrane of the male gametes. Male cells have been incubated in tertiary amine local anesthetics (procaine and tetracaine) prior to the addition of female gametes. Dark field photomicrographs show that anesthetic-treated male gametes have lost their organized swimming pattern (Fig. 12). The circular patterns seen are very reminiscent of those that occur during rapid mixing with pheromone (Fig. 9). When female gametes are added, the male cells are unable to orient to the pheromone gradient or to the female gametes (Fig. 13) and they often pass by the female cells without any change in behavior. If the gametes are washed in DS, to remove the anesthetic, attraction is restored. The two anesthetics are known to displace Ca^{2+} from membranes (Low *et al.*, 1979) and since Machlis (1973a) has shown that Ca^{2+} is required for attraction, these results provide evidence for the importance of metal cations to the transduction process. The anesthetics also can interfere with membrane potentials (Seeman, 1974) so the possibility exists that changes in the electrical nature of the plasma membrane are important to effective attraction.

Experiments are being carried out to examine the role of macromolecular structures in the orientation and movement of the male gametes to the female cells. Observations of male gamete motility in the presence of the microtubule inhibitor colchicine show no change from attraction in the absence of the drug. It is known that the nine sets of triplet cytoplasmic microtubules in the motile cells of *Allomyces* are resistant to the action of colchicine (Olson, 1972). Dark field

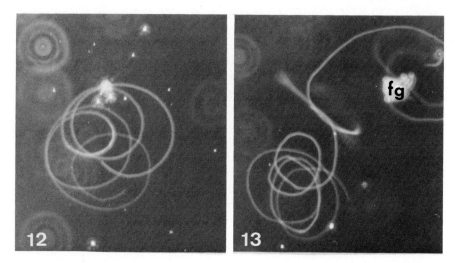

Figs. 12 and 13. Male motility in the presence of procaine ($10^{-2}M$). Fig. 12: The organized male motility run has been lost and the cells swim in circles. ×200. Fig. 13: Male attraction to the female gamete (fg) does not occur and the male gamete continues to swim in a circular pattern. ×200.

analysis of attraction in the presence of the microfilament inhibitor cytochalasin B also does not cause any change in turning or attraction. Filaments measuring 6–7 nm have been found in the male and female gametes of *A. macrogynus,* and cytochalasin B will block fertilization and gamete fusion by specifically interacting with the female gametes (Pommerville, in preparation). Either these filaments in the male gamete are not affected by the drug or they play no role in the attraction process.

E. Conclusion

By way of conclusion I would like to discuss the importance of the pheromone to *Allomyces*. Obviously, from what has been discussed and reported here, the pheromone is very important for fertilization and therefore genetic recombination. Without the attractant efficient fertilization would not occur (fertilization efficiency in *A. macrogynus* is essentially 100%) as demonstrated by the inability of the male gametes to orient to female cells in the presence of the local anesthetics. The male gametes of *A. macrogynus,* by constantly changing direction, increase their chances of finding a pheromone gradient and female gametes. Furthermore, the actual female gametes play an important role in male orientation and attraction. The female gametes are sluggish swimmers and do not move over as large an area as the male gametes (Pommerville, 1978a). By remaining relatively immobile, they act essentially as point sources of pheromone secretion and the male gametes will have an increased chance of finding and mating with them. However, if all the female cells spread out so as to present many single point sources of pheromone secretion, male gametes would find it more difficult to orient to any one cell (see Pommerville, 1978a, Fig. 5). However, the female gametes tend to remain in various size clusters and do not move about or spread for a prolonged time. By remaining in clusters male attraction is optimized and the ensuing events of fertilization and zygote development are promoted.

IV. PERSPECTIVES

This chapter has described the current state of our understanding of chemical cell communication between gametes of the fungus *Allomyces*. Discovery and characterization of the pheromone and several properties of sirenin were discussed. Previously unpublished information on the timing of sirenin secretion and its sensitivity to RNA and protein synthesis were presented. The pheromone is generated and secreted only by the female gametes even when they have not yet been released from the gametangia. The major focus of the chapter was a description of male gamete attraction to the female cells, including how they orient themselves to the pheromone gradient. It was proposed that the male gametes use a temporal sensing system to determine their position in the gradient. Some preliminary experiments described possible transduction events required for processing of the pheromone signal.

The information gathered on cell communication and the chemosensory processes in *Allomyces* presents several areas for future research. It remains to be discovered where and how the male gametes inactivate the sirenin molecule. It has been shown by an agar well bioassay (Pommerville, 1977, 1978c) that female gametes can be attracted to male cells. At the time this information was published it was suggested that the male gametes generate and secrete a pheromone that is responsible for female attraction. Although this proposal cannot be ruled out, it is also possible that female attraction is caused by the inactivation product of sirenin. After male gametes inactivate the pheromone, it could diffuse back through the bioassay and cause attraction of the female cells. If this is so, it represents a bioassay for the sirenin inactivation product. Once the inactivation product has been isolated it should be assayed by attempting to observe female attraction. The ability of keerosin to attract female gametes also should be tried.

The location where the pheromone is inactivated could be determined by using radioactive pheromone in conjunction with radioautography or scintillation counting. These experiments could determine if the attractant is taken up into the male cells. If the pheromone does not get into the male cytoplasm inactivation must occur at the plasma membrane. Such results would require isolation and characterization of the male gamete plasma membrane and experiments designed to determine the ability of membrane preparations to inactivate the pheromone. Equilibrium and kinetic methods should be employed to measure binding to possible pheromone receptors and if specific binding sites exist, they may be isolated by gel or affinity chromatography. The localization of receptors using labeled pheromone may prove difficult since the dissociation rate of a pheromone–receptor interaction may be so high that functional localization would not be possible. From the discussion of the temporal sensing system it should be clear that receptors need not be localized at the cell surface of male gametes.

In terms of sensory transduction, many experiments need to be done. The role of the membrane potential to klinokinesis and the interaction with Ca^{2+} require more study. Intracellular events responsible for the inhibition of directional changes have not been studied at all and interactions between flagellar motility and attraction should be examined.

Results obtained from current future experiments should go a long way toward better understanding the mechanism of cell communication in *Allomyces*. These results would be important not only to the *Allomyces* system and to other fungi, but to the general areas of plant and animal cell communication. Often it has been the simpler organisms, as the fungi and algae, that have provided the physiological and biochemical knowledge with which to understand the higher, more complex forms of behavior. The work with *Allomyces* has the potential for contributing to this understanding.

REFERENCES

Bean, B. (1979). Chemotaxis in unicellular eukaryotes. *In* "Encyclopedia of Plant Physiology" (W. Haupt and M. E. Feinleib, eds.), New Series, Vol. 7, pp. 335–354. Springer-Verlag, Berlin and New York.

Bhalerao, U. T., Plattner, J. J., and Rapoport, H. (1970). Synthesis of *dl*-sirenin and *dl*-isosirenin. *J. Am. Chem. Soc.* **92**, 3429–3433.

Carlile, M. J., and Machlis, L. (1965a). The response of male gametes of *Allomyces* to the sexual hormone sirenin. *Am. J. Bot.* **52**, 478–483.

3. Pheromones in *Allomyces*

Carlile, M. J., and Machlis, L. (1965b). A comparative study of the chemotaxis of the motile phases of *Allomyces*. *Am. J. Bot.* 52, 484–486.

Emerson, R. (1941). An experimental study of the life cycles and taxonomy of *Allomyces*. *Lloydia* 4, 77–144.

Emerson, R. (1954). The biology of water molds. *In* "Aspects of Synthesis and Order in Growth" (D. Rudnick, ed.), pp. 171–208. Princeton Univ. Press, Princeton, New Jersey.

Emerson, R., and Fox, D. L. (1940). Carotene in the sexual phase of the aquatic fungus *Allomyces*. *Proc. R. Soc. London, Ser. B* 128, 275–293.

Emerson, R., and Wilson, C. M. (1954). Interspecific hybrids and the cytogenetics and cytotaxonomy of Euallomyces. *Mycologia* 46, 393–434.

Fähnrich, P. (1974). Untersuchungen zur Entwicklung des Phycomyceten *Allomyces arbuscula*. II. Einfluss von Inhibitoren der Protein- und Nucleinsäuresynthese auf die Gametogenese. *Arch. Microbiol.* 99, 147–153.

Greenham, L. W., and Poste, G. (1971). The role of lysosomes in virus-induced cell fusion. 1. Cytochemical studies. *Microbios* 3, 97–104.

Jaenicke, L. (1972). "Sexuallockstoff im Pflanzenreich," Rep. No. 217. Rheinisch-Westfälischen Akad. Wiss. Westdeutscher Verlag, Opladen.

Karlson, P., and Luscher, M. (1959). "Pheromones": A new term for a class of biologically active substances. *Nature (London)* 183, 55–56.

Klapper, B. F., and Klapper, M. H. (1977). A natural inhibitor of sexual attraction in the water mold *Allomyces*. *Exp. Mycol.* 1, 352–355.

Kochert, G. (1978). Sexual pheromones in algae and fungi. *Annu. Rev. Plant Physiol.* 29, 461–486.

Koshland, D. E., Jr. (1974). Chemotaxis as a model for sensory sytems. *FEBS Lett.* 40, S3–S9.

Koshland, D. E., Jr. (1977). A response regulator model in a simple sensory system. *Science* 196, 1055–1063.

Loomis, W. F. (1979). Biochemistry of aggregation in *Dictyostelium*. *Dev. Biol.* 70, 1–12.

Low, P. S., Lloyd, D. H., Stein, T. M., and Rogers, J. A. (1979). Calcium displacement by local anesthetics. Dependence on pH and anesthetic charge. *J. Biol. Chem.* 254, 4119–4125.

Machlis, L. (1958a). Evidence for a sexual hormone in *Allomyces*. *Physiol. Plant.* 11, 181–192.

Machlis, L. (1958b). A study of sirenin, the chemotactic sexual hormone from the watermold *Allomyces*. *Physiol. Plant.* 11, 845–854.

Machlis, L. (1973a). Factors affecting the stability and accuracy of the bioassay for the sperm attractant sirenin. *Plant Physiol.* 52, 524–526.

Machlis, L. (1973b). The chemotactic activity of various sirenins and analogues and the uptake of sirenin by the sperm of *Allomyces*. *Plant Physiol.* 52, 527–530.

Machlis, L., Nutting, W. H., William, M. W., and Rapoport, H. (1966). Production, isolation, and characterization of sirenin. *Biochemistry* 5, 2147–2152.

Machlis, L., Nutting, W. H., and Rapoport, H. (1968). The structure of sirenin. *J. Am. Chem. Soc.* 90, 1674–1676.

Macnab, R. M. (1978). Bacterial motility and chemotaxis: The molecular biology of a behavioral system. *CRC Crit. Rev. Biochem.* 5, 291–341.

Morrison, P. J. (1977). Gametangial development in *Allomyces macrogynus*. I. The ultrastructure of early stages of development. *Arch. Microbiol.* 113, 163–172.

Nutting, W. H., Rapoport, H., and Machlis, L. (1968). The structure of sirenin. *J. Am. Chem. Soc.* 90, 6434–6438.

Olson, L. W. (1972). Colchicine and the mitotic spindle of the aquatic phycomycete *Allomyces*. *Arch. Mikrobiol.* 84, 327–338.

Plattner, J. J., and Rapoport, H. (1971). The synthesis of d- and l-sirenin and their absolute configurations. *J. Am. Chem. Soc.* 93, 1758–1761.

Plattner, J. J., Bhalerao, U. T., and Rapoport, H. (1969). Synthesis of dl-sirenin. *J. Am. Chem. Soc.* 91, 4933.

Pommerville, J. (1977). Chemotaxis of *Allomyces* gametes. *Exp. Cell Res.* 109, 43–51.

Pommerville, J. (1978a). Analysis of gamete and zygote motility in *Allomyces*. *Exp. Cell Res.* 113, 161–172.

Pommerville, J. (1978b). Gametes and fertilization. *In* "Lower Fungi in the Laboratory" (M. S. Fuller, ed.), pp. 47–48. Dep. Bot., Univ. of Georgia, Athens.

Pommerville, J. (1978c). Chemotaxis of motile spores (*Allomyces, Phytophthora*, and *Pythium* spp.). *In* "Lower Fungi in the Laboratory" (M. S. Fuller, ed.), pp. 185–186. Dep. Bot., Univ. of Georgia, Athens.

Pommerville, J., and Fuller, M. S. (1976). The cytology of the gametes and fertilization of *Allomyces macrogynus*. *Arch. Microbiol.* 109, 21–30.

Rønne, M., and Olson, L. W. (1976). Isolation of male strains of the aquatic phycomycete *Allomyces macrogynus*. *Hereditas* 83, 191–202.

Seeman, P. (1974). The membrane expansion theory of anesthesia: Direct evidence using ethanol and a high-precision density meter. *Experientia* 30, 759–760.

Turian, G. (1969). "Différenciation fongique." Masson, Paris.

4

Sexual Pheromones in *Volvox* Development

GARY KOCHERT

I.	Introduction and Survey of the Literature	73
II.	Current Research	81
	A. Characterization of the Sexual Pheromones of *Volvox*	81
	B. Mode of Action of *Volvox* Pheromones	84
III.	Perspectives	88
	References	92

I. INTRODUCTION AND SURVEY OF THE LITERATURE

The several species of the green alga *Volvox* present a fascinating variety of sexual interactions in their relatively simple life cycles (Starr, 1970; Kochert, 1975). The genus is cosmopolitan in distribution and is a common resident of temperate freshwater habitats. The basic structure of the organism is that of a spheroidal aggregate of 0.5 to 2.0 mm in diameter with the cells embedded in a transparent gelatinous matrix (Fig. 1). The number of cells per organism varies with the species from about 1000 to more than 10,000, and each organism contains both somatic and reproductive cells. Somatic cells each have two flagella which project through the matrix into the surrounding medium and provide the motive force for the active swimming exhibited by the organism. Reproductive cells function in either sexual or asexual reproduction as outlined below.

Although each individual of *Volvox* is called a "colony" in the traditional literature, it is clear that this term is inappropriate (Kochert,

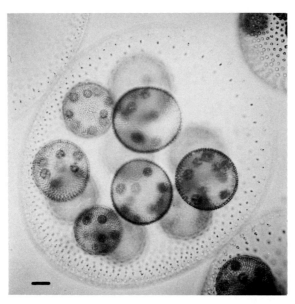

Fig. 1. An asexual spheroid of *Volvox carteri* f. *nagariensis*. This individual contains 12 nearly mature juvenile spheroids, each of which contains 8–10 gonidia. The anterior end of the organism is toward the upper right. Scale, 50 μm.

1975). Each individual is an organism with differentiated cell types and a definite polarity which is expressed in several ways. Each organism exhibits definite anterior and posterior poles. When swimming the anterior pole is directed forward and the organism rotates in one direction about its longitudinal axis in a most stately fashion. No dorsoventral axis is present and each organism is radially symmetrical about the longitudinal axis. This radial symmetry extends even to the detailed placement of the somatic cells in the matrix material. Each somatic cell is arranged so that the plane made by its flagellar bases would be bisected by a meridian line from the anterior to the posterior pole. Further, each cell is so placed so that the chloroplast eyespot is found to be on the side of the cell furthest from the anterior pole of the organism (Fig. 2). This exacting cell placement is necessary for the translation of the independent beating of each individual somatic cell's flagella into the regular motion exhibited by the whole organism. No communication between the somatic cells is necessary to ensure a seemingly coordinated swimming pattern in this type of system. Indeed *Volvox* has been proposed to fit a "Roman galley" model where the seemingly

4. Pheromones in *Volvox*   75

coordinated motion of a ship results not from communication between the rowers, but from individual responses of rowers to a common stimulus (Kochert, 1973). Direct evidence for this model is also found in mutants isolated by R. J. Huskey (personal communication). In these mutants the pattern of cell placement has been perturbed, and mutant organisms cannot swim even though they have full power of flagellar movement. Polarity is also exhibited in the fact that reproductive cells are commonly localized in the posterior portion of the organism.

Reproductive cells are of three types in *Volvox*. There are asexual reproductive cells (gonidia), which can form the next generation of organisms by a process of division and cell differentiation; androgonidia, which form bundles of sperm; and eggs, which can be fertilized and will form a resistant zygospore. In the genus *Volvox* these few cell types form the theme on which an amazing set of variations in sexual patterns and interactions has evolved.

Each species of *Volvox* is generally assumed to be haploid during the entirety of its life cycle with the exception of the zygospore. This is the sole diploid cell in the life cycle and it does not propagate itself by mitosis but undergoes meiosis after a period of maturation. Three products of meiosis abort (Starr, 1975) so that each fertilization results in but one recombinant product. All species of *Volvox* also exhibit asexual reproduction by gonidial division and a generalized *Volvox* life cycle is presented in Fig. 3. It should be realized, however, that varia-

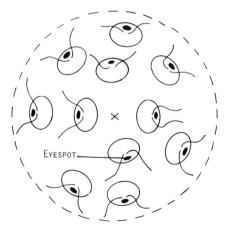

Fig. 2. A diagrammatic view of a portion of the anterior pole of a *Volvox* spheroid to illustrate placement of flagella and eyespots. X, anterior pole.

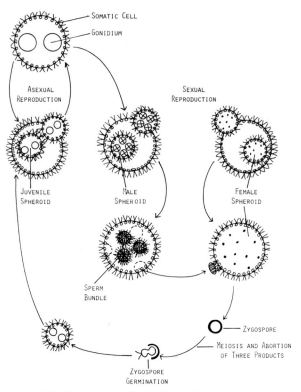

Fig. 3. A generalized *Volvox* life cycle.

tion in the details of asexual and sexual reproduction make it impossible to present a single detailed life cycle representative of the genus *Volvox* as a whole.

Asexual reproduction is accomplished by division and differentiation of the gonidia to form juvenile spheroids. These are formed inside the parent and released when mature. Juvenile spheroid formation has been examined thoroughly in some species at the light microscope level (Janet, 1923; Kochert, 1968; Starr, 1969), and is now being investigated in detail in *V. carteri* with transmission and scanning electron microscopy (D. L. Kirk, personal communication). The process has common features in all species in that a number of divisions occur resulting in a single layer of cells arranged in a spherical, hollow aggregate. The division process is such that cells are formed with their flagellar ends directed toward the inside of the juvenile spheroid. An inversion process then follows in which the developing juvenile spheroid turns itself inside out by a series of shape changes elegantly detailed by Kirk and associates (Viamontes and Kirk, 1977; Viamontes et al., 1979). No more cells are formed after inversion, but formation of extracellular matrix between the cells, which begins at this stage, causes the organism to become larger since the cells are forced apart as a result of matrix deposition. It was the clear differentiation of somatic and reproductive cell types which occurs during this process that interested early developmental biologists in the potential of *Volvox* as a model system.

Two basic types of gonidial division are present in the genus. In one of these, gonidia begin to divide when they are quite small. Mitosis and cytokinesis occur along with growth, and the developing juvenile spheroid gradually grows in size. It is typical of this process that visual differentiation of reproductive from somatic cells occurs late in juvenile spheroid formation, and one cannot distinguish the somatic cells from the reproductive cells until after gonidial division and inversion and shortly before daughter spheroid release. In the other pattern of asexual reproduction, gonidia go through a period of enlargement and the onset of divisions is delayed until the gonidia are quite large. A true cleavage then ensues with the cleavage products becoming smaller at each stage. In some of the species exhibiting this pattern the differentiation between gonidia and somatic cells occurs very early in the division process. In *V. carteri*, for example, gonidia are differentiated from somatic cells by unequal cleavages at the 16- or 32-celled stage of development. The two cell types are thus clearly distinguishable through subsequent cleavage. inversion, and expansion of the juvenile spheroid. This early differentiation into "soma" and "germ" lines was commented on by Weismann (1889) long before it was possible to grow and experimentally manipulate the organisms in laboratory culture.

Several patterns of sexual reproduction are exhibited by the various species of the genus. Some species are monoclonic, that is, sexual reproduction occurs within a single clone; in the others, two clones must be mixed for sexual reproduction to occur (diclonic). All species form morphologically differentiated sperm, but egg formation varies among the species. In some species no visibly differentiated eggs or female spheroids are formed and gonidia can be fertilized to form zygospores; if not fertilized they divide to form juvenile spheroids in the usual pattern. In other species eggs are formed which are morphologically different from the gonidia. In some cases they will die if not fertilized; in other cases unfertilized eggs will divide to form juvenile spheroids (i.e., they will function as gonidia).

Further classification of species and strains can be made on the basis of whether sperm and eggs are formed in separate spheroids (dioecious) or the same spheroid (monoecious) (for review, see Kochert, 1975). All diclonic species thus far described are dioecious and produce male and female spheroids in separate clones. Monoclonic species may be either monoecious, dioecious, produce a mixture of single-sexed and mixed spheroids, or produce male spheroids and no morphologically distinguishable females. In the latter case gonidia apparently function as eggs as outlined above. Figure 4 diagrams some of these patterns of sexual reproduction.

In many species sexual pheromones have been shown to be involved in the coordination of sexual reproduction. The first of these was discovered and described by Darden (1966) who showed that in *V. aureus* (a monoclonic, monoecious species without morphologically distinct females) male spheroids release a pheromone into the culture medium which induces the production of more males when added to an asexually growing culture. In this case the pheromone is inducing gonidia to form male rather than asexual juvenile spheroids. Sexual pheromones were next discovered in *V. carteri,* a diclonic, dioecious species (Kochert, 1968; Starr, 1969). In this system male spheroids release a pheromone which both induces the formation of males in the male clone and females in the female clone. As in *V. aureus* the action of the pheromone is to induce gonidia (which would, in the absence of pheromone, cleave to form asexual spheroids) to cleave to form sexual spheroids. Subsequently sexual pheromones have been discovered to be operative in several other species of *Volvox* (Starr, 1970; Kochert, 1975) including at least one monoclonic, monoecious species (*V. capensis*) (R. C. Starr, personal communication). In the latter case the

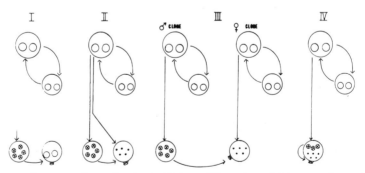

Fig. 4. Patterns of reproduction in *Volvox*. All species exhibit asexual reproduction as indicated by the cycle in the upper part of the diagram for each type. Reproductive cell types: ○, gonidium; ⊗, sperm bundle; ●, egg. I, Monoclonic, dioecious (no specialized females); II, monoclonic, dioecious; III, diclonic, dioecious; IV, monoclonic, monoecious.

pheromone induces the formation of spheroids containing both sperm and eggs.

Each of the sexual induction systems described above seems to fit into one of two general types (McCracken and Starr, 1970). In one of these, characterized by *V. carteri*, the pheromone acts on a gonidium and in some fashion alters its cleavage pattern so that sexual rather than asexual reproductive cells are differentiated, and the resultant juvenile spheroid is male or female rather than asexual. *Volvox carteri*, it will be recalled, shows early differentiation of somatic and reproductive cells during the rapid cleavage of a large gonidium. To be effective in this species the sexual pheromone must be present when the gonidium is small and be present during the enlargement of the gonidium prior to cleavage (Kochert, 1968). If the pheromone is added to mature gonidia or to gonidia which have begun cleavage, no effect is seen. In this type of induction, then, the pheromone acts on a gonidium to alter the cleavage pattern and the type of reproductive cell produced.

The other type of induction system is typified by *V. rousseletii* (diclonic, dioecious). In this case the pheromone appears to act directly on the young uncommitted reproductive cells and determines whether they will become gonidia, somatic cells, eggs, or androgonidia. *Volvox rousseletii* is one of the species in which visible differentiation of reproductive from somatic cells does not occur until after inversion. Juvenile spheroids are released from the parent with three cell types evident: several thousand somatic cells, 5-20 large cells, and 100-300 medium-sized cells. If no pheromone is present the medium-sized cells become somatic cells and are soon indistinguishable from the other somatic cells. The large cells function as gonidia and divide to form juvenile spheroids. If the pheromone is present, however, the middle-sized and large cells have a different fate. In spheroids from male clones both cell types will cleave to form bundles of sperm. In female clones both cell types will differentiate without cleavage into eggs, which will die if not fertilized. In this induction pattern juvenile spheroids remain uncommitted to the asexual or the sexual pathway until very late and in some cases the juvenile spheroid is uncommitted until after release from the parent.

Sexual pheromones have been purified and partially characterized in some species. In all cases the pheromones appear to be species-specific and interspecies induction does not occur (Starr, 1970). The early work in *V. aureus* and *V. carteri* indicated that the pheromones were nondialyzable, and susceptible to loss of biological activity when treated with certain proteolytic enzymes (Darden, 1966; Kochert, 1968; Starr, 1969). Subsequently the pheromones were independently purified from two isolates of *V. carteri:* one from Japan (*V. carteri* f. *nagarenisis*), and one from the United States (*V. carteri* f. *weismannia*). Both of these were reported to be glycoproteins of about 30,000 molecular weight and they do not cross-induce (Kochert and Yates, 1974; Starr and Jaenicke, 1974). Pheromones in all other systems so far investigated also appear on preliminary evidence to be proteins or glycoproteins (Kochert, 1975). The one exception is *V. capensis* where from preliminary work it appears that the formation of the monoecious spheroids typical of this species can be induced by glutamic acid (R. C. Starr, personal communication). This is an interesting finding and has general implications in terms of the mechanism of action of the *Volvox* pheromones. A list of *Volvox* isolates known to produce sexual pheromones is included as Table I.

Stocks of nearly all *Volvox* species are maintained in the Culture Collection of Algae at the University of Texas. A complete catalogue has been published and cultures of these organisms are obtainable for a small fee (Starr, 1978). Stock *Volvox* cultures and cultures used for pheromone production and experimentation are maintained axenically in *Volvox* medium buffered at pH 8.0 rather than 7.0 as originally described (Provasoli and Pintner, 1959). Stocks can also be maintained axenically or bacterized in soil–water medium (Starr, 1978) and will require less frequent transfer in this medium. Culturing conditions for pheromone production have been published (see references cited in Table I) and will vary somewhat with each species or isolate. Pheromone bioassay can be conveniently accomplished in culture tubes containing 9 ml of *Volvox* medium. The bioassay must be per-

TABLE I

Sexual Pheromones in *Volvox*

Organism	Chemical nature	Reference
V. carteri f. *nagarensis*	Glycoprotein, MW 30,000, pI 10.5	Starr and Jaenicke (1974)
V. carteri f. *weismannia*	Glycoprotein, MW 32,000, pI 10.4	Kochert and Yates (1974)
V. aureus	Glycoprotein?	Darden (1966)
V. rousseletii	Protein? 10,000 MW?	McCracken and Starr (1970)
V. gigas	Protein? 20,000 MW?, autoclavable	Vande Berg and Starr (1971)
V. obversus	?	Karn and Starr (1974)
V. dissipatrix	?	Starr (1970)
V. capensis	Glutamic acid?	R. C. Starr (personal communication)

formed under axenic conditions. One ml of the material to be assayed is added to the first tube, the contents are filter sterilized if necessary, and 1 ml is axenically transferred to another tube to begin a series of 1-ml serial dilutions. Asexual spheroids are then added to each tube, and after a suitable period of time, counts are made and the results expressed as percent sexual colonies.

Under ideal conditions the tubes containing the higher concentrations of pheromone (i.e., where pheromone is not limiting) will give 95–100% sexual spheroids. When the limiting concentration of pheromone is reached, one tube will give less than 100% and the rest of the tubes will show no sexual spheroids. Results can be expressed in terms of the limiting dilution where induction was seen. Several conditions must be observed to obtain reproducible results. These include the following: (1) The organisms selected as inoculum must be from uncrowded, rapidly growing cultures and must all be at the same developmental stage, preferably from the same synchronized culture. The developmental stage best used as inoculum will vary with the species or isolate and the original references must be consulted. (2) Light and temperature levels should be constant and should be those which give rapid growth. (3) Many components of common buffers and many reagents will kill the inoculum organisms or inhibit their growth. These include Tris buffer, phosphate, sucrose, and salts of various sorts. Since the organism responds to very small pheromone concentrations, one can usually get useful results even if the first few tubes of a serial dilution are ruined by such interfering substances. Light of about 2000 lx from cool-white fluorescent tubes is sufficient to maintain stock cultures of *Volvox*. A light-dark cycle of about 16:8 is necessary for long-term maintenance of stocks. For pheromone production and bioassay, light at the highest levels compatible with healthy growth should be used. "Power Groove" fluorescent bulbs (General Electric) have been especially useful to generate higher light intensities. Maximum temperatures tolerated will again vary with the *Volvox* isolate, but will usually be in the range of 25°–30°C.

II. CURRENT RESEARCH

A. Characterization of the Sexual Pheromones of *Volvox*

Present work in my laboratory is predominantly in two areas: characterization of sexual pheromones and investigation of their mode of action. The *Volvox* sexual pheromones are interesting molecules. They are produced in fairly large amounts, and with one exception they are produced only in male cultures. Only two have been purified to near homogeneity and these are both from isolates of *V. carteri* as outlined above. The general properties of the pheromones from both these sources seem to be similar. Both were reported to be glycoproteins of about 30,000 molecular weight with a pI in the range of 10.2–10.4. Both these pheromones seem to be extremely stable molecules. They accumulate in concentrations in the range of 10^6 times above the threshold level needed for biological activity. We have performed experiments in which we added an aliquot of purified pheromone to a young culture of nonproducer targer organisms, took aliquots of the medium periodically for more than a month until the culture was senescent, and bioassayed the aliquots (Hagen and Kochert, unpublished observations). Essentially full biological activity was retained and there was no evidence that the organisms degraded or metabolized the pheromone. The *V. carteri* pheromones appear to be stable for years at room temperature when maintained axenically. They are also very resistant to repeated cycles of freezing and thawing, heating, and pH changes from pH 3.0 to pH 12.0. After 2 min of boiling in sodium dodecyl sulfate (SDS) and electrophoresis on SDS gels, considerable biological activity is retained (Kochert and Yates, 1974).

It was interesting to us that, although the two pheromones appear to be generally similar, they showed no cross-induction. We have thus initiated studies to compare the pheromones produced by the Japanese and American isolates of *V. carteri*. We have coelectrophoresced the two pheromones and we find they form separate bands on SDS gels with the pheromone from the American isolate being slightly larger in apparent molecular weight (Fig. 5). It is also of interest that the pheromone from the American isolate reveals a minor band of slightly smaller molecular weight on SDS gels. We believe this to be a slight modification of the main band on the basis of preliminary peptide mapping results, but more data are needed.

The pheromones also appear to differ somewhat in carbohydrate content. The pheromone from the Japanese isolate was reported to have a carbohydrate content of 40% (Starr and Jaenicke, 1974) while the American counterpart was only about 20% carbohydrate (Kochert and Yates, 1974). The pheromone from the Japanese isolate was reported to have the following carbohydrate composition: xylose (25.5%), mannose (15.6%), arabinose (6.6%), galactose (4.6%), glucose (32.5%), unknown peak (11.3%), and *N*-acetylglucosamine (3.9%) (Starr and Jaenicke, 1974). We find galactose, mannose, xylose, arabinose, and glucosamine to be the main sugar components of the pheromone from the American isolate. We also find a variable amount of glucose, which we believe to be a contaminant in our samples.

4. Pheromones in Volvox

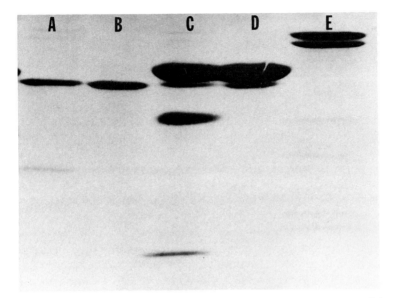

Fig. 5. Coelectrophoresis of purified pheromones isolated from Japanese and American isolates of *Volvox carteri* and treated 1 hr at 37°C with 25 μg/ml of *Staphylococcus aureus* V8 protease. (A). Pheromone from the Japanese isolate treated with V8 protease; (B) pheromone from the Japanese isolate; (C) pheromone from the American isolate treated with V8 protease; (D) pheromone from the American isolate; (E) V8 protease.

We have recently begun a comparison of the two pheromones on the basis of protease sensitivity and peptide maps using the procedures outlined by Cleveland *et al.* (1977). Under these conditions, both pheromones are resistant to trypsin. The pheromone from the American isolate is resistant to chymotrypsin, but the pheromone from the Japanese isolate is not. Both are cleaved by *Staphylococcus aureus* V8 protease, by protease K, and by protease Type VI from *Streptomyces griseus*. In the latter cases the peptide maps generated are quite different for each enzyme, and for each enzyme our preliminary experiments indicate the peptide maps are different for the two pheromones. Cyanogen bromide peptide maps are also quite different for the two pheromones. We have raised antibodies against the pheromone from the American isolate in rabbits. These antibodies do not show a precipition reaction in standard Ouchterlony double-diffusion plates against the pheromone from the Japanese isolate.

It would appear, then, that the pheromones produced by these two forms of the same species are quite similar in general chemical structure, but they differ in protease sensitivity and do not cross-induce or cross-react antigenically, in the tests we have so far applied. The total carbohydrate content varies although the sugar components appear qualitatively similar.

B. Mode of Action of *Volvox* Pheromones

The other main aspect of our current work on sexual pheromones involves investigation of their mode of action. Several factors are important to note here. First, the pheromones are active in very low concentrations. Biological activity is commonly seen at about 10^{-14} M (Kochert and Yates, 1974; Starr and Jaenicke, 1974). If one calculates on the basis of numbers of molecules required per target cell it is evident that only a few hundred molecules will suffice. The actual number may be even less since these numbers were derived from standard assay tubes where the density of target organisms is very low and no attempt was made to maximize the number of organisms. Such low numbers of molecules per cell eliciting a biological reaction would not be unprecedented. Diphtheria toxin appears to be able to kill cells at a concentration approaching one molecule per cell (Yamaizumi *et al.*, 1978). These low levels make it enormously difficult to perform certain sorts of experiments since technology is only now appearing which will allow one to deal with such small numbers of molecules.

Another interesting fact about *V. carteri* pheromone action is that its biological activity appears to require long contact and appears to be reversible right up to the time when its morphological effects are evident. We have investigated this most extensively in the American isolate (Kochert and Crump, in press). In this case the pheromone must be added when gonidia are quite small and must be continuously present through the period of gonidial enlargement and the beginning of cleavage for full biological activity. One can wash away the pheromone with culture medium and partially reverse biological activity (i.e., a lower percentage of gonidia will be induced) as late as early cleavage of the target gonidium. Interestingly, a short pulse of ultraviolet (uv) irradiation from a germicidal lamp will also reverse biological activity. In these experiments, gonidia are placed in pheromone-containing medium when they are small, allowed to enlarge in the presence of the pheromone, removed to the surface of agar plates and subjected to a 5-10 sec dose of uv, and then returned to pheromone-containing medium. The majority of such irradiated gonidia do not form females, although they are not visibly damaged by the treatment and will form females in the next generation. Furthermore, the sensitivity to uv is life cycle stage dependent. The same dose given early in the period of gonidial enlargement has very little effect on biological activity. It would appear that either the uv-sensitive component is not present at the early stages or the target cells have time to "recover," perhaps by synthesizing more of the component in question.

The most striking parallel to *Volvox* pheromones to be found in other biological systems are animal gondadotropic hormones. These molecules are glycoproteins of about the same molecular weight as the pheromones from *V. carteri*. Their biological function is also parallel since they induce the differentiation of gametes in both males and females. Extensive data have been gathered (Bahl, 1977; Catt and Dufau, 1978) to support the hypothesis that gonadotropic hormones (and other protein, peptide, and catacholamine hormones) exert their biological activity through the following series of steps: (1) interaction of the hormone with a cell surface, specific receptor, (2) activation of adenyl cyclase by some form of coupling to the hormone receptor complex, (3) increase in the cellular cAMP level, (4) activation of protein kinases by cAMP, and (5) phosphorylation of enzymes leading to activation or repression of key pathways thereby changing the phenotype of the target cell (Fig. 6).

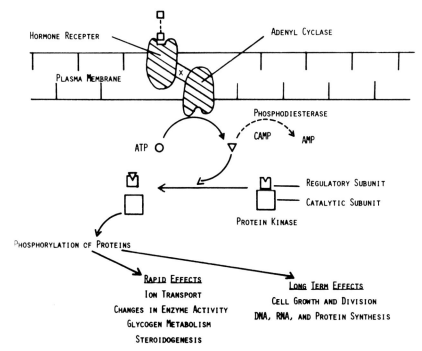

Fig. 6. Diagrammatic representation of the mode of action of a gonadotropic hormone in a mammalian system.

Because of the parallels noted previously it is difficult to escape the hypothesis that *Volvox* pheromones act by the same general mechanism as that described for gonadotropic hormones. Accordingly, we have for some time been conducting experiments which parallel those done in animal systems in an attempt to demonstrate specific pheromone receptors in *V. carteri* f. *weismannia* (Noland and Kochert, in press).

To perform the requisite binding experiments we have radiolabeled the pheromone with ^{125}I. With the chloramine T procedure (Hunter and Greenwood, 1962) one can achieve specific activities in the range of 10^7 cpm/μg of pheromone. By bioassay the labeled product appears to be as potent as the native molecule. It is difficult to be sure of this conclusion, however, given the small number of molecules required for activity, the nature of the bioassay, and the difficulty of accurately quantitating the small amounts of pheromone labeled at one time. It is apparent, moreover, that labeled preparations are less stable than the native molecule and break down to a specific set of smaller components over a period of about 2 weeks.

With these limitations in mind we have performed a variety of pheromone binding experiments. In one series we have investigated binding of pheromone to whole asexual spheroids. Time course studies reveal that binding was relatively rapid and reached equilibrium by about 30 min under our conditions. In dose–response experiments increasing amounts of equilibrium binding could be achieved when progressively larger amounts of labeled pheromone added (Fig. 7). A large portion of the bound radioactivity could be eluted by incubation in fresh medium. No saturation of the binding could be achieved even though the amount of pheromone added was increased to a level more than 1000 times greater than that needed for biological activity. Competition experiments were also conducted with whole organisms. In these we added an excess of unlabeled pheromone to see if the amount of labeled pheromone binding would be reduced. Only small amounts of competition were achieved.

We believe the results of these binding experiments indicate that the binding we observe to whole organisms is largely nonspecific. We also believe that this binding represents an ionic interaction between the very basic pheromone molecules and the acidic extracellular matrix. Previous work has established that the matrix has an overall negative charge, probably due to sulfated polysaccharide groups (Burr and McCracken, 1973; McCracken and Barcelona, 1976). It has also been shown that *Volvox* spheroids will readily adhere to polylysine-coated microscope slides by such an ionic interaction (Viamontes *et al.*, 1979).

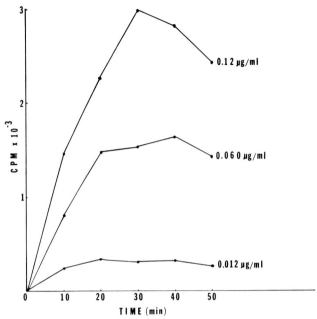

Fig. 7. Time course of binding of different concentrations of ^{125}I-labeled pheromone purified from American isolates of *Volvox carteri*.

Several other lines of evidence support this hypothesis. If one performs binding experiments in elevated salt concentrations (which would be expected to decrease ionic interactions), binding is decreased. Other labeled proteins will also bind to whole *Volvox* spheroids and the more basic protein chymotrypsinogen will bind more than bovine serum albumin. The labeled pheromone from *V. carteri* f. *weismannia* also binds to spheroids of other *Volvox* isolates, presumably also by ionic interaction. The amount of pheromone bound per spheroid varies with the *Volvox* species, and spheroids of some isolates (e.g., *V. carteri* f. *nagariensis*) bind as much pheromone as do those of *V. carteri* f. *weismannia*. We now believe this variation in pheromone binding is the result of a difference in makeup or amount of spheroid matrix rather than a true species-specific binding to specific receptors. Ionic binding of pheromone to matrix may have an important ecological role in "collecting" pheromone from the dilute concentration present in natural waters, but the large background binding generated complicates the detection of specific receptors.

To reduce background ionic binding we have performed additional experiments with isolated gonidia. We have used gonidia isolated from young spheroids recently released from the parent. Time course binding studies reveal a rapid achievement of equilibrium similar to that seen with whole spheroids. Data from the experiments involving addition of increasing amounts of pheromone reveal a multiphasic response. At low concentrations of pheromone, binding is proportional to the amount of pheromone added. In the range of 10^5 molecules of pheromone per target cell an apparent saturation is reached. Still larger doses of labeled pheromone bring about further increases in binding. Competion experiments with isolated gonidia indicate that about 50% of the bound, labeled pheromone is susceptible to competition by unlabeled pheromone. We interpret these results to mean that two sorts of binding sites are present on gonidia. One is nonsaturable and noncompetible, and the other can be saturated at relatively low pheromone concentrations and is competible by unlabeled pheromone. We believe the latter type of binding site may represent specific binding to a surface receptor comparable to that seen in the animal hormone systems described above. We also observe that at relatively low pheromone levels about 50% of the binding can be eliminated by pretreatment of gonidia with trypsin. This is compatible with a surface localization of a portion of the binding sites.

IV. PERSPECTIVES

The small amount we have learned about sexual differentiation in *Volvox* serves to emphasize the potential for future studies. Several areas would seem to provide good possibilities. One of these areas seems to be further characterization of pheromones from various species. A complete characterization of at least one pheromone will ultimately be necessary for a complete understanding of the mode of action. Characterization experiments will also be of interest in an evolutionary sense. There are more than 20 species of *Volvox* and one or more of the species are common residents of freshwater habitats all over the world. Many of these have been shown to produce sexual pheromones. In the cases thus far investigated absolute species specificity of pheromone action has been reported even though all the pheromones (except for *V. capensis* as noted above) appear to be glycoproteins. The molecular basis of the strict species specificity is unknown and differences between the pheromones may be major or relatively minor. Our comparative data on the pheromones from two isolates of *V. carteri* would indicate they are quite different. The main evidence for this conclusion comes from peptide mapping, however, and the applicibility of these techniques to glycoproteins with different sugar content is not firmly established. Much more work needs to be done in the isolation and characterization of *Volvox* pheromones. Amino acid sequence studies will be of interest to determine whether the differences between the pheromones represent major differences or minor amino acid sequence differences in the peptide backbone.

Peptide and glycoprotein hormones appear to be highly conserved in animal systems. Quite a lot of work has been done with gonadotropic hormones, and these hormones from various species have been shown to have extensive antigenic and biological cross-reactivity (Vaitukaitis *et al.*, 1976). Certainly such creatures as pigs and cows seem to be more distantly related to one another than we believe the various species of *Volvox* are to one another, yet the *Volvox* pheromones show no interspecies activity while the gonadotropic hormones do. If it turns out that the *Volvox* pheromones are quite different from one species to another it would have interesting evolutionary consequences. When proteins in different species are found to be very similar, it is taken as evidence of a common origin. If the molecules are very different in closely related organisms it could be taken as evidence for independent origin of the molecule in question. In other words, different molecules for the same function were recruited out of the repertoire of possible gene products in the two different organisms. It seems hardly credible that each *Volvox* species would have separately evolved a pheromone signaling system to coordinate sexual reproduction, yet the pheromones are definitely distinct enough so that they do not cross-react.

It should be recognized, however, that the *Volvox* pheromones have a greater "need" to be highly species specific than do such hormones as gonadotropic hormones in vertebrates. *Volvox* pheromones are released into the surrounding medium. This would in nature be a body of water containing many other organisms and many other types of molecules including perhaps pheromones from other *Volvox* species since these organisms often occur in mixed populations. It would seem, then, to be of considerable advantage for the organism to be able to respond quite specifically only to its own pheromone. This is different in the case of animal hormones secreted into a circulatory system where, in effect, each organism is a closed system. A high degree of specificity must be maintained in this case between different types of hormones in the same organism but not between the same hormone in different organisms. The fact that hormones cross-react is not likely to cause problems since gonadotropins from a foreign species can not be introduced into the circulatory system to interfere with normal reproductive processes. Indeed, one of the main operational differences between pheromones and hormones is that pheromones must be more species-specific than hormones since they serve as a vital communication mechanism through an external medium between individuals of the same species.

Continued work on characterization may also yield valuable information on synthesis of the pheromone. One possibility often mentioned is that the pheromone is not a specific synthetic product, but is a degradation product of some structural component of the organism. Kinetics of production would indicate that the major portion of pheromone production occurs in cultures containing males following sperm bundle production and breakdown. Certainly a lot of degradation is going on in cultures at this stage of development. After the sperm bundles from male spheroids are released, the remaining "hulk" composed of somatic cells and surrounding matrix soon disintegrates. In addition the sperm bundles themselves are ephemeral and soon disintegrate. Sperm bundles appear to produce a protease during the normal fertilization reaction (Hutt and Kochert, 1971; Coggin *et al.,* 1979) and this protease can degrade the matrix material as evidenced by fertilization pore formation. It is very possible that the pheromone is a degradation product produced by the sperm protease as a result of enzymatic attack on the spheroid matrix. This would explain the stability of the pheromone since it would be derived from an extracellular, stable structural component of the spheroid. There is little evidence to support or refute this hypothesis. Negative evidence includes the observations that (1) the protein portion of the extracellular matrix contains a high proportion of hydroxyproline while the purified pheromone contains little or none, and (2) the matrix has an overall negative charge while the pheromone is highly positive. However, analysis of the sugar components indicates that the same sugars are present in both the pheromone and isolated extracellular matrix (Kochert, unpublished observations). Clearly much more work is needed in characterization of both phermones and matrix to solve this problem.

It should be evident also that we know very little about the mode of action of any *Volvox* pheromone. One of the problems which must be overcome is that we know very little about primary effects of the pheromones. We add the pheromone to a *V. carteri* culture when the target gonidia are small and the only indication of response we presently recognize is many hours later when the gonidia cleave to form females. A lot of growth, development, cell division, and differentiation have occurred in the intervening period. We know of no biochemical parameter to measure which gives an indication of pheromone primary effects. As mentioned above we are presently working on the hypothesis that the *Volvox* pheromone acts by the same general mechanism as animal peptide hormones. This involves to binding a specific surface receptor followed by a rise in intracellular cAMP concentration, activation of protein kinase, and phosphorylation of key enzymes. We have some evidence that there is a specific receptor for the *V. carteri* f. *weismannia* pheromone. We also know that neither cyclic AMP, dibutyrl-cAMP, nor a variety of inhibitors or activators of the cAMP system will mimic the biological effect of the pheromone when added to target cultures. Since we know nothing about permeability of any of these compounds, negative results of this sort are not very conclusive, however. By using a cAMP radioimmunoassay procedure we have detected cAMP in *Volvox* gonidia. We have also detected a phosphodiesterase activity that may be part of a cAMP-based control system. Beyond this, very little is known.

It is important to realize also that our hypothesis about *Volvox* pheromone action may be totally off the mark. The mass of detailed and interesting work on animal hormonal mechanisms tends to dominate one's thinking and one can be led to try to "force" the *Volvox* system into one of the existing animal hormone models. These well-established mechanisms must of course be tested in *Volvox*, but it is necessary to keep in mind that *Volvox* pheromones may operate by totally different and as yet undescribed mechanisms.

A variety of approaches are needed to describe in detail the mechanism of *Volvox* sexual induction. The *V. carteri* system is presently being investigated extensively and many of these approaches are under way in various laboratories. We need to know exactly what is happening morphologically and ultrastructurally during enlargement and cleavage of induced and control gonidia. What is the cellular basis for the unequal cleavage that differentiates somatic cells from gonidia and how is this pattern changed to produce eggs rather than gonidia when pheromone is present? The kind of detailed investigation carried out by Kirk and associates on the morphology and mechanisms of inversion (Viamontes *et al.*, 1979) and presently being applied to cleavage will provide important information.

The genetics of sexual induction also provide a very exciting and productive avenue of approach. Pioneering work on genetics of *Volvox* sexuality was done by Starr (1970). Many sorts of fascinating mutants have been described in detail by Starr (1970) and by Huskey and colleagues who have systematically isolated and characterized many mutants (Sessons and Huskey, 1973). These include mutants which form sexual spheroids in the absence of the pheromone, those which do not respond to pheromone, and many sorts of pattern mutants which have altered numbers of locations of reproductive cells. The very nature of the genetic control of sexual type in *Volvox*, or indeed in any other alga, is very imperfectly known. In some species, it will be recalled, both sperm and eggs are formed in the same spheroid, while other species form sperm and eggs only in separate clonal cultures. In the diclonic species, sexual type is inherited in a 1:1 fashion and the nature of this locus is not understood. Studies which further our knowledge in these areas will be crucial to our understanding of pheromone action.

Volvox has many advantages as a model system for studies of pheromone action. These have been outlined before, and it is sufficient here to note that these characteristics make possible the kind of combined approach to the problem described above. It is not unreasonable to expect that the type of morphological, genetic, and biochemical approaches described above will one day converge to yield a detailed mechanism of pheromone action at all levels of cellular organization. *Volvox* will then finally achieve a measure of that potential to further aid our understanding of mechanisms of cellular differentiation which it so clearly proclaims to its every observer.

REFERENCES

Bahl, O. P. (1977). Human chorionic gonadotropin, its receptor and mechanism of action. *Fed. Proc., Fed. Am. Soc. Exp. Biol.* **36**, 2119–2127.

Burr, F. A., and McCracken, M. D. (1973). Existence of a surface layer on the sheath of *Volvox*. *J. Phycol.* **9**, 345–346.

Catt, K. J., and Dufau, M. L. (1978). Gonadotropin receptors and regulation of interstitial cell function in the testis. *In* "Receptors and Hormone Action" (B. W. O'Malley and L. Birnbaumer, eds.), Vol. 3, pp. 291–339. Academic Press, New York.

Cleveland, D. W., Fischer, S. G., Kirschner, M. W., and Laemmli, U. K. (1977). Peptide mapping by limited proteolysis in sodium dodecyl sulfate and analysis by gel electrophoresis. *J. Biol. Chem.* **252**, 1102–1106.

Coggin, S. J., Hutt, W., and Kochert, G. (1979). Sperm-bundle female somatic cell interaction in the fertilization process of *Volvox carteri* f. *weismannia* (Chlorophyta). *J. Phycol.* **15**, 247–251.

Darden, W. H. (1966). Sexual differentiation in *Volvox aureus*. *J. Protozool.* **13**, 239–255.

Hunter, W. M., and Greenwood, F. C. (1962). Preparation of iodine-131 human growth hormone of high specific activity. *Nature (London)* **194**, 495–496.

Hutt, W., and Kochert, G. (1971). Effects of some protein and nucleic acid synthesis inhibitors on fertilization in *Volvox carteri*. *J. Phycol.* **7**, 316–320.

Janet, C. (1923). "Le Volvox. Troisième Mémoire." Protat Frères, Macon.

Karn, R. C., and Starr, R. C. (1974). Sexual and asexual differentiation in *Volvox obversus* (Shaw) Printz, strains WD3 and WD7. *Arch. Protistenkd.* 116, 142-148.

Kochert, G. (1968). Differentiation of reproductive cells in *Volvox carteri*. *J. Protozool.* 15, 438-452.

Kochert, G. (1973). Colony differentiation in green algae. *In* "Developmental Regulation" (S. J. Coward, ed.), pp. 155-167. Academic Press, New York.

Kochert, G. (1975). Developmental mechanisms in *Volvox* reproduction. *Symp. Soc. Dev. Biol.* 33, 55-90.

Kochert, G., and Yates, I. (1974). Purification and partial characterization of a glycoprotein sexual inducer from *Volvox carteri*. *Proc. Natl. Acad. Sci. U.S.A.*, 71, 1211-1214.

McCracken, M. D., and Barcelona, W. J. (1976). Electron histochemistry and ultrastructural localization of carbohydrate-containing substances in the sheath of *Volvox*. *J. Histochem. Cytochem.* 24, 668-673.

McCracken, M. D., and Starr, R. C. (1970). Induction and development of reproductive cells in the K-32 strains of *Volvox rousseletii*. *Arch. Protistenkd.* 112, 262-282.

Provasoli, L., and Pintner, I. J. (1959). Artificial media for freshwater algae: Problems and suggestions. *In* "The Ecology of the Algae" (C. A. Tryon and R. T. Hartman, eds.), pp. 84-96. Pymatuning Lab. Field Biol., Univ. of Pittsburgh, Pittsburgh, Pennsylvania.

Sessons, A., and Huskey, R. J. (1973). Genetic control of development in *Volvox*: Isolation and characterization of morphogenetic mutants. *Proc. Natl. Acad. Sci. U.S.A.*, 70, 1335-1338.

Starr, R. C. (1969). Structure, reproduction and differentiation in *Volvox carteri* f. *nagariensis* Iyengar, strains HK9 and HK10. *Arch. Protistenkd.* 111, 204-222.

Starr, R. C. (1970). Control of differentiation in *Volvox*. *Dev. Biol.* 4, Suppl. 59-100.

Starr, R. C. (1975). Meiosis in *Volvox carteri* f. *nagariensis*. *Arch. Protistenkd.* 117, 187-191.

Starr, R. C. (1978). The culture collection of algae at the University of Texas at Austin. *J. Phycol.* 14, Suppl., 47-100.

Starr, R. C., and Jaenicke, L. (1974). Purification and characterization of the hormone initiating sexual morphogenesis in *Volvox carteri* f. *nagariensis*. *Proc. Natl. Acad. Sci. U.S.A.* 71, 1050-1054.

Vaitukaitis, J. L., Ross, G. T., Braunstein, G. D., and Rayford, P. L. (1976). Gonadotropins and their subunits: Basic and clinical studies. *Recent Prog. Horm. Res.* 32, 289-331.

Vande Berg, W. J., and Starr, R. C. (1971). Structure, reproduction, and differentiation in *Volvox gigas* and *Volvox powersii*. *Arch. Protistenkd.* 113, 25-31.

Viamontes, G. I., and Kirk, D. L. (1977). Cell shape changes and the mechanism of inversion in *Volvox*. *J. Cell Biol.* 75, 719-730.

Viamontes, G. I., Fochtmann, L. J., and Kirk, D. L. (1979). Morphogensis in *Volvox*: Analysis of critical variables. *Cell* 17, 537-550.

Weismann, A. (1889). "Essays Upon Heredity" (E. B. Poulton, S. Schönland, and A. E. Shipley, transl. and eds.). Clarendon, Oxford.

Yamaizumi, M., Mekada, E., Uchida, T., and Okada, Y. (1978). One molecule of diphtheria toxin fragment A introduced into a cell can kill the cell. *Cell* 15, 245-250.

5

Cell Interaction by Gamones in *Blepharisma*

AKIO MIYAKE

I.	Introduction	95
II.	Survey of the Literature	100
III.	Current Research	101
	A. Mating Types	101
	B. General Scheme of Preconjugant Interaction	101
	C. Isolation and Characterization of Gamones	103
	D. Biosynthesis of Gamones	106
	E. Taxis to Gamone 2	108
	F. Induction of Cell Union by Gamone	109
	G. Gamones in Five Species of *Blepharisma*	118
	H. Gamone Receptors	118
	I. Molecular Scheme of Preconjugant Interaction	120
	J. Initiation of Meiosis	121
IV.	Perspectives	123
	References	126

I. INTRODUCTION

Blepharisma is a genus of ciliate characterized by the possession of a red pigment, blepharismin (formerly zoopurpurin) and a set of well-developed membranelles which flutter like eyelashes. Blepharismin is a hypericin-type pigment contained in numerous minute granules which are usually located just under the cell membrane giving a red coloration to the cell (for review, see Giese, 1973). The albino mutant which has much less pigment appears white (Chunosoff et al., 1965).

Although the physiological function of this pigment is largely unknown, the color difference between the wild-type and albino cells greatly helps identifying cell types in studies of cell interaction (Fig. 1).

The genus *Blepharisma* belongs to class Polyhymenophora, order Heterotrichida (for recent discussion about classification, see Corliss, 1979; Hirshfield *et al.*, 1973). It is one of the most widely investigated ciliate genera (Giese, 1973). It is also the first and, so far, the only ciliate in which sexual pheromones have been isolated and characterized (see below). Dozens of species are currently recognized in *Blepharisma*, but pheromonal interaction has been mainly investigated in *B. japonicum* v. *intermedium* (formerly *B. intermedium*), the largest and most brightly colored species in the genus.

Like other ciliates, blepharismas have many fine surface projections consisting of either single or compound cilia. In spite of these cilia *Blepharisma* (particularly *B. japonicum*) is sluggish as compared to many other ciliates. In addition to locomotive and feeding functions,

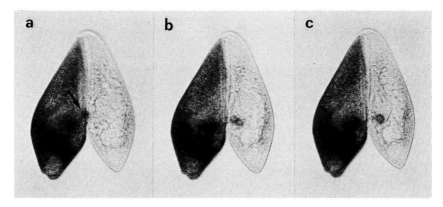

Fig. 1. Conjugant pairs of *Blepharisma japonicum* at the time of karyogamy. Left: Red-pigmented (wild-type) cell, mating type II. Right: albino cell, mating type I. Cells unite along the whole length of the peristomial floor. At this stage of conjugation, the red pigment tends to accumulate at the posterior end of the cell and also at the vicinity of the micronuclei, making gametic nuclei visible without staining. (a) The migratory gametic nucleus of the red cell (arrow) is at the border of the united cells. Gametic nuclei in the albino cell are invisible because of the lack of pigment. (b) The same pair as in (a), 31 min later. The migratory gametic nucleus of the red cell has entered the albino cell. This nucleus has brought in pigment, which also surrounds the stationary gametic nucleus of the albino cell. (c) The same pair as in (b), 6 min later. The two gametic nuclei have fused to form a fertilization nucleus. Photographs of living, unstained cells. ×200. (From Miyake, 1978.)

5. Gamones in *Blepharisma*

these surface projections play important roles in sexual interaction (also see Chapter by Ricci on *Oxytricha* and Chapter 14 by Hiwatashi on *Paramecium*).

Blepharisma has another ciliate characteristic, nuclear dimorphism: each cell has two types of nuclei. The micronucleus is diploid and functions as a carrier of nuclear genes through both asexual and sexual processes. The macronucleus usually contains much more DNA than the micronucleus, controls most of the cellular phenotypes, and is retained through asexual processes. In sexual processes, it is replaced by a new macronucleus that develops from the synkaryon which is micronuclear in origin. A typical vegetative cell of *B. japonicum* has one long macronucleus stretching along most of the cell length and about 15 spherical micronuclei (diameter 1–2 μm), the number of which differs from cell to cell.

The main developmental cycles in the life history of *Blepharisma* are binary fission, physiological reorganization, encystment–excystment, and gamontogamy (Fig. 2).

In binary fission, a cytoplasmic furrow transversely divides the cell into two daughter cells of equal sizes. The macronucleus condenses into a spherical mass and then stretches to divide into two halves each entering a daughter cell. Prior to cytokinesis some of the micronuclei undergo mitosis.

In physiological reorganization, the macronucleus condenses into a spherical mass and then stretches as in binary fission but does not divide. Although cytoplasmic division does not occur some of the micronuclei undergo mitosis.

In the encystment–excystment cycle, each cell rounds up, shrinks, and produces a cyst wall around itself. The dormant cyst of *Blepharisma* probably provides protection against adverse conditions. Under certain conditions, the cell emerges from the cyst wall and resumes the vegetative form. Usually no nuclear division occurs during the cycle.

In gamontogamy, vegetative cells develop into preconjugants, i.e., cells which can undergo sexual interaction. Preconjugants of complementary mating types interact by excreted gamones (pheromones), gain the capacity to unite, and form conjugant pairs (preconjugant interaction). In each of the united cells (conjugants) about two-thirds of the micronuclei complete meiosis I. One of the division products completes meiosis II to produce two haploid gametic nuclei. One of the nuclei is exchanged between the conjugants of a pair. Fertilization then follows (Fig. 1). The synkaryon divides mitotically and produces

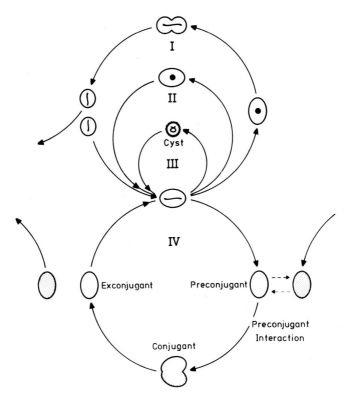

Fig. 2. Life cycles of *Blepharisma japonicum*. I, Binary fission; II, physiological reorganization; III, encystment–excystment; IV, gamontogamy. The macronucleus is shown only in asexual cycles (I, II, and III). The micronucleus is not shown. Cells of complementary mating types are differently shaded.

new micro- and macronuclei. Meanwhile the pair dissociates into exconjugants, the old macronucleus degenerates, and is replaced by a new one as the exconjugant develops into a new vegetative cell.

Blepharisma is usually cultured on lettuce or cerophyl medium inoculated with food bacteria such as *Enterobacter aerogenes* (formerly *Aerobacter aerogenes*) or *Pseudomonas ovalis,* using essentially the same method used for *Paramecium* (Sonneborn, 1970). An axenic culture method exists for *B. japonicum* (Smith and Giese, 1967), but a defined medium does not exist.

For studies on the gamones, a modified lettuce juice medium (Miyake and Beyer, 1973) inoculated with *E. aerogenes* has been mainly used. In order to maintain a high yield of cells and reproducible experimental results, cultures should be kept under a strictly monoxenic condition (for details, see Kubota et al., 1977). Contamination by undesirable bacteria is revealed by persistent turbidity. Cells are grown at 24°–27°C in cotton-plugged Erlenmeyer flasks. The fission rate can be controlled to about one fission per day by doubling of the culture volume every day with fresh medium. To maintain an adequate oxygen supply, the amount of the culture is limited to one half of the volume of the flask. Successful cultures of *B. japonicum* yield 500–1000 cells per ml or about 0.5 ml packed cells per liter.

Blepharisma can also be cultured on a small ciliate, *Saprophilus* sp., in the lettuce medium containing *E. aerogenes* (Miyake et al., 1979b). In these cultures, contamination is reduced probably because *Saprophilus* eat bacteria less discriminately than *Blepharisma*.

The continuous supply of large amount of cells is achieved by using a culture device consisting of two 20-liter bottles which are connected by a silicon tube. The first bottle contains only the lettuce medium with *E. aerogenes* and the second contains a complete *Blepharisma* culture. The gentle aeration of the *Blepharisma*-containing bottle with sterilized air is done with candle filters or similar devices which produce fine air bubbles. As *Blepharisma* cells are harvested, the lost volume is compensated by adding bacterized medium from the first bottle which is in turn compensated for the lost volume by the addition of fresh lettuce medium. This relatively simple, inexpensive device contributed a great deal in supplying large amount of cells needed for the isolation and identification of gamones.

As in many other ciliates, the transformation of vegetative cells into preconjugants is controlled by both internal and external factors. The main internal factor is the "maturity." After conjugation, cells enter a period of immaturity during which they cannot become preconjugants. This period, which is usually measured by the number of fissions, lasts for 0–100 fissions in many ciliates (for review, see Bleyman, 1971; Nanney, 1974). In *B. japonicum,* Bleyman (1975) reported that by 50 fissions after conjugation 84% of the clones were mature. Once mature, they remain mature for the rest of the clone's life which may last for several hundred fissions or even longer (A. Miyake, unpublished). The regulation of the immaturity phase in ciliates is discussed in detail in Chapter 14 by Hiwatashi.

The main external condition required for preconjugant differentiation is the deprivation of food. In the laboratory, this is achieved as follows. Two days after the last feeding, cells are harvested, washed with, and suspended in a synthetic medium: SMB I (1.5 mM NaCl, 0.05 mM KCl, 0.4 mM CaCl$_2$, 0.1 mM MgCl$_2$, 2×10^{-3} mM EDTA, 2 mM sodium phosphate buffer pH 6.8) (Miyake, 1968); SMB II (same as SMB

I except that $MgCl_2$ is replaced by $MgSO_4$) (Miyake and Beyer, 1973); or SMB III (same as SMB I except that the half of $MgCl_2$ is replaced by $MgSO_4$) (A. Miyake, unpublished). The cell density of the suspension is about 5×10^3 cells per ml or about 1 ml packed cells in 200 ml. Debris in the culture is removed, if necessary, by filtering (Miyake and Honda, 1976; Kubota et al., 1977). Most cells transform into highly reactive preconjugants within 1 day, if mature cells are used. Cells remain as preconjugants for a week or longer if they are transferred into fresh SMB every 2–3 days. Thus, unlike many other ciliates, even very starved cells can be preconjugants in *B. japonicum*.

Strains of various species of *Blepharisma* are maintained by I. R. Isquith, Department of Biology, Fairleigh Dickinson University, Teaneck, New Jersey; H. I. Hirschfield, Department of Biology, New York University, New York, New York; Y. Takagi, Department of Zoology, Nara Women's University, Kitauoya-Nishimachi, Nara 630, Japan, and by the present author.

II. SURVEY OF THE LITERATURE

Isquith and Hirshfield (1966, 1968) first described complementarity for conjugation between a wild-type (red) and a mutant (albino) strain in *B. japonicum* (then *B. intermedium*). Miyake (1968) found that these two strains interact by means of gamones and began isolating and characterizing these gamones. Miyake and Beyer (1973) then demonstrated that the interaction occurs between complementary mating types I and II, which may have the same pigmentation, and they established the basic scheme of sexual interaction by gamones. Bleyman (1975) has reported on the genetics of mating types. Inaba (1965) has reported on interspecific conjugation.

The gamone of mating type II (gamone 2) was isolated, characterized (Kubota et al., 1973), and chemically synthesized (Tokoroyama et al., 1973). The gamone of mating type I (gamone 1) has also been isolated and characterized (Miyake and Beyer, 1974; Braun and Miyake, 1975).

Cytological aspects of conjugation in *B. japonicum* were reinvestigated by Miyake et al. (1979a). Unlike Suzuki (1957) who reported three micronuclear pregamic divisions including a peculiar unequal one, they found that the pregamic divisions of this species consist of only two divisions which are meiotic.

Sexual interaction in *Blepharisma* has been briefly reviewed and discussed in relation to other systems of interacting cells (Crandall,

1977; Honda, 1979; Miyake, 1974a,b, 1978, 1981; Nanney, 1977, 1980; Sonneborn, 1978). Kubota *et al.* (1977) described details for the purification procedures for gamones 1 and 2. The book "Blepharisma" by Giese (1973) is a rich collection of previous works on this genus but does not contain any work on the sexual pheromones.

III. CURRENT RESEARCH

A. Mating Types

Clones of *Blepharisma* established from cells collected in nature are mostly selfers, i.e., cultures in which conjugation occurs spontaneously. This condition greatly hindered the discovery of mating types in this genus. Thus, the sexual complementarity in *Blepharisma* was first shown between an albino mutant and a red wild-type culture of *B. japonicum* (Isquith and Hirshfield, 1966, 1968). These cultures were both selfers, but the affinity for conjugation between albino and red cells was detected based on the relative abundance of albino–red pairs over albino–albino and red–red pairs. Fortunately, it was later found in *B. japonicum* that clones pure for each of the complementary mating types I and II can be obtained by careful selection (for review, see Miyake, 1981). It was also found that cells of this species can change their mating type in both directions (type I → type II and type II → type I) even during asexual reproduction (Miyake and Beyer, 1973; Bleyman, 1975). When this occurs in a clone, the clone becomes a selfer. Little is known about the mechanism of this switchover of cell types.

Mating types are more elusive in *B. americanum, B. musculus, B. stoltei,* and *B. tropicum,* because selfing is even more abundant than in *B. japonicum,* but at least one pair of complementary mating types, I and II, have been detected in each of them (Miyake and Bleyman, 1976; Miyake and Mancini, 1978, and unpublished).

B. General Scheme of Preconjugant Interaction

That complementary mating types I and II of *B. japonicum* interact sexually by means of gamones is demonstrated by placing preconjugants of one mating type in the cell-free fluid of preconjugants of the other mating type. Cells treated in this way gain the capacity to unite in 1–2 hr and form pairs between themselves. In these pairs cells adhere by their peristomes (face-to-face pair). Since the appearance of such pairs among single cells can be easily detected, this cellular response is used to detect and assay the gamone activity. A unit of activity (U) is defined as the smallest amount of gamone activity that can induce at least one face-to-face pair in 500–1000 cells suspended in 1 ml SMB (Miyake and Beyer, 1973). To measure the activity, a sample is serially diluted by the factor of 2 and mixed with the same volume of cell suspension. If the pair is induced up to the 2^n times dilution, the sample is assumed to contain 2^{n+1} U of the gamone activity. It should be noted that activity measured in this way (U) tends to be lower than the real activity which should be between U and 2U. By using this assay method it was found that the excretion of gamone by one mating type is induced or enhanced by gamone of the other mating type (Section III,D).

Based on these findings, Miyake and Beyer (1973) constructed the basic scheme of preconjugant interaction consisting of 7 steps (Fig. 3). Type I cells autonomously excrete gamone 1 (step 1). This gamone specifically reacts with type II cells (step 2) and transforms them so that they can form cell union (step 3) and at the same time induces the production and excretion of gamone 2 (step 4). Gamone 2 specifically reacts with type I cells (step 5) and transforms them so that they can undergo cell union (step 6). Transformed cells unite in pairs upon contact (step 7). If the gamone 1 production of mating type I is lowered by a prolonged starvation, the production is enhanced by gamone 2 (Section III,D,1) making the reaction chain consisting of steps 1, 2, 4, and 5 in Fig. 3 a loop of positive feedback.

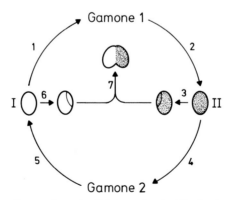

Fig. 3. Diagram of preconjugant cell interaction in *Blepharisma japonicum*. I, Mating type I cell; II, mating type II cell. (From Miyake, 1978.)

C. Isolation and Characterization of Gamones

Preliminary investigations indicated that gamones 1 and 2 (then factors A and R) are a heat-sensitive nondialyzable substance and a heat-resistant dialyzable substance, respectively (Miyake, 1968). Since this result suggested that gamone 2 was easier to handle, purification of this gamone was carried out first.

1. Gamone 2 (Blepharismone)

Mating type II cells were suspended (5 ml packed cells per liter) in a 9:1 mixture of SMB I and cell-free fluid of mating type I cells. The latter, prepared by suspending newly grown type I cells in SMB I (5 ml packed cells per liter) for 1 day, provided gamone 1 which served to induce and maintain the maximum production of gamone 2. After 1 day at 25°C, the cell-free fluid was obtained by removing cells by mild centrifugation. The cells were resuspended as described above and the process was repeated for several days. Harvested cell-free fluid was stored frozen and used as the starting material for gamone 2 purification.

Gamone 2 in this sample was concentrated *in vacuo* (<50°C), extracted with ethanol, dried, dissolved in 75% *n*-propanol and applied to cellulose column chromatography (eluting medium, 75% *n*-propanol). Gamone 2 was eluted in a single peak. Fractions in this peak were pooled, dried, dissolved in water, and chromatographed on paper (developing solvent, 75% *n*-propanol). Gamone 2 activity was localized in a single fraction fluorescing in pale blue under uv irradiation. This fraction was extracted with water. Gamone 2 crystallized from water as pale yellow prisms with the activity of 10^6 U/mg (for more details, see Kubota *et al.*, 1973, 1977). In a typical experiment, 3.68 mg of pure gamone 2 was obtained from 48 liters of the cell-free fluid (average activity 80 U/ml). This very high recovery rate (96%) must be taken with reserve, however, because the measured activity of gamone is the subject to the error amounting up to 100% (Section III,B), and because the variation in the sensitivity of tester cells may also cause an equivalent error.

Gamone 2 was identified as 3-(2'-formylamino-5'-hydroxybenzoyl)-lactate (see Fig. 12 for the chemical structure) by using X-ray crystallography and other methods (Kubota *et al.*, 1973). Although gamone 2 was originally called blepharismin (Kubota *et al.*, 1973), the name was changed to blepharismone (Miyake, 1974a) to avoid confusion since the red pigment, zoopurpurin, was renamed as blepharismin (Giese, 1973). Pure blepharismone dried and kept at $< -10°C$ retains the same activity for years. The stability is perhaps due to the association of calcium which stabilizes the β-carboxy-β'-hydroxyketone structure as pointed out by Kubota et al. (1973). Indeed, if blepharismone is eluted through a P-2 (Bio-Gel) column with pure water, most of the activity is lost within a few days under the same conditions (Miyake, 1981). Blepharismone is more effective in acidic medium (Miyake, 1981) suggesting that the protonated form of blepharismone might be more biologically active.

Racemic blepharismone was chemically synthesized starting from 3-hydroxyacetophenone via 2-formylamino-5-hydroxyacetophenone (Tokoroyama et al., 1973). Synthetic blepharismone is approximately half as active as the natural compound. This result is explained by assuming that one of the enantiometric forms (L) is more biologically active than the other form (D). This assumption is supported by the finding that L isomers of gamone 2 inhibitors are more effective than their D isomers (Section III,H).

2. Gamone 1 (Blepharmone)

In the isolation of unstable gamone 1, the finding that serum albumin is a potent protector of this gamone played a critical role. Gamone 1 in cell-free fluid from an SMB II suspension of type I cells loses ½ or more of its activity in 1 day at 25°C and ¾ or more of its activity after freezing and thawing. However, if 0.01% bovine serum albumin (described simply as albumin below) is added, the activity changes little under these conditions (Miyake and Beyer, 1973). The fact that the gamone 1 activity is measured about 4 times higher under the presence of 0.01 % albumin (Miyake and Beyer, 1973) is probably due to this protection effect of albumin.

Mating type I albino cells were suspended (10 ml packed cells/liter) in SMB II containing 32 U/ml gamone 2 and 0.01% albumin. Gamone 2 was used to induce maximum production of gamone 1. After 1 day at 25°C, the cell-free fluid was obtained by removing cells by mild centrifugation. The cells were resuspended as described above, and the process was repeated for several days. Cell-free fluid was stored frozen and used as the starting material for gamone 1 purification.

The cell-free fluid was concentrated *in vacuo* ($< 25°C$), lyophilized, dissolved in water, and centrifuged (10,000 g, 30 min). Gamone 1 in the supernatant was purified by two successive gel filtrations (P-150, Bio-Gel) and carboxymethyl cellulose chromatography (Fig. 4). The latter was carried out so that gamone 1 was only weakly retained, eluting as a peak shortly after the front. In all these column chromatographies, 0.01% albumin was added to the elution medium.

5. Gamones in *Blepharisma*

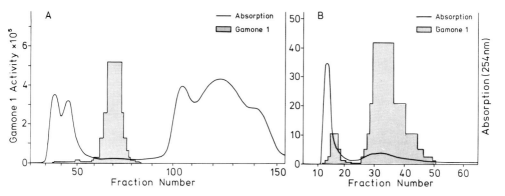

Fig. 4. Elution patterns of gamone 1 of *Blepharisma japonicum* in the purification procedure. Chromatography at 4°C; the fraction size was 6 ml. (A) Bio-Gel P-150 column (5 × 40 cm) chromatography. The elution medium: SMB II plus 0.01% bovine serum albumin. (B) Carboxymethyl cellulose column (2.5 × 35 cm) chromatography. The equilibrium medium: SMB II, pH 5.6 (the NaCl was omitted, and the phosphate buffer was increased to 4 mM). The elution medium: the equilibrium medium plus 0.01 % bovine serum albumin. Before the sample application, the column was eluted with 2 liters of elution medium. (From Miyake and Beyer, 1974. Copyright 1974 by the American Association for the Advancement of Science.)

The last step of purification was DEAE-cellulose column chromatography in which gamone 1 eluted at the front whereas albumin was retained. Lyophilized gamone 1 was white fluffy material with the activity of 1.6×10^7 U/mg. In a typical experiment, 2.0 mg of pure gamone 1 was obtained from 20 liters of the cell-free fluid mentioned above (Miyake and Beyer, 1974).

The purity of the gamone 1 preparation thus obtained was tested by acrylamide gel electrophoresis and isoelectric focusing. In each a single band was stained by protein-staining dyes which coincided with the gamone 1 activity. The results of isoelectric focusing indicated that the isoelectric point of gamone 1 is pH 7.5. If the staining after electrophoresis was carried out by the periodic acid–Schiff's reagent (PAS) for the determination of carbohydrate, a single band was seen at the position stained by amido black. Comparing the intensity of the PAS reaction of the band with that of orosomucoid and β_2-glycoprotein 1, the carbohydrate content of gamone 1 was estimated to be 5% of the protein. Based on these results, Miyake and Beyer (1974) concluded that gamone 1 is a slightly basic glycoprotein and named this glycoprotein blepharmone.

Amino acid and sugar analysis of the purified sample of gamone 1 indicates that it contains 175 amino acids excluding tryptophan (Lys_7,

His$_1$, Arg$_4$, Asp$_{26}$, Thr$_{17}$, Ser$_{19}$, Glu$_7$, Pro$_8$, Gly$_{13}$, Ala$_{13}$, Cys$_4$, Val$_{11}$, Met$_6$, Ile$_8$, Leu$_{12}$, Tyr$_{13}$, Phe$_6$) and 6 sugars (glucosamine$_3$, mannose$_3$) (Braun and Miyake, 1975). Thus, blepharmone may be identified by its characteristic amino acid composition, i.e., an unusually high content of tyrosine and a low content of glutamic acid.

The molecular weight of gamone 1 was estimated to be 20,000 and 30,000 by gel filtration and SDS acrylamide gel electrophoresis, respectively. The result of membrane filtration (Miyake and Beyer, 1974) cannot distinguish these two values, but amino acid and sugar analysis mentioned above strongly support the former value. The discrepancy might be due to the fact that the mobility of some glycoproteins lacking sialic acid is unusually low in SDS acrylamide gel electrophoresis as suggested by Braun and Miyake (1975).

Among the enzymes tested (Pronase, trypsin, papain, nagarse, leucine aminopeptidase, DNase, RNase, neuraminidase, α-amylase, β-galactosidase, hyaluronidase), only Pronase clearly destroyed the gamone 1 activity. Gamone 1 activity was rapidly lost in 8 M urea and by heating to 50°C (Miyake and Beyer, 1973). These results are consistent with the conclusion that gamone 1 is a protein.

D. Biosynthesis of Gamones

Well-fed cells of *B. japonicum* neither excrete gamones nor have the capacity to respond to gamone of the other mating type. If such cells are washed with and suspended in SMB, the gamone production and the responsiveness to gamone both appear within several hours as they transform into preconjugants. The transformation is reversible. If fed, the gamone production stops and the responsiveness to gamone disappears within a few hours. Such a switch of the cellular activity by the nutritive condition might have an important bearing on the basic mechanism of differentiation (Wolfe, 1973). The other important factor for gamone biosynthesis is the presence of gamone of the other mating type.

1. Gamone 1

If rapidly growing type I cells are suspended in SMB II, the gamone excretion increases from 0 to 0.01 U/hr/cell in 1 day and then decreases. The change of the intracellular gamone activity follows a similar pattern. If cells in the fourth day are exposed to gamone 2 (32 U/ml), gamone 1 excretion starts increasing within 2 hr reaching the level of the excretion by 1-day-old cells. The intracellular gamone 1 activity doubles at the same time. Gamone 2 has no such stimulating effect on 1-day-old cells (Miyake and Beyer, 1973). The threshold concentration of gamone 2 for induction of gamone 1 is at least one order lower than that for the induction of cell union (Miyake, 1981).

2. Gamone 2

Type II cells are either in the "autonomously gamone excreting" (*augex*) form or in the "*non-augex*" form. The latter excretes gamone 2 only under the presence of gamone 1. These two forms are interconvertible during asexual reproduction.

If *non-augex* type II cells are incubated with 10^4 U/ml gamone 1, both intra- and extracellular gamone 2 activity are first detected after 120 min indicating that these cells produce gamone 2 in about 2 hr in the presence of gamone 1 and that produced gamone 2 is immediately excreted (Miyake and Beyer, 1973). If the gamone 1 treatment of type II cells is discontinued after 1, 1.5, or 2 hr by washing the cells, gamone 2 production begins at nearly the same time and continues for 3–4 hr at the same rate as in the control, which is continuously exposed to gamone 1. Induction is achieved also by a shorter treatment, though the gamone 2 production never reaches the level of the control (Miyake, 1978). These results indicate that gamone 1 induces a mechanism for gamone 2 biosynthesis in 1–2 hr and that the mechanism, once induced, continues functioning for hours even if gamone 1 is removed from the medium. This suggests that gamone 1 induces an enzyme or an enzyme system participating in gamone 2 biosynthesis. The threshold concentration of gamone 1 for the gamone 2 induction is at least one order lower than that for the induction of cell union (Miyake, 1981).

The induction of gamone 2 by gamone 1 in the *non-augex* form appears to be amenable to biochemical analysis, because (1) the system consists of a single type of cells and a known signal substance (gamone 1); (2) what is induced is a stable signal substance (gamone 2) which can be easily determined by its potent biological activity; (3) gamone 2 production rises from 0 to the detectable level within a few hours; and (4) cells are starved and hence their general biochemical processes are likely to be reduced to the minimal level.

Type II cells of *augex* autonomously produce gamone 2. Like gamone 1 biosynthesis in type I cells, gamone 2 biosynthesis of *augex* decreases as cells starve. At this time of reduced gamone 2 production, gamone 1 enhances the production.

The chemical structure of gamone 2 suggests that tryptophan is the

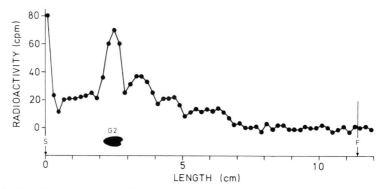

Fig. 5. Biosynthesis of radioactive gamone 2 by *Blepharisma japonicum*, mating type II, incubated with [^{14}C]tryptophan. Preconjugants of type II cells (3.5 ml packed cells) were suspended in 1 liter SMB II containing gamone 1 (10^3 U/ml) plus L-[3-^{14}C]tryptophan (0.04 μCi/ml) and incubated for 17 hr at 25°C. Gamone 2 in the cell-free fluid was isolated as described in the text. However, cellulose column chromatography was omitted and the alcohol extract was directly applied to paper chromatography (developing solvent, *n*-propanol: conc. NH$_4$OH:H$_2$O 75:15:10). The fraction containing gamone 2 was extracted with water and one fiftieth of it was applied to cellulose thin layer chromatography (developing solvent as above). Fractions, each 2 mm wide, were extracted 3 times with 0.2 ml water. Radioactivity of each fraction was then counted with 9.4 ml of counting solution (9 gm butyl-PBD in 1 liter dioxane). S, Position of sample application; F, elution front; G2, gamone 2 spot revealed by uv irradiation. Other conditions and methods were as described in Miyake and Honda (1976).

precursor for its biosynthesis (Kubota *et al.*, 1973). This assumption is supported by the fact that type II cells incubated with [^{14}C]tryptophan produce ^{14}C-labeled gamone 2 (Fig. 5).

E. Taxis to Gamone 2

Gamone 2 is a potent attractant of type I cells: gamone 2 activity of 10^{-3} U or even lower attracts type I cells under the experimental conditions used (Honda and Miyake, 1975) (Fig. 6).

Well-fed cells which are not responsive to gamone 2 by cell union are also not attracted by gamone 2. If such cells are starved in SMB, the ability to be attracted by gamone 2 appears after about 3 hr at nearly the same time at which the ability to respond to gamone 2 by cell union appears. 5-Hydroxy-L-tryptophan, a competitive inhibitor of gamone 2 in the induction of cell union (Section III,H) is also an inhibitor of taxis to gamone 2 (Honda and Miyake, 1975). The kinetics of these two inhibitions are very similar (Honda, 1979).

These results suggest that both taxis and cell union are mediated by the same gamone receptor. However, the treatment by cycloheximide (20 µg/ml, 2.5 hr), which strongly inhibits protein synthesis and induction of cell union (Section III,F), has no effect on the attraction to gamone 2 (Honda and Miyake, 1975). The result suggests that, contrary to cell union, taxis does not require the direct participation of protein synthesis.

The details of the behavioral response which results the attraction are still to be investigated. In the sessile ciliate *Tokophrya*, the only other ciliate in which taxis to sexual pheromone has been reported, cells of complementary mating types orient and stretch toward each other if they are placed within a certain distance (Sonneborn, 1978). Whether such a directed orientation participates in taxis to gamone 2 in *B. japonicum* is still unknown. Taxis to gamone 1 was not detected (Honda and Miyake, 1975).

F. Induction of Cell Union by Gamone

1. Process of Cell Union Formation

When preconjugants of mating types I and II of *B. japonicum* are mixed together or when cells of one mating type are treated by gamone of the other mating type, conjugant pairs are formed in several hours (Fig. 7). The process is divided in three stages, preunion, ciliary union, and conjugant union, based on morphological characteristics of cellular association.

a. Preunion Stage. In this stage, which lasts for 1–2 hr, cells do not yet associate in any spatially fixed way. Even irregular transient cell contact is hardly visible except at the beginning and the end of the stage. If mating types I and II, both stayed in SMB for 1 day, are mixed together, a very weak sticking together of cells occurs in all combinations of mating types for a few minutes and then disappears. Toward the end of this stage, a stronger sticking appears. Cells then form face-to-face pairs to enter the next stage (A. Miyake, unpublished). If cells are resuspended in SMB and then mixed, these stickings are hardly detectable because, if resuspended in SMB, even unmixed cells lightly stick together for a few hours (Miyake and Beyer, 1973). Since the behavior of cells is considerably affected by subtle stimuli such as the resuspension in SMB and the illumination for observation, more careful studies are needed to evaluate the significance of the sticking together of cells in this stage, particularly the very weak one at the time of mixing.

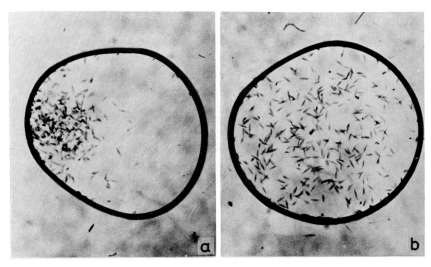

Fig. 6. Taxis of *Blepharisma japonicum*, mating type I to gamone 2. 10 µl of cell suspension and 1 µl of cell-free fluid of the same cell suspension containing gamone 2 (10^3 U/ml) (a) or none (b) were placed on a microscope slide and covered by a cover slip lifted 0.16 mm by spacers. The two drops were arranged so that, after covering, the smaller, satellite drop was situated very close to the left side of the larger, cell-containing drop. The two drops were then merged together by slightly moving the cover slip. Photographed after 30 min. Diameters of the disks were approximately 5 mm. (From Honda and Miyake, 1975. Reproduced by permission of Macmillan Journals Ltd.)

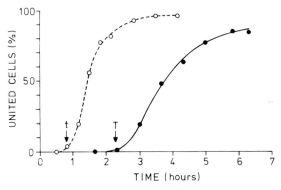

Fig. 7. Induction of homotypic cell unions in *Blepharisma japonicum*, and the transition from ciliary to conjugant union. Gamone 2 (800 U/ml) was added at time 0 to mating type I (10^3 cells/ml) at 25°C. (○), Ciliary union + conjugant union; (●), conjugant union; t, appearance of ciliary union; T, appearance of conjugant union. (From Miyake and Honda, 1976.)

5. Gamones in Blepharisma

The preunion stage can be shortened to some extent by using highly sensitive cells and increasing gamone concentrations, but it can never be totally eliminated indicating that this stage is needed for cells to prepare for the next two morphogenetic stages.

b. Stage of Ciliary Union. This stage, which also lasts for 1–2 hr, begins with the formation of the face-to-face pair. In this pair, each cell takes a fixed position to the other facing to each other with their peristomal floors (Fig. 8). Throughout this stage, the pair is held by the ciliary adhesion. The union is easily broken up by mechanical stirring, but, if not disturbed, the pair persists until it is transformed into a conjugant pair.

The face-to-face pair is formed in all combinations of mating types. Thus, both heterotypic (type I–type II) and homotypic (type I–typeI, type II–type II) pairs are formed.

c. Stage of Conjugant Union. Conjugant union is formed by a direct contact of cell bodies, not by ciliary adhesion (Miyake, 1974a). In *B. japonicum*, conjugant union begins at the anterior end of the peristomial floor and extends posteriorly. Except at the very beginning, cells in a conjugant pair (Fig. 1) are so firmly united that a surgical operation is needed to separate them.

The difference between the heterotypic and homotypic unions becomes apparent in this stage. In heterotypic pairs, the united region reaches the posterior end of the peristomial floor, whereas in homotypic pairs the extension of the union stops halfway. More important, meiosis and other nuclear changes of conjugation occur only in heterotypic pairs. Heterotypic pairs persist at least until karyogamy which takes place about 1 day after the pair formation at 24°C. The

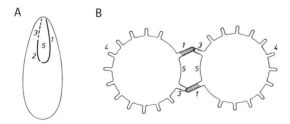

Fig. 8. Diagrammatic illustrations of (A) the peristome and (B) a pair of cells attached by ciliary union (cross section) in *Blepharisma japonicum*. 1, Adoral zone of membranelle; 2, undulating membrane (UM); 3, cilia anterior to UM; 4, somatic cilia; 5, peristomial floor. (A, modified from Bedini *et al.*, 1978, by permission of Company of Biologists Ltd.; B, modified from Honda and Miyake, 1976.)

duration of homotypic pairs is more variable ranging from a few hours to several days depending on the availability of gamone of the other mating type.

2. Three Local Surface Changes for Cell Union

Ciliary and conjugant unions are both formed at the peristomial area. The peristome of *B. japonicum* consists of the adoral zone of membranelles (AZM), the undulating membrane (UM), a row of cilia anterior to UM (antUMC), and a stripe of nonciliated cell surface (peristomial floor) which is surrounded by them (Fig. 8A). Both AZM and UM are made of compound cilia. Cytostome opens at the posterior end of the peristomial floor. The remainder of the cell surface is covered by single cilia (somatic cilia).

During the stage of ciliary union, cells are held together by specific adhesion between AZM of one cell and antUMC of the other (Honda and Miyake, 1976). For the development of this complementarity, first AZM and then antUMC must change their surfaces as shown by the experiment in Fig. 9.

Because AZM and antUMC run nearly parallel at the left and the right margins of the peristomial floor, respectively, cells in ciliary union are held together by two sets of AZM-antUMC bondings facing to each other with their peristomial floors (Fig. 8B). In homotypic union the united cells are of the same mating type. Thus, the homotypic ciliary union provides an interesting example of cell union in which two identical cells are held together in a regular spatial arrangement by locally differentiated complementary cell surfaces.

For the occurrence of conjugant union, the third area of the cell surface, the peristomial floor, must gain the capacity to unite. Unlike AZM and antUMC, peristomial floors appear to be changed in such a way that they can unite between themselves upon their contact. Since cells in ciliary union are already facing each other with their peristomial floors, conjugant union is formed without changing the face-to-face arrangement of cells in ciliary union. By the time this takes place, the peristomial floor has been considerably bulged suggesting a profound change occurring under it.

Local degeneration of cilia, briefly reported by Miyake and Beyer (1973) to occur before conjugant union is formed, was not confirmed by Bedini *et al.* (1978) in their electron microscopic observation. It is likely that Miyake and Beyer, who used only optical microscopes, misinterpreted the prominence of cilia-lacking surface, which is due to the bulging of peristomial floors, as the indication of cilia degeneration.

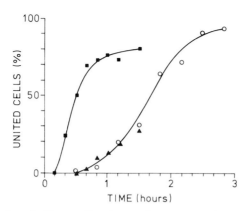

Fig. 9. Formation of ciliary cell union in the mixture of gamone-pretreated and untreated cells in *Blepharisma japonicum*. Mating type II cells (red) were treated by gamone 1 (10^3 U/ml). After 2.5 hr, when many cells had formed ciliary union, they were mechanically stirred to separate most of the united cells and mixed with mating type I cells (albino) in a 1:13 ratio. Pairs were formed immediately after mixing, but they were all red–red pairs. At 16 min after mixing, albino cells started uniting, but only with red cells. These red–albino pairs were united by a single AZM–antUMC bonding formed between albino AZM and red antUMC indicating that AZM but not antUMC of albino cells had acquired the ability to unite. At 40 min after mixing, albino–albino pairs started appearing indicating that now antUMC of albino cells can also unite. (■), Percentage of red cells united with albino cells (excluding red cells in red–red pairs); (▲), percentage of albino cells united with albino cells. The curve of the induction of homotypic unions of albino cells by gamone 2 (○) is shown for comparison. The result indicates that the development of the complementarity between AZM and antUMC for ciliary union is achieved by changing both AZM and antUMC in this order. Essentially the same result was obtained when mating type I cells (red), pretreated by gamone 2, were mixed with untreated mating type I cells (albino). (From Honda and Miyake, 1976.)

Cilia degeneration before the formation of conjugant union does occur in another ciliate, *Paramecium,* and is discussed in Chapter 14 by Hiwatashi.

3. Roles of Ciliary Union

The formation of ciliary union can be prevented by mechanical stirring. If the stirring of gamone-treated cells is stopped at the time at which cells in the control (without stirring) start forming conjugant pairs, conjugant pairs are formed unaffected (Honda and Miyake, 1976). Therefore, the continuous occurrence of ciliary union during the stage of ciliary union is not indispensable for the formation of conjugant union. It appears that the main roles of ciliary union are (1) acceleration of preconjugant interaction by holding together preconjugants of complementary mating types, and (2) preparation for conjugant union by holding two cells in such a way that they face each other by the future sites of conjugant union, the peristomial floors.

4. Homotypic Union as an Experimental Tool

Homotypic union is unique in that it lacks meiosis and other nuclear changes of conjugation. This characteristic, however, becomes apparent only in later stages of cell union. In early stages, heterotypic and homotypic unions are similar in many respects: (1) both unions are induced by gamone (Miyake, 1968); (2) they look alike under the optical microscope (Miyake and Beyer, 1973) as well as the electron microscope (Bedini et al., 1978) during the first few hours; (3) in both of them cells start uniting by ciliary adhesion (Miyake and Beyer, 1973) between the AZM of one cell and the antUMC of the other (Honda and Miyake, 1976); and (4) their resistance to Pronase increases in the same way (Miyake and Beyer, 1973). It may be concluded that these two unions are formed by the same mechanism. The difference develops only thereafter as heterotypic and homotypic cell contacts lead to different consequences. Therefore, in homotypic union, it is as if the process of conjugation is stopped at the stage of cell union.

This situation provides at least two investigative opportunities. First, in homotypic unions, the mechanism of cell union can be investigated under the condition that it is not complicated by the occurrence of nuclear changes of conjugation. Second, the homotypic union provides an excellent control in the investigation of the problem how the heterotypic cell contact induces meiosis.

5. Protein Synthesis

Induction of homotypic union by gamone is completely inhibited by 10 μg/ml cycloheximide applied at the beginning of gamone treatment. Cycloheximide at this concentration does not inhibit taxis to gamone 2 (Section III,E), but it rapidly stops most of the [^{14}C]lysine incorporation into 10% TCA insoluble material. At lower concentrations, inhibition of the incorporation and inhibition of cell union are closely correlated. Based on these results, Miyake and Honda (1976) concluded that cycloheximide inhibits cell union by inhibiting protein synthesis and therefore that protein synthesis is needed for the induction of cell union by gamone.

Gamone also increases amino acid incorporation. The increased incorporation begins about 5 min after the beginning of gamone treatment and continues for about 2 hr (Fig. 10). The end of the increased incorporation roughly coincides with the beginning of the stage of

5. Gamones in *Blepharisma*

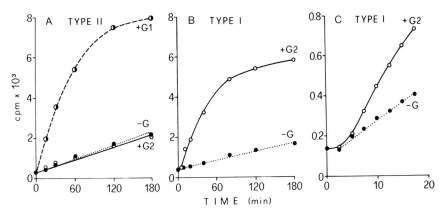

Fig. 10. Gamone-induced amino acid incorporation in *Blepharisma japonicum*. Gamones (10^4 U/ml for gamone 1, 1.6×10^2 U/ml for gamone 2) were added at time 0. [^{14}C]Amino acid was added at time 0 in (A) and (B), 30 min before time 0 in (B). In order to reduce the "background incorporation" which may overshadow the induced incorporation, very starved skinny cells (but still having the strong ability to respond to gamone and unite) were used. (A) Mating type II cells, [^{14}C]leucine. (B) Mating type I cells, [^{14}C]leucine. (C) Mating type I cells, [^{14}C]lysine. +G1, gamone 1; +G2, gamone 2; −G, no gamone. (From Miyake and Honda, 1976.)

ciliary union. The overall separation pattern by SDS electrophoresis of the proteins of the gamone-treated cells labeled during the first 2 hr is distinctly different from that of the nontreated cells. Miyake and Honda (1976) therefore concluded that gamone induces protein synthesis. Because the induced cell union is the homotypic union in which little or no further changes of conjugation occur, it is likely that the induced proteins mainly participate in cell union. These results suggest that cells synthesize proteins needed for cell union during the most part of the preunion stage.

Gamone-induced proteins are, however, of various kinds and many of them also appear to be synthesized in nontreated cells though in smaller amounts. The results of the density gradient centrifugation and agarose chromatography of cell fractions of labeled cells suggest that the bulk of the gamone-induced proteins are associated with lipid-containing particles larger than 10^6 daltons. It appears therefore that most of the gamone-induced proteins are membrane proteins (Miyake and Honda, 1976).

If 10 μg/ml cycloheximide is added shortly after some cells start forming ciliary union, new pairs continue to be formed in spite of the fact that protein synthesis is almost completely blocked and it is not until 4 hr later that all pairs separate. On the other hand, if gamone is removed at a comparable time, new pairs are not formed and those already present all separate within half an hour in spite of the fact that the accumulated gamone-induced protein decreases only slightly during this time. These results suggest that in addition to the accumulation of induced protein, a cycloheximide-insensitive process, which might be the transportation of the synthesized proteins to the site of cell union, also participates in the induction of cell union by gamone (Miyake and Honda, 1976).

6. Ultrastructure

In conjugant pairs, cell membranes at the contact area are juxtaposed with a distance of about 200 nm. This regular arrangement is interrupted in some places by cytoplasmic bridges and vacuoles (Bedini et al., 1978).

During the first few hours in the stage of conjugant union, the cytoplasm of the united region is characterized by the presence of PACM (perpendicularly associated with the cell membrane) microtubules and the abundance of saccules. PACM microtubules, first observed by Ototake (1969), originate from the cell membrane or its vicinity, run perpendicularly or obliquely (and seldom in parallel) to the cell membrane. In heterotypic pairs, they disappear in a few hours after formation of conjugant union, whereas in homotypic pairs they persist for 18 hr or even longer. Saccules are present all through the cytoplasm and all the time, but they tend to be more abundant near the united surfaces in early stages of conjugant union (Bedini et al., 1978).

If isolated cells are singly treated by gamone, PACM microtubules are formed at the presumptive site of cell union and saccules become abundant there. In later stages, many small bodies apparently derived from the saccules are also seen outside the cell at this site. If a doublet, a morphological mutant with two peristomes and hence with the capacity to unite with two cells, unites homotypically with only one cell, the free peristomial floor also shows similar structural changes [Miyake, Nobili, Lanfranchi, and Bedini, cited in Miyake and Honda (1976) and Miyake (1978)] (Fig. 11). Therefore, the formation of PACM microtubules and the accumulation of saccules are both responses of cells to gamone treatment, not the result of the cell union.

These results strongly suggest that PACM microtubules and saccules are devices for cell union, particularly conjugant union. Saccules might contain gamone-induced proteins and might be transported to the site of cell union by PACM microtubules as suggested by Miyake and Honda (1976).

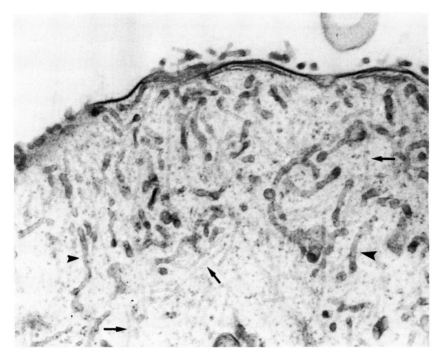

Fig. 11. Presumptive site of cell union of *Blepharisma japonicum*, mating type II, treated by gamone 1 (10^4 U/ml) for 7 hr. Cross section at the anterior part of a free peristomial floor of the doublet cell at the terminus of a homotypic chain consisting of four doublet cells. Arrow, PACM microtubule; arrowhead, one of the saccules supposedly containing gamone-induced proteins. Electron microscopy was carried out as described by Bedini *et al.* (1978). Other conditions and methods were as described by Miyake *et al.* (1977). ×34,000. (Courtesy of A. Miyake, R. Nobili, A. Lanfranchi, and C. Bedini.)

7. Hypothetical Mechanism

Based on the results described above, a hypothetical mechanism, with which a chemical signal of gamone is translated into a specific pattern of cellular arrangement in conjugant pairs, is now constructed. During the preunion stage, gamone induces in its target cells synthesis of proteins, mostly membrane proteins, which are needed for cell union. Some of the early synthesized proteins provide the complementarity between AZM and antUMC so that they can stick to each other upon their contact. This brings about the ciliary union in the face-to-face arrangement. Most of the synthesized proteins are packed in saccules and are transported to the peristomial floor by PACM microtubules which are also induced by gamone and emanate from the peristomial floor. These saccules increase the surface of peristomial floor by incorporating themselves into it and at the same time provide the peristomial floor the capacity to unite with a similarly transformed peristomial floor of the other cell.

Although this hypothesis is based only on *Blepharisma* works, it is essentially the same as previous hypotheses on the mechanism of conjugant-union formation (Miyake and Honda, 1976; Miyake, 1978) in which the results in other ciliates are also considered. Thus the present hypothesis is consistent with such works as the blocking experiments of conjugant-union formation carried out in various ciliates by using enzymes and protein synthesis inhibitors and the studies on the accumulable factor for conjugant-union formation in *Paramecium* and *Tetrahymena* (for reviews of these works, see Crandall, 1977; Miyake, 1974a, 1978, 1981).

G. Gamones in Five Species of *Blepharisma*

In *B. americanum, B. musculus, B. stoltei,* and *B. tropicum,* each of the complementary mating types I and II excretes a gamone which induces homotypic union in the other mating type. Gamone of mating type I (gamone 1) is nondialyzable and heat sensitive like gamone 1 of *B. japonicum.* Gamone of mating type II (gamone 2) is dialyzable and heat resistant like gamone 2 of *B. japonicum.* Thus the gamone system in these five species of *Blepharisma* appears to be similar (Miyake and Bleyman, 1976; Miyake and Mancini, 1978, and unpublished).

Type I cells of *B. americanum, B. musculus, B. stoltei,* and *B. tropicum* all respond to gamone 2 of *B. japonicum* by cell union. Gamones of type II cells in the former four species all induce cell union in type I cells of *B. japonicum* (Miyake and Bleyman, 1976; Miyake and Mancini, 1978, and unpublished). Gamone 2 of *B. americanum* (strain Berlin) is not separable from gamone 2 of *B. japonicum* by cellulose thin layer chromatography (n-propanol:conc.NH_4OH:H_2O; 75:10:15) (Miyake, 1981). Therefore, gamone 2 of these five species appear to be the same molecule, blepharismone. On the other hand, mating type I of each species has one of the blepharismones (A, J, M, etc.) which are more or less specific to species or syngen.

H. Gamone Receptors

As the material basis of gamone recognition, a pair of receptors, receptors 1 and 2, each specific for gamones 1 and 2, respectively, were postulated (Miyake and Beyer, 1973; Miyake, 1974a, 1981). These receptors were supposed to identify the extracellular signal (gamone) and in turn to produce an intracellular signal which switches on cellular mechanism of conjugation.

Revoltella et al. (1976) demonstrated that ^{125}I-labeled gamone 2 binds to type I cells fixed with formaldehyde and glutaraldehyde and that this binding is inhibited by unlabeled gamone 2. About 3×10^6 molecules of gamone 2 are bound per cell under the same condition. The mating type specificity of this binding was not tested, however, by comparing the binding capacities of type I and type II cells. Ricci et al. (1976) showed that phytohemagglutinin W, concanavalin A, and antitubulin antibodies inhibit the induction of homotypic union by gamone 2 only if suboptimal concentrations of gamone 2 are used. They concluded that the receptor site for gamone 2 is distinct from the binding site of these ligands. The same conclusion was also reached by Revoltella et al. (1976).

The induction of homotypic union in type I cells by gamone 2 is competitively inhibited by tryptophan and 5-hydroxytryptophan. The relative activity of 0.54 µg/ml gamone 2 solution with 0.0, 0.5, 1.0, 2.0, 4.0 and 8.0 mM of these inhibitors is 1 · 1/2 · 1/4 · 1/8 · 1/16 · 1/32 for L-5-hydroxytryptophan, 1 · 1 · 1/2 · 1/2 · 1/4 · 1/8 for D-5-hydroxytryptophan, 1 · 1/2 · 1/4 · 1/4 · 1/8 · 1/16 for L-tryptophan and 1 · 1 · 1 · 1/2 · 1/4 · 1/4 for D-tryptophan. Thus L isomers are two to four times more effective than D isomers in this inhibition. L-Leucine is a weak inhibitor, but glycine, L-alanine, L-serine, and L-valine have no inhibiting effect at 0.25–8.00 mM (Miyake, 1974a, and unpublished; Beyer and Miyake, 1973). In the induction of homotypic union in type II cells by gamone 1, L-tryptophan and 5-hydroxy-L-tryptophan have no inhibiting effect at 0.25–4.00 mM (Miyake, 1978, and unpublished). These results indicate that 5-hydroxytryptophan, tryptophan, and possibly also leucine compete with gamone 2 over the receptor of gamone 2 because of their structural similarities to gamone 2. The fact that L isomers are more potent inhibitors than D isomers suggests that L gamone 2 is more effective than D gamone 2.

Other compounds structurally related to gamone 2, 3-indoleacetic acid (~0.25 mM), 5-hydroxyindoleacetic acid (~0.25 mM), serotonin creatinine sulfate (~2 mM), N-formyl-L-kynurenine (~0.5 mM), and β-(indole-3)-DL-lactic acid (ILA) (~1 mM), were also tested. Except 5-hydroxyindoleacetic acid, which was slightly inhibiting, they all increased gamone 2 activity to some extent (A. Miyake, unpublished; Beyer and Miyake, 1973). However, the result requires reinvestigation, because the pH of the medium, an important factor for gamone 2 activity (Miyake, 1981), was not carefully controlled in these experiments. If the pH of the medium was kept at 6.8, the promoting effect of ILA decreased to the barely detectable level (A. Miyake, unpublished).

The direct interaction between gamones 1 and 2 is not detectable in the *in vitro* experiment: the activities of solutions of gamone 1 (7.5 × 10^3 U/ml) and of gamone 2 (1.6 × 10^4 U/ml) are not appreciably changed by mixing with the same amount of solutions of gamone 2 (1.6 × 10^4 U/ml) and gamone 1 (10^6 U/ml), respectively (Miyake, 1978, 1981, and unpublished).

I. Molecular Scheme of Preconjugant Interaction

Based on the results described above, a molecular scheme of preconjugant interaction will now be constructed. The diagram of preconjugant interaction presented by Miyake (1974a) as a working hypothesis (Fig. 12) may serve to illustrate the scheme.

Type I cells excrete gamone 1, a glycoprotein. This molecule reacts with a receptor (hatched area in Fig. 12) on type II cells and induces protein synthesis. Most of the synthesized proteins participate in cell union by transforming three localities of the cell surface, AZM, ant-

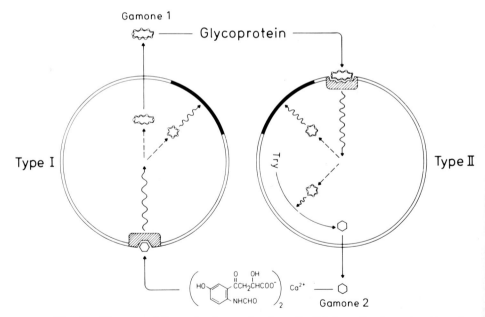

Fig. 12. Diagram of the molecular mechanism of cell interaction in conjugation of *Blepharisma japonicum*. Circles rimmed with a waved line represent proteins. See text for details. (From Miyake, 1974b. Reproduced by permission of Springer-Verlag.)

UMC, and peristomial floor. AZM and antUMC first gain the capacity to adhere with each other thus making possible the formation of face-to-face pairs by ciliary union. Then the peristomial floors are transformed so that they can unite with each other to form conjugant pairs (blackened area in Fig. 12). The transformation of peristomial floor is achieved by accumulating gamone-induced proteins at this area by PACM microtubules which are also induced by gamone–receptor reaction and radiate from the cell membrane at the peristomial floor (not shown in Fig. 12). The other proteins are enzymes that transform tryptophan to gamone 2, calcium- 3- (2'- formylamino- 5'- hydroxybenzoyl)lactate. This molecule is excreted, reacts with a receptor on type I cells, and induces protein synthesis. One of the synthesized proteins is gamone 1, which is excreted and stimulates type II cells as described above, thus closing the loop of positive feedback of gamone induction. The others are proteins for cell union and type I cells gain the capacity to unite just as in type II cells. At the same time, the reaction between gamone 2 and its receptor causes a behavioral change in type I cells so that they move toward the higher concentration of gamone 2 and hence toward type II cells (not shown).

J. Initiation of Meiosis

Preconjugant cell interaction ends with the formation of conjugant union. In the conjugant cell interaction which follows, the role of gamones has not yet been established. Therefore, in this chapter, which deals with cell interaction by gamones, the treatment of conjugant interaction is limited to the earliest stage, initiation of meiosis and other nuclear changes of conjugation, in which possible roles of gamones are being investigated.

In heterotypically united cells of *B. japonicum,* micronuclei start swelling within a few hours and enter meiosis I (Miyake *et al.,* 1979a). Prior to this, by 1–2 hr after the onset of ciliary union, cells are determined to undergo meiosis and other nuclear changes of conjugation. This was demonstrated as follows. Miyake *et al.* (1979b) surgically separated heterotypically united cells at various times after they formed a face-to-face pair by ciliary union. If the occurrence of nuclear changes was later ascertained in an isolated cell, the cell was considered to be activated, i.e., determined for meiosis and other nuclear changes of conjugation, at the time of the operation. Activated cells appear after 0.8 hr and their number steadily increases reaching 100% after 1.8 hr. The time at which 50% of the cells are activated is 1.0 hr and this is regarded as the activation time. Thus, cells are activated by 1 hour of the heterotypic cell contact.

On the other hand, homotypically united cells never undergo such nuclear changes even if they are kept united for days and stay side by side with heterotypic pairs. Therefore, some intracellular signal, which is present in heterotypically united cells but not present in homotypically united cells, must turn on the cellular mechanism for meiosis by 1–2 hr after the onset of heterotypic union.

The role of protein synthesis during this critical period was studied by Miyake et al. (1979b). If pairs are continuously exposed to 10 μg/ml cycloheximide, no cells are activated at least for 5 hr. However, if such cycloheximide-treated pairs are washed and incubated in an inhibitor-free medium, activation occurs with a delay corresponding to the duration of cycloheximide treatment. Based on these results they concluded that heterotypic cell contact induces and maintains the synthesis of a protein, whose accumulation to a certain threshold is required for activation.

The intracellular signal for activation was also investigated by using chains of homotypically united doublet cells. A doublet is a morphological mutant possessing two peristomes. Since conjugant union occurs at the peristome, each doublet cell can unite with two cells. If doublet cells are treated with gamone of the complementary mating type, they unite side by side in chains. Like homotypic pairs, they do not undergo the nuclear changes. However, if even a single cell of the complementary mating type unites at one end of such a chain, the nuclear changes occur there and then propagate through the chain (Miyake, 1975).

Miyake et al. (1977) prepared homotypic chains of doublet cells of various lengths and induced a heterotypic union at one end of each chain. After various times, they surgically separated all cells and observed the occurrence of nuclear changes in isolated cells. In this way they investigated how the activation time of a cell is affected by its position in the chain and by the length of the chain to which it belongs. Analyzing the results, they found that the propagation of activation is slower in longer chains and that the propagation slows down as it proceeds in the chain. These results indicate that the signal for activation is something that is "diluted" along the chain, thus providing evidence that the signal is a substance that is transferable through united cells and requires a certain concentration to work. It appears, therefore, that the heterotypic cell contact induces an intracellular signal substance which turns on the synthesis of protein needed for activation. But it is also possible that the signal and the protein are identical.

Since the problem how heterotypic cell contact induces the hypothetical signal for activation is still to be solved, the possibility that gamones directly participate in this process deserves consideration. Gamone evokes various cellular responses such as chemoattraction, gamone production, and cell union, each requiring a specific threshold concentration of gamone. Although gamone does not activate the cell at concentrations so far used to induce these responses, it might do so at higher concentrations which may be easily achieved if gamone is excreted into a narrow intercellular space of paired cells. This hypothesis is supported by Metz's (1947) finding, that in *P. aurelia* the mating reaction alone is sufficient in inducing the nuclear changes of conjugation. In preliminary experiments, however, highly concentrated solutions of gamone 1 (10^6 U/ml) and gamone 2 (10^5 U/ml) failed to induce any nuclear changes in isolated cells of mating types II and I, respectively (Miyake, 1981). The other possibility that activation is achieved by a mutual exchange of gamones between the united cells through cytoplasmic bridges formed at the beginning of conjugant union has not yet been critically tested.

Currently, the intracellular signal for the initiation of meiosis is being investigated by Miyake and Heckmann (unpublished) and by Santangelo, Nobili, and Salvini (unpublished) using the microinjection technique for free-living ciliates first developed for *Paramecium* (Knowles, 1974; Koizumi, 1974; Koizumi and Preer, 1966). Since this technique makes it possible to inject an intracellular signal into individual cells, it may also be used to investigate other intracellular signals such as the one produced by a gamone–receptor interaction in preconjugant interaction.

IV. PERSPECTIVES

Preconjugant interaction of *B. japonicum* is a simple system of interacting cells in the sense that it consists of only two types of cells, each cell type excreting only one signal substance. However, what actually occurs in this interaction is a complex sequence of events. The complexity is mainly due to two facts: (1) the cellular response to a single gamone is multiple including behavioral (chemotaxis), biochemical (gamone induction), and morphogenetic (cell union) responses; and (2) the interaction triggers a cascade of developmental events including meiosis, fertilization, divisions of syncaryon, nuclear differentiation, reorganization of cell surface patterns, etc. Therefore, the success of future molecular analysis of sexual interaction in this ciliate largely depends on how effectively this integral system is dissected into component processes which can be individually investigated.

The dissection has already been carried out to some extent. First, both of the signal molecules, gamones 1 and 2, have been isolated, purified, and gamone 2 has even been chemically synthesized. This allows to investigate the effect of each gamone in the simplest conceivable system consisting of a single type of cell and a purified gamone. Second, the fact that each cellular response to gamone requires a different threshold of gamone concentration makes it possible to investigate one or two particular cellular responses while the other cellular responses are absent or greatly reduced. Third, it was found that the whole process of conjugation occurs only in heterotypic pairs with the process stopping at the stage of cell union in homotypic pairs. This fortunate situation allows one to investigate some of the component processes of preconjugant interaction uncomplicated by the occurrence of the later processes of conjugation. A potentially more powerful method, dissection of the system by using genetic mutants, may also be exploited in future.

Many of the unit processes thus dissected can be investigated as a specific case of a biological process of general importance. Examples are (1) taxis to gamone 2 as the directed movement of cells; (2) gamone induction by gamone as the recognition of a specific chemical signal, and also as the regulation of biochemical synthesis of a specific substance; (3) gamone-induced homotypic union as the cell union, and more specifically as the uniting of same type of cells, and also as the local differentiation of cell surfaces; and (4) induction of nuclear changes of conjugation by heterotypic cell contact as the initiation of meiosis, and also as the induction of developmental processes by the contact between two cell types.

Perhaps the single accomplishment most needed at this moment is the isolation and characterization of gamone receptors. This would allow the molecular analysis of the crucial problem of how the reaction between gamone and receptor evokes specific cellular responses. In this way, studies on gamones may join current researches on how a signal molecule evokes a specific response in its target cells.

In this respect, blepharismone is particularly interesting because of its similarity to a neurotransmitter, serotonin. They are chemically alike both being a tryptophan derivative. Functionally, they are both intercellular molecular messengers of specific information. This raises a question whether blepharismone and other gamones of small molecular weight, which would be found in future in other ciliates, would be neurotransmitters or hormones in multicellular organisms (Miyake, 1978). The number of known ciliate species is already as large as 7200 and is constantly increasing (Corliss, 1979), suggesting that an enormous number of different kinds of molecules are being used as signals in their sexual interaction. During the long history of evolution, perhaps they have already exploited most of the biochemical processes to produce various kinds of small molecules which can be used as signals. If this is the case, a systematic survey on low molecular weight gamones in ciliates, particularly in heterotrichous ciliates many of which appear to be gamone-excreting species, might turn out to be a rich excavation of hitherto unknown neurotransmitters and hormones.

In addition to the similarity of gamone 2 to serotonin, gamones in *Blepharisma* have two more striking features which deserve comment. One is the chemical difference between gamones 1 and 2. Gamone 1 is a glycoprotein of 20,000 molecular weight, whereas gamone 2 is a much smaller molecule, a tryptophan derivative. This molecular asymmetry in a symmetrical system of interacting cells tempted Nanney (1977) to speculate that gamones 1 and 2 might interact in a key and lock fashion. However, this is unlikely because the two gamones mixed *in vitro* do not neutralize each other (Section III,H) and because type II cells of each species of *Blepharisma* can distinguish species-specific gamone 1 while apparently having the same molecule, blepharismone, as gamone 2 (Section III,G). On the other hand, such a molecular difference between a pair of signals in interacting cells is seen also in other systems. For example, the interaction of negative feedback between the thyroid gland and the anterior pituitary is mediated by thyroxine (amino acid) and thyrotropic hormone (glycoprotein).

A possible clue for understanding the significance of such a molecular difference between signals in cell interaction is obtained by considering another conspicuous feature of *Blepharisma* gamones. As described in Section III,G, blepharismone is probably generic, whereas blepharmone is species-specific. This fact suggests that the origin of blepharismone is as old as the origin of the genus *Blepharisma* or possibly even older, and that species-specific blepharmones appeared later, one by one, as new species were born in this genus. This suggests also that *Blepharisma* or its ancestors had a period during which they had only blepharismone as a conjugation signal. At that time there were no complementary mating types; each cell had blepharismone and its receptor. Therefore conjugation was always selfing. Blepharismone, the production of which was controlled by both internal and external factors, was then used as a signal to herald the time of conjugation (Miyake and Bleyman, 1976). Irrespective of the validity of this inference, a comparative study of ciliate gamones should help in revealing phylogenic relationships among ciliates.

Finally, it should be pointed out that the preconjugant interactions of three ciliates described in this volume, *Blepharisma, Oxytricha,* and *Paramecium,* are quite different from each other. Indeed, sexual interaction in ciliates is subject to great diversity (Nanney, 1977; Miyake, 1978, 1981). Some of them use excreted signals, others use cell bound signals, and still others appear to use signals loosely bound to the cell; positive feedback in signal production was found in some ciliates; in order to exchange signals, cells may collide with, agglutinate with, chemotactically approach or stretch toward target cells. These interactions are so variable that they appear to cover most types of cell interactions hitherto described or postulated in different organisms (Miyake, 1981). Yet, still many more different types of interaction are expected to be found because only a handful of ciliates have been investigated so far in respect to sexual interaction.

Such a diversity is not surprising, however, because it is a general rule that sexual interaction of each species is meticulously programmed so that it fits its own way of life. Indeed, the mode of sexual interaction is almost as variable as the mode of life, as it is seen, for example, in the infinite variety of sexual interaction in metazoa. Thus, in the long history of evolution, ciliates must have developed a tremendous number of different types of cell interaction; they might already have exploited most of the available cellular mechanisms to meet what their sexual interactions demand. A comparable situation is found in multicellular organisms which have developed an enormous number of different types of cell interactions to meet what their complex daily life and embryonic development require.

Therefore, it may be profitable to investigate the sexual interaction in ciliates in the light of the results obtained in the study of cell interaction in multicellular organisms and vice versa.

ACKNOWLEDGMENTS

This work was supported by the Deutsche Forschungsgemeinschaft.

REFERENCES

Bedini, C., Lanfranchi, A., Nobili, R., and Miyake, A. (1978). Ultrastructure of meiosis-inducing (heterotypic) and non-inducing (homotypic) cell unions in conjugation of *Blepharisma. J. Cell Sci.* 32, 31–43.

5. Gamones in *Blepharisma*

Beyer, J., and Miyake, A. (1973). On the molecular mechanism of gamone-induced conjugation in *Blepharisma intermedium*. *Prog. Protozool.*, Int. Congr. Protozool., 4th, Clermont-Ferrand. Abstr. p. 42.

Bleyman, L. K. (1971). Temporal patterns in the ciliated Protozoa. *In* "Developmental Aspects of the Cell Cycle" (I. L. Cameron, G. M. Padilla, and A. M. Zimmerman, eds.), pp. 67-91. Academic Press, New York.

Bleyman, L. K. (1975). Mating types and sexual maturation in *Blepharisma*. *Genetics* 80, S14-S15.

Braun, V., and Miyake, A. (1975). Composition of blepharmone, a conjugation-inducing glycoprotein of the ciliate *Blepharisma*. *FEBS Lett.* 53, 131-134.

Chunosoff, L., Isquith, I. R., and Hirshfield, H. I. (1965). An albino strain of *Blepharisma*. *J. Protozool.* 12, 459-464.

Corliss, J. O. (1979). "The Ciliated Protozoa." Pergamon, Oxford.

Crandall, M. (1977). Mating type interactions in microorganisms. *In* "Receptors and Recognition" (P. Cuatrecasas and M. F. Greaves, eds.), Series A, Vol. 3, pp. 45-100. Chapman & Hall, London.

Giese, A. C. (1973). *"Blepharisma."* Stanford Univ. Press, Stanford, California.

Hirshfield, H. I., Isquith, I. R., and DiLorenzo, A. M. (1973). Classification, Distribution and Evolution. *In "Blepharisma"* (A. C. Giese, ed.), pp. 304-332. Stanford Univ. Press, Stanford, California.

Honda, H. (1979). Cell attraction and cell-to-cell contact. *Seibutsu Butsuri* 18, 19-26. (In Jpn.; Engl. sum.)

Honda, H., and Miyake, A. (1975). Taxis to a conjugation-inducing substance in the ciliate *Blepharisma*. *Nature (London)* 257, 678-680.

Honda, H., and Miyake, A. (1976). Cell-to-cell contact by locally differentiated surfaces in conjugation of *Blepharisma*. *Dev. Biol.* 52, 221-230.

Inaba, F. (1965). Conjugation between two species of *Blepharisma*. *J. Protozool.* 12, 146-151.

Isquith, I. R., and Hirshfield, H. I. (1966). Non-Mendelian traits in *Blepharisma*. *J. Protozool.* 13, Suppl., p. 27.

Isquith, I. R., and Hirshfield, H. I. (1968). Non-Mendelian inheritance in *Blepharisma intermedium*. *J. Protozool.* 15, 513-516.

Knowles, J. K. C. (1974). An improved microinjection technique in *Paramecium aurelia*. *Exp. Cell Res.* 88, 79-87.

Koizumi, S. (1974). Microinjection and transfer of cytoplasm in *Paramecium*. *Exp. Cell Res.* 88, 74-78.

Koizumi, S., and Preer, J. R., Jr. (1966). Transfer of cytoplasm by microinjection in *Paramecium aurelia*. *J. Protozool.* 13, Suppl., p. 27.

Kubota, T., Tokoroyama, T., Tsukuda, Y., Koyama, H., and Miyake, A. (1973). Isolation and structure determination of blepharismin, a conjugation initiating gamone in the ciliate *Blepharisma*. *Science* 179, 400-402.

Kubota, T., Miyake, A., and Tokoroyama, T. (1977). Isolation and purification of conjugation-inducing substances in *Blepharisma*. *In* "Experimental Methods on Natural Organic Compounds—Extraction and Isolation of Physiologically Active Substances" (S. Natori, N. Ikekawa, and M. Suzuki, eds.), pp. 485-498. Kodansha, Tokyo. (In Jpn.)

Metz, C. B. (1947). Induction of "pseudoselfing" and meiosis in *Paramecium aurelia* by formalin killed animals of opposite mating type. *J. Exp. Zool.* 105, 115-139.

Miyake, A. (1968). Induction of conjugating union by cell-free fluid in the ciliate *Blepharisma*. *Proc. Jpn. Acad.* 44, 837-841.

Miyake, A. (1974a). Cell interaction in conjugation of ciliates. *Curr. Top. Microbiol. Immunol.* **64,** 49–77.
Miyake, A. (1974b). Conjugation of the ciliate *Blepharisma:* A possible model system for biochemistry of sensory mechanisms. *In* "Biochemistry of Sensory Functions" (L. Jaenicke, ed.), pp. 299–305. Springer-Verlag, Berlin and New York.
Miyake, A. (1975). Control factor of nuclear cycles in ciliate conjugation: Cell-to-cell transfer in multicellular complexes. *Science* **189,** 53–55.
Miyake, A. (1978). Cell communication, cell union, and initiation of meiosis in ciliate conjugation. *Curr. Top. Dev. Biol.* **12,** 37–82.
Miyake, A. (1981). Physiology and biochemistry of conjugation in ciliates. *In* "Biochemistry and Physiology of Protozoa" (M. Levandowsky and S. H. Hutner, eds.), Vol. 4, pp. 125–198. Academic Press, New York.
Miyake, A., and Beyer, J. (1973). Cell interaction by means of soluble factors (gamones) in conjugation of *Blepharisma intermedium. Exp. Cell Res.* **76,** 15–24.
Miyake, A., and Beyer, J. (1974). Blepharmone: A conjugation-inducing glycoprotein in the ciliate *Blepharisma. Science* **185,** 621–623.
Miyake, A., and Bleyman, L. K. (1976). Gamones and mating types in the genus *Blepharisma* and their possible taxonomic application. *Genet. Res.* **27,** 267–275.
Miyake, A., and Honda, H. (1976). Cell union and protein synthesis in conjugation of *Blepharisma. Exp. Cell Res.* **100,** 31–40.
Miyake, A., and Mancini, P. (1978). Pathway of mating information in the genus *Blepharisma* and species relationships. *Boll. Zool.* **45,** 227.
Miyake, A., Maffei, M., and Nobili, R. (1977). Propagation of meiosis and other nuclear changes in multicellular complexes of *Blepharisma. Exp. Cell Res.* **108,** 245–251.
Miyake, A., Heckmann, K., and Görtz, H.-D. (1979a). Meiosis in *Blepharisma japonicum. Protistologica* **15,** 473–486.
Miyake, A., Tulli, M., and Nobili, R. (1979b). Requirement of protein synthesis in the initiation of meiosis and other nuclear changes in conjugation of *Blepharisma. Exp. Cell Res.* **120,** 87–93.
Nanney, D. L. (1974). Aging and long-term temporal regulation in ciliated Protozoa. A critical review. *Mech. Ageing Dev.* **3,** 81–105.
Nanney, D. L. (1977). Cell–cell interactions in ciliates: Evolutionary and genetic constraints. *In* "Microbial Interactions" (J. L. Reissig, ed.), Receptors and Recognition, Series B., Vol. 3, pp. 353–397. Chapman & Hall, London.
Nanney, D. L. (1980). "Experimental Ciliatology." Wiley, New York.
Ototake, Y. (1969). Electronmicroscopy of the cortical structure in *Blepharisma intermedium* during conjugation. *Biol. J. Nara Women's Univ.* **19,** 45–47 (In Jpn.; Engl. sum.).
Revoltella, R., Ricci, N., Esposito, F., and Nobili, R. (1976). Cell surface control in *Blepharisma intermedium* Bhandary (Protozoa Ciliata) 1. Distinctive membrane receptor-site in mating type I cells for various interacting ligands. *Monit. Zool. Ital.* **10,** 279–292.
Ricci, N., Esposito, F., Nobili, R., and Revoltella, R. (1976). Cell surface control in *Blepharisma intermedium* (Protozoa Ciliata). Blocking of gamone II-induced homotypic pairing by different membrane interacting ligands. *J. Cell. Physiol.* **88,** 363–370.
Smith, S. G., and Giese, A. C. (1967). Axenic media for *Blepharisma intermedium. J. Protozool.* **14,** 649–654.
Sonneborn, T. M. (1970). Methods in *Paramecium* research. *Methods Cell Physiol.* **4,** 241–339.

Sonneborn, T. M. (1978). Genetics of cell–cell interactions in ciliates. *Birth Defects, Orig. Artic. Ser.* 14(2), 417–427.
Suzuki, S. (1957). Conjugation in *Blepharisma undulans japonicum* Suzuki with special reference to the nuclear phenomena. *Bull. Yamagata Univ. Nat. Sci.* 4, 43–68.
Tokoroyama, T., Hori, S., and Kubota, T. (1973). Synthesis of blepharismone, a conjugation inducing gamone in ciliate *Blepharisma*. *Proc. Jpn. Acad.* 49, 461–463.
Wolfe, J. (1973). Conjugation in *Tetrahymena:* The relationship between the division cycle and cell pairing. *Dev. Biol.* 35, 221–231.

6

Sex Pheromones in *Neurospora crassa*

M. S. ISLAM

I.	Introduction	131
II.	Survey of the Literature	134
III.	Current Research	138
	A. Pheromone Extraction	139
	B. Biological Characterization	140
	C. Biochemical Characterizations	147
IV.	Perspectives	150
	References	157

I. INTRODUCTION

Neurospora, the so-called "red bread mold," has been extensively used during the last 40 years for the study of genetics, particularly at the biochemical level. The advantage of working with this organism is that it exhibits orthodox genetic behavior with easily manageable sexual crosses and can be grown very conveniently in inexpensive, chemically defined media. Thus, it has been possible to induce, isolate, and examine mutants affecting subtle biochemical characteristics. Recently, this organism has been shown to be a very convenient model system for studying the hormonal regulation of sex gene expression in eukaryotes.

The genus *Neurospora* was originally known as *Monilia* before its sexual stages were discovered by Shear and Dodge in 1927. Initially, most of the cytological and cytogenetical studies were done with

N. tetrasperma and *N. sitophila* but later *N. crassa* was most extensively utilized for genetic studies.

Neurospora crassa exists as two mating types, *A* and *a*, which are determined by a difference in a single gene. Each of the mating types is normally capable of producing primordial reproductive organs or protoperithecia which are small, dark brown bodies, formed by a thick weaving of hyphal cloak around a coiled branch called the ascogonium [Fig. 1(5)]. From the tips of the ascogonium develop long, receptive hyphae called trichogynes. A conidium or mycelial fragment from the opposite mating type is first received by the trichogynes [Fig. 1(6)] and the nucleus migrates down to the body of the protoperithecium (ascogonium) where it continues to divide along with the female nucleus [Fig. 1(7)]. Nuclear division at this time is synchronous. Individually, the newly formed nuclei enter projections in the ascogonium [proasci, Fig. 1(8)] where they fuse with the haploid maternal nuclei of the opposite mating type. The first two divisions [Fig. 1(9, 10)] of the diploid zygote nucleus are meiotic, resulting in the formation of four nuclei with the coincident reduction of the chromosome number from 14 (diploid) to 7 (haploid). An additional mitotic division [Fig. 1(11)] results in eight nuclei which are arranged in a row in order of their formation during meiosis. During ascus development, the hyphal "mantle" of the protoperithecium enlarges and takes on the typical form of perithecium [Fig. 1(13)], which becomes black through the accumulation of melanin pigment. [Details of chromosome behavior in the ascus have been reported by McClintock (1945) and Singleton (1953) and are beyond the scope of this review.] Later, cell walls develop around each nucleus [Fig. 1(12)] and oval sexual spores, ascospores [Fig. 1(1)] (28 μm long), are produced. During maturation the ascospores blacken and each comes to possess two nuclei by an additional mitotic division. After being heat activated, each ascospore germinates and can then either propagate by itself by vegetative reproduction or mate with the opposite mating type. Each ascospore forms a haploid, genetically homogeneous mycelium [Fig. 1(2)] of one or the other mating type. In *Neurospora* any part of the mycelium can act as a reproductive unit.

The vegetative mycelium of *Neurospora* is coenocytic because of incomplete septation. The young mycelium, on a solid substrate, is whitish but after about 7 days of growth becomes pink-red with the formation of asexual reproductive bodies, i.e., the conidia. Conidia [macro, Fig. 1(3)] are formed by constriction of the aerial hyphae and are usually present in rows. The conidia are oval, multinucleate structures which normally measure 7 μm long and contain 4–6 nuclei.

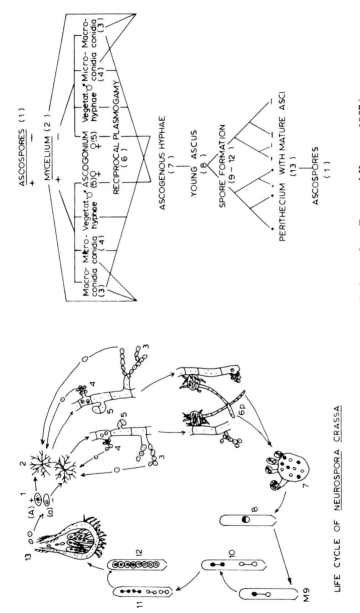

Fig. 1. Life cycle of *Neurospora crassa.* (Redrawn from Esser and Kuenen, 1967.)

Another type of asexual spore is the microconidium [Fig. 1(4)]. Microconidia are spherical, binucleate structures, about 3 μm in diameter. They are produced endogenously in constricted cells. In submerged cultures the formation of conidia is suppressed and coenocytic mycelia are produced.

Neurospora can be conveniently grown in a minimal synthetic medium (Medium N; Vogel, 1956). This medium is widely used for the vegetative growth of wild as well as mutant strains. For the growth of "auxotrophs" this medium is supplemented with the required biochemical (amino acids, vitamins, nucleic acid metabolites, etc.) at an appropriate concentration, usually 5–10 mg/100 ml. A slightly different medium known as "complete medium" may also be utilized for the culture of biochemically deficient strains. To prepare complete medium, Medium N is supplemented with a carbon source, 0.5% yeast extract, and 0.5% casein hydrolysate (N-Z-case, Sheffield). For mating studies a synthetic crossing medium (SC) is used (Westergaard and Mitchell, 1947). Presently, the biotin concentration of medium N is usually increased 100-fold and phosphate is supplied as K_2HPO_4 and KH_2PO_4 rather than as the monobasic salt alone. The pH of the medium is 6.5. (For a ready reference of culture media and methods, see Stanford *Neurospora* Methods, 1963.)

Although *Neurospora* was originally collected from pieces of bread in a bakery in France, it was found to be widely available in moist tropics where it appears following the burning of vegetation. A list and the characteristics of 675 *Neurospora* spp. collected from nature between 1968 and 1974 are available in the paper of Perkins *et al.* (1976). The results of their work led them to group the newly collected strains into four previously known species, *N. crassa, N. sitophila, N. intermedia,* and *N. tetrasperma*. The different wild-type strains of *Neurospora* and various mutants are available from the Fungal Genetics Stock Center, Humboldt, ARCATA, United States.

II. SURVEY OF THE LITERATURE

The hormonal regulation of sexual development in fungi was first proposed by Sachs and DeBary in the 1880's but Burgeff (1924) provided the first experimental evidence for the existence of such substances (for review, see Raper, 1957, 1960). His observations were based on the fact that the plus (+) and minus (−) strains of *Mucor mucedo* could produce "zygophores" which grew toward each other even when separated by a celloidin membrane. Since then the presence of diffusible substances which function in the sexual cycle of fungi has been demonstrated in several organisms, for example, *Achlya* (Raper, 1939a,b, 1951, 1957, 1960; Barksdale, 1960, 1963, 1969); *Allomyces* (Machlis, 1958a,b); *Ascobolus* (Bistis, 1956, 1957; Bistis and Raper, 1963); *Mucor* (Banbury, 1954, 1955; Plempel, 1957; Plempel and Braunitzer, 1958); *Saccharomyces* (Levi, 1956; Yanagishima, 1969, 1971); *Bombardia* (Zickler, 1952); and *Glomerella* (Markert, 1949; Driver and Wheeler, 1955).

In *Neurospora* the presence and activity of "sexual hormone(s)" have been the subject of much controversy. This has been reviewed extensively by Raper (1952). Although their existence has been demonstrated by some investigators, others have not been able to corroborate these results. In 1931 Moreau and Moruzi claimed to have demonstrated "sexual hormones" in *Neurospora (Monilia)*. In their experiments a U-tube filled with medium was inoculated at one end with one mating type strain (M) and the other end was inoculated with an opposite mating type strain (N). Although single-strain cultures produced only small protoperithecia, in the U-tube culture the arm containing the strain N showed the development of large-size perithecia (sclerotia) while in the arm containing the strain M, normal, mature perithecia (sclerotia) were observed. Small portions of the agar contained in the horizontal section of the U-tube were cut out and incubated under humid conditions. Since no growth was observed, it was concluded that no hyphal growth had occurred between the two vertical ends of the tube. In the view of these authors, the production of perithecia was due to the action of diffusible hormones from one mycelium to the other through the solidified medium. However, in the same year, Dodge (1931) repeated the U-tube experiments, using a conidial strain and a nonconidial strain. Perithecia appeared in the ends of the tubes only when air spaces were formed by the shrinking agar medium and therefore were present throughout the greater portion of the length of the horizontal part of the U-tube. In the cases where perithecia were initiated, it was shown that both strains, conidial and nonconidial, were present in the medium in the immediate vicinity of the perithecia. Thus, Dodge (1931) showed that in the U-tube experiments the strains actually crossed into the other culture. In a subsequent paper, Moreau and Moruzi (1932) claimed to provide further evidence for hormonal action in the induction of perithecia. Their new observations pointed to the presence of a wide "restraint area" in the region where compatible mycelia intermingle, together with the occurrence of perithecia on either side of these areas.

Aronescu (1933, 1934) subjected the hormonal theory to intensive genetic analysis. Using compatible conidial and nonconidial strains and the U-tube technique, she isolated asci from perithecia wherever they appeared, and determined the sexual and conidial characteristics of the mycelia derived from the eight separated spores of each. Theoretically, three possibilities exist: (a) segregation due to syngamy of both factors in a perfectly normal Mendelian ratio, (b) no segregation for both factors, all spores alike and identical to the parent strain (selfing of a single strain under the hormonal influence of the other), and (c) segregation for the sex factor but conidial character always that of the "sole" parent. In an analysis of about 50 asci, spores of each parental type were formed, providing conclusive evidence that the distance between the two ends of the tube had been bridged by hyphal growth and that compatible nuclei had been brought together. The same results were again obtained when the work was repeated using the same strains as used previously by Moreau. Aronescu also disputed Moreau's claim that two different mycelia (but of the same mating type), when grown together on a single plate, produce perithecia. In many trials her results were uniformly negative. Thus, a more accurate method of separating hyphae was required to prove the existence of sex pheromones.

Lindegren (1936) claimed to have confirmed the existence of sex hormones in *Neurospora*. A bisexual, heterokaryotic, self-sterile strain (containing f^+ and f^- nuclei) was mated to a highly fertile strain (F^+). Two kinds of zygote resulted: (a) f^+/f^- and (b) F^+/f^-. According to Lindegren, hormones from the F^+ strain had enabled the heterokaryotic, sterile strains to copulate. Although in their initial investigation Moreau and Moruzi (1931) assumed that their wild-type *Neurospora* strains were homokaryotic and that hormones had stimulated them to reproduce parthenogetically, in the light of Lindegren's work their strains were probably bisexual and self-sterile, and hormones possibly supplied an essential substance for fruiting. Commenting on the work of Dodge and Aronescu, Lindegren (1936) stated, "In Dodge and Aronescu's experiments homokaryotic (+) and (−) strains were used in two arms of the U-tube. Such an experiment cannot test the production of hormones. They also used single conidium cultures of the Moreau and Moruzi wild strain in an attempt to reproduce Moreau and Moruzi experiments." According to Lindegren (1936), such an experiment is "particularly liable to fail because by making single conidium isolates, the heterokaryon is easily separated."

In 1937 Moreau and Moreau described experiments in which they introduced further evidence in favor of the existence of sex hormones in *Neurospora*. In a cross of *N. sitophila* with *N. tetrasperma,* some asci were detected which contained spores of the *tetrasperma* type. It was therefore concluded that the origin of these spores stemmed from the "selfing" of *tetrasperma*-type nuclei under the influence of "sex hormones."

A subsequent paper by Moreau and Moreau (1938a) provided more convincing evidence for hormonal activity in *Neurospora*. Mycelia of one sexual strain growing on an agar plate were killed by heating for 10 min in steam at atmospheric pressure. Blocks 1 cm^2 of agar were cut out from these plates and placed on the surface of a living mycelium of a strain of opposite mating type. Perithecia developed on this compatible strain. A negative result was obtained if the temperature was raised to 110°C. Later, Moreau and Moreau (1938b) treated the mycelium by ether or chloroform vapor and obtained similar results. They concluded that their results indicated the presence of hormones in the medium of the killed thallus, which initiated perithecia development in the compatible strain. However, a genetic analysis of the sex and conidial characteristics of the progeny from "induced" perithecia was not carried out.

Backus (1939), while studying the mechanics of ascogonial fertilization in *N. sitophila*, observed two phenomena which suggested the presence of a sex hormone. The germination of conidia was completely inhibited in regions previously overgrown by mycelia of the opposite compatibility strain. This effect was interpreted as being due to an inhibiting substance provided by the opposite mating type. In addition, it was observed that the long, sparsely branched trichogynes did not produce short, lateral branches in the absence of conidia, microconidia, or hyphal fragments of the opposite compatibility strain in the immediate vicinity. In the presence of potential fertilizing elements, one to several lateral branches were regularly observed to arise upon the trichogyne and grow directly toward the source of the fertilizing nuclei.

Hirsch (1954) and Barbesgaard and Wagner (1959) later provided some information about environmental and biochemical factors involved in protoperithecia formation and postulated that the incidence of protoperithecia development is directly proportional to the melanin content of the fungus.

Meanwhile, Ito (1956) had reported the effect of culture filtrate on the induction of protoperithecia formation in *Neurospora*. He found that the filtrate of the medium containing mycelia of a single strain of either mating type was less effective in stimulating perithecia formation than the filtrate of medium containing mycelia of both mating types. The effect of mixed-culture filtrate was found to surpass the additive effect of two single-culture filtrates of each mating type. This observation was taken to indicate the presence of a cooperative but unknown second factor in the mixed-culture filtrate. In fact, the single-culture filtrate of one mating type stimulated the formation of (proto)perithecia on mycelia of opposite mating type, although those (proto)perithecia failed to produced asci and ascospores. Ito (1956) concluded that the formation of perithecia on the haploid vegetative thallus of one mating type was stimulated by a diffusible substance secreted by mycelia of the opposite mating type, quite irrespective of sexual fushion of both mating types. Finally, Ahmad et al. (1967) reported that filter-sterilized mycelial extract from a mated culture increased the fertility of semisterile mutants of *lys-5* locus. Mycelial extracts of single-strain cultures (either Em a or Leu-1A) and culture filtrates (Em a, Leu-1A, and Em a × Leu-1A) did not have positive effects on the restoration of the weakly fertile mutants of the *lys-5* locus.

As a result of the controversy over the existence of sex pheromones in *Neurospora,* the author, under the influence and encouragement of his supervisor, Prof. J. Weijer, decided to embark upon a thorough and renewed investigation of this phenomenon with the ultimate objective of isolating and characterizing such substances. The initial plan of this study was a corroborative one with Dr. N. V. Vigfusson who had begun studies on the genetics of the sexual cycle of *Neurospora crassa.* Although the genetic control of the sexual cycle had been extensively analysed in three other homothallic ascomycetes, *Sordaria macrospora* (Esser and Straub, 1956, 1958), *Sordaria fimicola* (Olive, 1956), and *Glomerella cingulata* (Wheeler and McGahen, 1952; Wheeler, 1954), an analysis had not been done for *Neurospora* (see Vigfusson et al., 1971). Vigfusson initiated the work, for the first time, under the guidance of Prof. J. Weijer, at the University of Alberta, Canada, and when I joined them in 1967 I decided to investigate the hormonal control mechanism, utilizing primarily some of their induced mutants (uv) blocked at specific phases in the sexual development.

III. CURRENT RESEARCH

Before describing our efforts in elucidating and analyzing sex pheromones (sex and fertility-inducing substances) in *Neurospora crassa,* I will mention some initial unsuccessful attempts which might be of interest. Initial experiments were started in an attempt to induce perithecia formation in wild crosses of *Neurospora* (Em A × Em a) in submerged culture. Several experiments ranging from shaking to nonshaking, nonshaking to shaking, and constant shaking conditions, as well as forced aeration of the cross culture, led to no success. Similarly cross-feeding experiments with sterile (uv-induced) and wild crosses did not give any substantial evidence for the presence of pheromonal substances. In addition, U-tube experiments with some of the sterile strains obtained from Vigfusson were uniformly unsuccessful. It was only with an acetone extract of wild cross (Em A × Em a) from agar medium that some evidence for fertility enhancement in the sterile crosses was obtained. This was the initial point for the biochemical extraction of fertility-inducing substances. Subsequently, by adopting basically similar procedures to those used by Machlis (1958b), hormonal substances from *Neurospora* could be conveniently isolated.

A. Pheromone Extraction

Details of the extraction procedures have already been reported (Islam and Weijer, 1969; Vigfusson *et al.*, 1971). For the extraction of sex and fertility-inducing substances, 15 ml of sterilized liquid crossing medium were poured into a standard petri dish containing five sheets of sterilized filter paper (Whatman No. 1, 9 cm). Inocula of both Em A and Em a were made simultaneously on the filter paper close to one another and the culture was incubated at a temperature of $23°–26°C$ for three weeks, by which time free ascospores could be detected. In some experiments cultures were allowed to grow for only two weeks. Usually, 50 plates were grown in a single experiment. In single-strain cultures (Em A or Em a) the same procedures were followed except that the minimal medium was used.

A water extract was made after 2 or 3 weeks of incubation. Filter papers containing the mycelia were initially soaked in 3000 ml of sterilized, distilled water for 2 hr with occasional stirring. The water was decanted and the filter papers were soaked again in 3000 ml of sterilized, distilled water. The water extract (6000 ml) was then filtered through glass wool and activated charcoal (Norit-A) was then added to the water extract (5 gm/500 ml). The mixture was agitated for a few minutes, then left at room temperature for about 3 hr. At this time, the charcoal was collected by filtration and dried at room temperature. Three hundred ml of chloroform were added to the partially dried charcoal and agitated for a few minutes, then left for 3 hr with occasional agitation. The chloroform mixture was filtered twice through five layers of filter paper No. 1. The extraction process was repeated once more, yielding about 500 ml of chloroform extract which was then evaporated under vacuum. The residue was dissolved in 15 ml chloroform and stored in the refrigerator. For purification of the *Neurospora* extracts, the lipid purification method of Folch *et al.* (1957) was followed. Redistilled water (0.2 by volume) was added to the final extract and mixed thoroughly, and the mixture was allowed to separate into two layers for 15 to 30 min, after which the upper layer (water) was discarded. The process was repeated once more, after which the solvent layer was stored under refrigeration. The extract (0.1 ml) was added directly to each plate (20 ml) or was dried on a sterilized, triangular piece of filter paper before addition to the culture. Extraction of hormonal substances was also tried with chloroform-methanol mixture (2:1), acetone, and ethyl alcohol (95%). On preliminary testing of the different extracts on wild crosses, it was found that the chloroform, chloroform–methanol, and acetone extracts showed biological activity by enhancing the number of perithecia produced. The alcohol extract showed no effect. The chloroform extraction procedure was adopted as the standard extraction method for later experiments, although initially some experiments were also done with a chloroform–methanol extract (Islam, 1970).

The present standard practice is to grow the single strains in liquid minimal medium (10 ml/100 ml Erlenmeyer flask) for 15 days, after which water extraction is made (1 liter/20 cultures), followed by activated charcoal (5 gm/500 ml) and chloroform (100 ml) extraction. The final residue is collected in 10 ml of chloroform.

B. Biological Characterization

1. *Extract of the Cross Em A × Em a*

The biological characterization was initiated using an extract from an Em A × Em a culture with a culture of the same type (Em A × Em a). The objective was to see the hormonal effect, if any, on the wild cross (Em A × Em a) which is by itself less fertile. In these experiments either a chloroform, a chloroform–methanol, or an acetone extract was found to increase five- to tenfold the number of perithecia developed. The addition of extracts to the filter paper in the medium gave rise to a strong, localized response by the organism, in that most of the perithecia developed on the area in a congregated manner. In the control plates the perithecia developed evenly over the entire plate (Fig. 2).

The chloroform and chloroform–methanol extracts were then tested for their activity on the single strains (i.e., Em A or Em a) and on sterile, biochemical mutants of the A mating type. On visual examina-

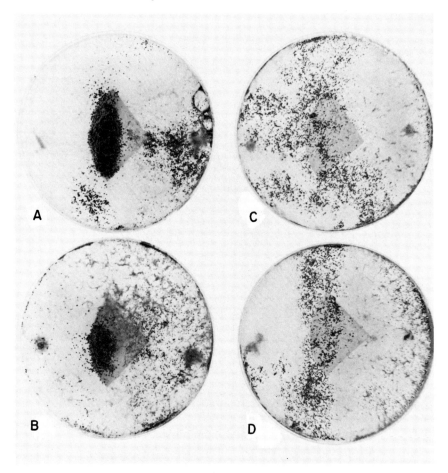

Fig. 2. Photographs showing the effect of the extract of the cross Em A × Em a on the cross Em A × Em a (wild-types), when compared with the untreated controls. (A, B) Cross extract; (C, D) controls.

tion of the treated plates it was concluded that the extract of the cross Em A × Em a increased the number of protoperithecia developed by the wild strain Em A but no induction of protoperithecia was noted when Em a was used as a tester strain. Using Em A as a tester strain, it was observed that an increase in the concentration of the extract yielded an increase in the number of visible protoperithecia and protoperithecia-like bodies per plate (Table I). In some instances, this experiment yielded fertile perithecia with mature ascospores (Islam and Weijer, 1972a). It was noted, however, that an increase in the

TABLE I

Dose Effect of the Extract of Cross Em A × Em a on Single Wild-Type Strain Cultures (Em A and Em a)[a]

Strain	Conc. of extract (ml)	7-day plates (development of protoperithecia)				15-day plates (no. of visible protoperithecia)				Avg.
		1	2	3	4	1	2	3	4	
Em A	0.1	+	+	+	+	35	38	32	38	35
	0.2	+	+	+	+	42	40	33	34	37
	0.3	+	+	+	+	45	40	42	38	41
	0.4	+	+	+	+	48	46	60	55	52
	0.5	+	+	+	+	68	65	52	54	60
	Control	−	−	−	−	−	−	−	−	0
Em a	0.1	−	−	−	−	−	−	−	−	0
	0.2	−	−	−	−	−	−	−	−	0
	0.3	−	−	−	−	−	−	−	−	0
	0.4	−	−	−	−	−	−	−	−	0
	0.5	−	−	−	−	−	−	−	−	0

[a] +, Development of visible protoperithecia and protoperithecia-like bodies; −, no development of visible protoperithecia.

concentration of the extract, even up to 0.5 ml/plate, did not have any visible effect on the development of protoperithecia and protoperithecia-like bodies by tester strain Em a. This extract was later found to be capable of rejuvenating the sexual potentiality of sterile isolate I-5 (obtained from the selfed progeny of Em A) when crossed to Em a (Islam, 1973) (Fig. 3). On a bisexual, self-sterile isolate I-28 the extract of Em A × Em a was found to have a suppressive effect in the development of visible protoperithecia and protoperithecia-like bodies (Islam and Weijer, 1971).

The effect of the extract of the cross Em A × Em a on sterile, biochemical mutants of A mating type was all the more interesting. Sterile mutants 5366-A, 10710-A, 7232-A, 9312-A, and 10402-A (St. L. background) were used as tester strains for determining the biological effect of the cross extract (Em A × Em a) on protoperithecia development and restoration of fertility. Of these strains, 5366-A was a pantothenic acid-deficient mutant whereas 7232-A, 10710-A, 9312-A, and 10402-A were leucine-deficient mutants. The extract proved to have positive biological effect on all biochemical mutants with respect to the enhancement of the number of protoperithecia developed by these strains. In addition, the extract improved the fertility of all the biochemical mutants except 5366-A, which remained completely

6. Sex Pheromones in *Neurospora crassa*

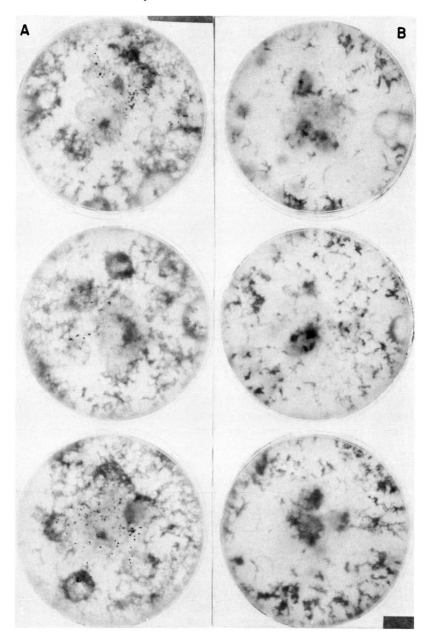

Fig. 3. Photographs showing the effect of the extract (A) of the cross Em *A* × Em *a* on a cross I-5 (sterile) × Em *a* (wild), when compared with the untreated control (B). I-5 (sterile) was an isolate from the "selfed" progeny of Em *A*.

sterile. 5366-*A* was characterized by complete male and female sterility and under control conditions this mutant produced an abundant number of small, brown protoperithecia and protoperithecia-like bodies when crossed to Em *a*. With the application of cross extract, an abundant number of perithecia-like bodies developed in a cross between 5366-*A* × Em *a*. But the asci were found to be immature and frequently without an ostiole. Upon dissecting these asci, no ascospores were observed. Although positive effects on the improvement of fertility were observed in the other four cases, in the cases of 7232-*A* and 10710-*A* the effect could be said to be marginal. The most prominent effects were in the cases of cross 9312-*A* × Em *a* and 10402-*A* × Em *a*. As such, the substances present in the extract were assumed to be more active on the later stages of sexual development (i.e., from karyogamy onward) (Vigfusson et al., 1971). The specificity of the extract of the cross Em *A* × Em *a* to the mating type *A* was also proved from an experiment with a semisterile mutant of *a* mating type (8455-*a*, St. L. background). Crosses between 8455-*a* and fluffy-*A* treated with the extract of Em *A* × Em *a* showed no increase in the number of perithecia when compared with control plates (untreated). Since 8455-*a* was a sterile male and a fertile female, fluffy-*A* was used as protoperithecial strain. In all other cases the crosses were "simultaneous."

2. *Single-Strain Extract (Em A, Em a)*

Since the biological property of the "cross extract" Em *A* × Em *a* was shown and since positive evidence for pheromonal activity had been obtained, it was decided to make single-strain extracts from Em *A* and Em *a* and to test their effects on the single, wild-type strains (Em *A*, Em *a*), on the wild-type cross (Em *A* × Em *a*), and on the sterile, biochemical mutants affected at different stages in the sexual cycle of *Neurospora crassa* (5366-*A*, 7232-*A*, 10710-*A*, 9312-*A*, and 10402-*A*). Extraction of the hormonal substances and the biological assay procedures were similar to those detailed above.

When the extracts of Em *A* or Em *a* were applied to the tester strain Em *A*, an increase in the number of visible protoperithecia and protoperithecia-like bodies initiated by Em *A* was observed. Tester strain Em *a*, on the other hand, did not respond to either extract. It was observed that, in comparing the results obtained with the extract of Em *A* and of Em *a* on tester strain Em *A*, the extract of Em *a* developed more protoperithecia and protoperithecia-like bodies than the extract of Em *A*.

Both the extracts Em *A* and Em *a* were found to have a positive effect on the fertility of the cross Em A × Em a. A four- to sixfold increase in the number of perithecia (when compared with untreated, control plates) was noted. In the treated plates the test organisms responded chemotactically to the extracts.

In the presence of extracts Em A and Em a, 5366-A developed a large number of perithecia-like bodies when crossed to Em a. The perithecia were mostly ill developed and small. When dissected, no ascospores were found. The effect of single-strain extracts on mutants 7232-A, 10710-A, 9312-A, and 10402-A when crossed to Em a consisted of a general increase in the production of protoperithecia and protoperithecia-like bodies as well as perithecia together with an improvement of fertility.

Both the single-strain extracts showed a positive effect on the cross I-5 (sterile) × Em a. Protoperithecia and protoperithecia-like bodies were developed in the treated plates (Islam, 1973). Upon dissection of the resulting perithecia, ascospores were detected. From the results obtained it appeared that the extract of Em a is more potent in its biological effect than the extract of Em A. In the control plates occasional (1–2) perithecium-like bodies (without ascospores) were observed.

The effect of single-strain extracts on the fertility of a semisterile mutant of a mating type (8455-a) was also tested but these proved to have no positive effect.

3. Futher Studies on the Nature of the Sex Pheromones

a. Biochemical Basis for "Sterility" of the Sterile Isolate I-5A. Isolate I-5 was from a "selfed" progeny of Em A, with selfing induced by the hormonal extract of cross Em A × Em a (Islam and Weijer, 1972a). This isolate was initially found to be completely male and female sterile but, with the application of pheromonal extracts, particularly from the cross Em A × Em a and from Em a alone, fertility could be restored in a cross with Em a (Islam, 1973). As a result, a more rigorous investigation of the biochemical basis for the sterility of this strain was carried out. The results indicated that the mycelial extract of I-5 possessed some fertility inhibitory principles. Later this strain was studied along with other uv-induced mutants for qualitative and quantitative variations in its ability to produce hormone. The results (based on spectrophotometric analysis) showed that this strain produces pheromonal substances (possibly fertility inhibitors) in excess, which could be the primary cause of its sterility (Shirin, 1974, p. 98). This was in agreement with our hypothesis.

b. Specificity of the Hormonal Extract of the Cross Em A × Em a. In earlier studies we followed the method of the "simultaneous crossing system" in which the two compatible strains were inoculated at two diagonally opposite sides of the petri dish, near the margin, thus allowing the growing mycelia to meet in the middle of the culture plates where the extracts were added. Thus, both the strains in a cross (sterile × fertile) acted both as male and as female parents. This gave no idea whether the inducing substances present in the extracts acted on the male sterility or on the female sterility (or on both) of A sterile strains, which were both male and female sterile. So it was decided to use the sterile strains both as male and female parents (reciprocal crossing system) and observe the effect of the extract of Em A × Em a on the improvement of fertility.

From the results obtained (Islam, 1975) it was evident that the sexual hormones present in the cross Em A × Em a are active on the male sterility of certain A strains, but not on the female sterility of the same mating type. The same extract enhanced the suppression of perithecia formation of a sterile strain.

c. Qualitative and Quantitative Differences of the Hormonal Extracts of Fertile and Sterile Strains. In an attempt to see the qualitative as well as quantitative differences between the fertile and sterile strains with respect to sex and fertility-inducing substances (sex pheromones), extracts were prepared from the homokaryons of Em A, Em a, St.L. A, St.L. a (wild types), and 7232-A, 9312-A, 10982-A, 8455-a, and 12042-a (sterility mutants belonging to specific genes). In addition, strain I-5A, which was a sterile isolate from the homokaryotic fruiting of Em A, was also utilized. These extracts were applied to the crosses of sterility mutants 7232-A, 9312-A, and 8455-a of *N. crassa*, using them as spermatial strain (male parent). The improvement of fertility of the extract-treated plates was compared to that of controls.

From the results obtained it was evident that the extracts of wild-type strains and more obviously of 7232-A had the potency of removing the sterility barrier of sterile strain 9312-A, in which case the block was thought to be at karyogamy and/or immediately afterward. For extracts of the other mutants such a conclusion was difficult to draw (Islam, 1980). From these results the presence of more of the active substances in the wild-type strains, particularly in Em A, was also ascertained.

From these studies it was also clear that the genetic background of the compatible strain was also a positive factor for the effective action of *Neurospora* sex pheromones (Islam, 1977).

6. Sex Pheromones in *Neurospora crassa*

d. Isolation of "Sterility" Mutants of *Neurospora crassa* with the Help of γ-Rays and Their Characterizations. Since some of the previously used, uv-induced mutants showed degenerate characteristics, new sterility (male-sterile) mutants were induced from the wild-type strain Em A using γ-rays. Twelve of these were selected for further study. Tests for female sterility were also done for these male-sterile mutants. They were grouped on the basis of their degree of fertility when crossed with the opposite mating type. The results are summarized in Table II.

A more rigorous study of the pheromonal regulation of the sex cycle in *Neurospora* will be carried out with these newly isolated mutants and work has already been done based on spectral (uv) analysis.

C. Biochemical Characterizations

Preliminary biochemical characterizations of the substances present in the extracts Em A × Em a, Em A and Em a were carried out using thin-layer chromatography followed by spectroscopical analysis using uv, ir, nmr, mass spectra, and microanalysis of the single-strain extracts (Em A and Em a) (Islam and Weijer, 1972b). Thin-layer chromatography revealed that the cross extract Em A × Em a con-

TABLE II

γ-Ray-Induced Sterility Mutants of *Neurospora crassa*[a]

Group	Male sterility mutants	Female sterility mutants	Characteristics[b]
A	366A, 516A, 575A	366A, 575A	Completely sterile, no development of perithecia
B	270A, 377A	270A, 342A	Many protoperithecia-like bodies; perithecia mostly ill developed, brown; spores very few or none.
C	136A, 166A, 278A, 342A, 361A, 363A	136A, 166A, 278A, 361A, 363A, 377A, 516A	Few to many developed perithecia with ostiole; spores very few to much less in number than control cross
D	242A	242A	Perithecia well developed but their number less than the control cross (semisterile)

[a] Strains 166A, 342A, and 361A later found to be infected and hence discarded.
[b] Fertility when crossed to opposite mating type.

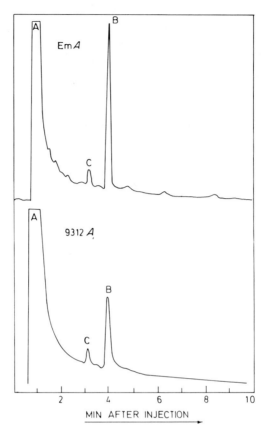

Fig. 4. Gas chromatograms of the pheromonal extracts of Em *A* (wild-type) and 9312-*A* (sterile). Column, SE 30; 200°C; gas flow, 15 lb; speed, 2 div. (17 mm)/min. (A) Solvent peak; (B) major peak of the pheromonal substances; (C) minor peak.

tained two substances, whereas each single-strain extract (Em *A* and Em *a*) contained only one. The physical appearance of the substances, their solubility in chloroform and chloroform–methanol solvent (lipid-specific) and their positive color test with iodine vapor and bromothymol blue solution suggested that the substances are lipoid in nature. The tendency of the substances to run with the solvent front when polar solvent such as benzene or 3:1 chloroform–benzene solution was used indicated that the substances are "hydrocarbon" in nature. The nonmobility of the substances on thin-layer chromatography with nonpolar solvents such as hexane and carbon tetrachloride indicated that the substances are unsaturated hydrocarbons.

All spectral analyses as well as analytical data of the biologically active components of the extracts of Em *A* and Em *a* suggest a chemical structure consisting of a long chain (straight or branched) of probably unsaturated hydrocarbons (Islam and Weijer, 1972b).

Both the biologically active compounds, from Em *A* and from Em *a*, exhibited similar behavior in thin-layer chromatograms and spectral analyses (uv, ir, nmr, and mass spectra), but differed to some extent in their molecular weights. The molecular weight of the biologically active compound present in the extract of Em *a* amounted to 354 to 372, whereas that from Em *A* ranged from 344 to 357. This small difference in molecular weight, together with the other chemical data, indicates that the two substances in question are closely related and probably differ in the length of hydrocarbon chain. Gas-chromatographic

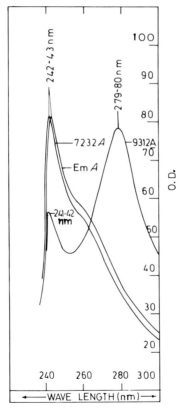

Fig. 5. uv-Spectrograms of the pheromonal extracts of Em *A* (wild-type), 9312-*A* (sterile), and 7232-*A* (sterile). Spectograms were recorded in an automatic recording Beckman Spectrophotometer (scale 2 A, 20 nm/inch, 1 inch/min).

analysis of the extracts of single strains, wild as well as sterility mutants, was later initiated with the aim of understanding the variations of substances present in these extracts, both qualitatively and quantitatively. The results from an analysis done in an SE-30 column at 200°C revealed the presence of a major component (represented by a prominent peak) in all the extracts, together with one and sometimes more minor components. The height of the major peak with respect to the minor one differed slightly from strain to strain, particularly in the case of 9312-A, where the height ratios seemed to be closer to each other (Fig. 4). Reisolation through thin-layer chromatography and subsequent analysis by glc indicated some degradation of the contained substances. From these studies it was determined that *Neurospora* sex pheromones degrade or volatilize even at low temperature.

Spectrophotometric analysis (uv) of the hormonal extracts of different wild-type strains and the sterility mutants is under current investigation. Ultraviolet spectrometry of Em A extracts showed an absorption peak (in chloroform) at 239–242 nm. Similar extracts were prepared from the cultures of different sterility mutants along with the wild-type strain (Em A) and their uv-absorption patterns were recorded. From the results obtained so far, a qualitative difference between the extracts of 9312-A and Em A could be conclusively proved (Fig. 5). Quantitative differences in pheromone level among the different wild-type strains as well as some of the sterility mutants have also been observed from these studies.

IV. PERSPECTIVES

Neurospora is an excellent organism for studying the hormonal regulation of the sexual cycle in eukaryotic microbes. The one-locus, two-allele incompatibility system coupled with an elaborate sexual morphogenetic cycle and the presence of an easily manipulated, orthodox genetic system make this organism most suitable for such study. The substances that we have been able to isolate have only limited biological function because they act only on the A mating type, on protoperithecia development, and in the enhancement of fertility of "late" sterile mutants (from karyogamy onward). The extract of the cross Em A × Em a has once again been shown to be specific to "male sterility" of the A mating type. So, there remain a number of stages in the sexual sequence of the A mating type itself for which the pheromonal regulatory mechanisms have yet to be explored. Raper (1957, 1960), working with *Achlya,* as reviewed by Horgen in Chapter

7, has demonstrated that sexual development in this fungus is a coordinated and sequentially developed process which is hormonally regulated at every step. It is possible that such a mechanism is also working in *Neurospora* but the system may be a more complex one because of its strict heterothallic nature.

From the early studies of Dodge (1931) and Lindegren (1936) it is evident that the sequence of morphological developmental steps with regard to sex in mating type A is identical to that in mating type a. From the present work it has been proved that in mating type A some of these steps are under hormonal control, whereas such hormonal control seems to be absent for sexual development in mating type a. The conclusion, however, that the sequence of events of sexual development (ability to initiate protoperithecia and to establish sex heterokaryons) in mating type a is *not* genetically controlled remains totally unacceptable. For this reason the present author is inclined to accept the possibility that the development of sex in mating type a involves a different gene product(s) than is employed for sexual development in mating type A. The present extraction procedures may not be sufficient for the isolation of these substances.

The biochemical identification of the hormonal substances has shown them to be hydrocarbons. Beyond that we have not been able to reveal their precise structure. This area will need active collaboration with chemists. With an opportunity being available, the author intends to embark on such studies.

REFERENCES

Ahmad, M., Das, A., Khan, M. R., and Huda, M. N. (1967). Improving fertility in crosses of *Neurospora crassa* lys-5 mutants. *Neurospora Newsl.* 11, 19.

Aronescu, A. (1933). Further studies in *Neurospora sitophila*. *Mycologia* 25, 43–54.

Aronescu, A. (1934). Further test for hormone action in *Neurospora*. *Mycologia* 26, 244–252.

Backus, M. P. (1939). The mechanics of conidial fertilization in *Neurospora sitophila*. *Bull. Torrey Bot. Club* 66, 63–76.

Banbury, G. H. (1954). Processes controlling zygophore formation and zygotrophism of *Mucor mucedo*. *Nature (London)* 173, 499.

Banbury, G. H. (1955). Physiological studies in the Mucorales. III. The zygotrophism of zygophores of *Mucor mucedo*. *J. Exp. Bot.* 6, 235–244.

Barbesgaard, P., and Wagner, S. (1959). Further studies on the biochemical basis of protoperithecia formation in *Neurospora crassa*. *Hereditas* 45, 564–572.

Barksdale, A. W. (1960). Interthallic sexual reactions in *Achlya*, a genus of the aquatic fungi. *Am. J. Bot.* 47, 14–23.

Barksdale, A. W. (1963). The role of hormone-A during sexual conjugation of *Achlya ambisexualis*. *Mycologia* 55, 627–632.

Barksdale, A. W. (1969). Sexual hormones of *Achlya* and other fungi. *Science* **166**, 831–837.
Bistis, G. (1956). Sexuality in *Ascobolus stercorarius*. I. Morphology of ascogonium; plasmogamy; evidence for a sexual hormonal mechanism. *Am. J. Bot.* **43**, 389–394.
Bistis, G. (1957). Sexuality in *Ascobolus stercorarius*. II. Preliminary experiments on various aspects of the sexual process. *Am. J. Bot.* **44**, 436–443.
Bistis, G., and Raper, J. R. (1963). Heterothallism and sexuality in *Ascobolus stercorarius*. *Am. J. Bot.* **50**, 880–891.
Burgeff, H. (1924). Untersuchungen über Sexualitat und Parasitismus bie Mucorineen. 1. *Bot. Abh. Heft* **4**, 5–135.
Dodge, B. O. (1931). Heterothallism and hypothetical hormone in *Neurospora*. *Bull. Torrey Bot. Club* **58**, 517–522.
Driver, C. H., and Wheeler, H. E. (1955). A sexual hormone in *Glomerella*. *Mycologia* **47**, 311–316.
Esser, K., and Kuenen, R. (1967). "Genetics of Fungi." Springer-Verlag, Berlin and New York.
Esser, K., and Straub, J. (1956). Fertilitat im Heterokaryon aus Zwei Sterilen Mutanten von *Sordaria macrospora*. *Z. Vererbungsl.* **87**, 625–626.
Esser, K., and Straub, J. (1958). Genitische untersuchungen an *Sordaria macrospora*. Kompensation und Induction bei Genbedingten Entwichlungsdefekten. *Z. Vererbungsl.* **89**, 729–746.
Folch, J., Less, M., and Stanley, G. H. S. (1957). A simple method for the isolation and purification of total lipides from animal tissue. *J. Biol. Chem.* **226**, 497–509.
Hirsch, H. M. (1954). Environmental factors influencing the differentiation of protoperithecia and their relation to tyrosinase and melanin formation in *Neurospora crassa*. *Physiol. Plant.* **7**, 72–97.
Islam, M. S. (1970). Sex hormones in *Neurospora crassa*. Ph.D. Thesis, Univ. of Alberta, Edmonton.
Islam, M. S. (1973). Sex hormones in *N. crassa*. Further studies on its biological properties. *Mycopathol. Mycol. Appl.* **51**, 87–97.
Islam, M. S. (1975). Sex hormones in *N. crassa*. Renewed evidence for their specificity. *Zentralbl. Bakteriol., Parasitenkd., Infektionskr. Hyg.* II **130**, 206–210.
Islam, M. S. (1977). Sex hormones in *N. crassa*. Hormone action is affected by genetic background. *Neurospora Newsl.* **24**, 5.
Islam, M. S. (1980). Sex hormones in *Neurospora crassa*. Differences in the biological property of the hormonal extracts of fertile and sterile strains. *Zentralbl. Bakteriol., Parasitenkd. Infektionskr. Hyg.* (II) (in press).
Islam, M. S., and Weijer, J. (1969). Sex hormones in *Neurospora crassa*. *Neurospora Newsl.* **15**, 24–25.
Islam, M. S., and Weijer, J. (1971). Sex hormones in *Neurospora crassa*. The effect of the extract of the cross Em A × Em a on a bisexual self-sterile strain. *Microb. Genet. Bull.* **33**, 4–5.
Islam, M. S., and Weijer, J. (1972a). Development of fertile fruit-bodies in the single strain culture of *Neurospora crassa*. *Folia Microbiol. (Prague)* **17**, 316–319.
Islam, M. S., and Weijer, J. (1972b). Identification of sexual hormones in *N. crassa*. *Indian J. Biochem. Biophys.* **9**, 345–349.
Ito, T. (1956). Fruit-body formation in red bread mold *Neurospora crassa*. Effect of culture filtrate in perithecial formation. *Bot. Mag.* **67**, 369–372.
Levi, J. D. (1956). Mating reaction in yeast. *Nature (London)* **177**, 753–759.
Lindegren, C. C. (1936). Heterokaryosis and hormones in *Neurospora*. *Am. Nat.* **70**, 404–406.

McClintock, B. (1945). *Neurospora.* I. Preliminary observations of the chromosomes of *Neurospora crassa. Am. J. Bot.* 32, 671-678.
Machlis, L. (1958a). Evidence for a sexual hormone in *Allomyces. Physiol. Plant.* 11, 181-192.
Machlis, L. (1958b). The study of sirenin the chemotactic sexual hormone from the water mold *Allomyces. Physiol. Plant.* 11, 845-854.
Markert, C. E. (1949). Sexuality in the fungus, *Glomerella. Am. Nat.* 83, 227-231.
Moreau, F., and Moruzi, C. (1931). Recherches experimentales sur la formation des peritheces chez les *Neurospora. C. R. Acad. Sci.* 192, 1475-1478.
Moreau, F., and Moruzi, C. (1932). Sur quelques aspects remarquables des cultures dispermes des Ascomycetes du genre *Neurospora. C. R. Soc. Biol.* 111, 864-865.
Moreau, F., and Moreau, Mme. F. (1937). Sur la theorie hormonale de la formation des peritheces des Ascomycetes. *Bull. Soc. Mycol. Fr.* 53, 297-300.
Moreau, F., and Moreau, Mme. F. (1938a). La formation hormonale des peritheces chez les *Neurospora. C. R. Acad. Sci.* 206, 369-370.
Moreau, F., and Moreau, Mme. F. (1938b). Nouvelles observations sur la formation des peritheces chez les *Neurospora. C. R. Acad. Sci.* 206, 1315-1316.
Olive, L. S. (1956). Genetics of *Sordaria fimicola.* I. Ascospore color mutants. *Am. J. Bot.* 43, 97.
Perkins, D. D., Turner, B. C., and Barry, E. G. (1976). Strains of *Neurospora* collected from nature. *Evolution* 30, 281-313.
Plempel, M. (1957). Die Sexualstoffe der Mucoraceae. *Arch. Mikrobiol.* 26, 154-174.
Plempel, M., and Braunitzer, G. (1958). Die isolierung der Mucorineen Sexualstoffe. I. *Z. Naturforsch, Teil B* 13, 302-305.
Raper, J. R. (1939a). Role of hormones in sexual reactions of heterothallic *Achlyas. Science* 89, 321-322.
Raper, J. R. (1939b). Sexual hormones in *Achlya.* I. Indicating evidence for a hormonal coordinating mechanism. *Am. J. Bot.* 26, 639-650.
Raper, J. R. (1951). Sexual hormones in *Achlya. Am. Sci.* 39, 110-121.
Raper, J. R. (1952). Chemical regulation of sexual processes in the Thallophytes. *Bot. Rev.* 18, 447-545.
Raper, J. R. (1957). Hormones and sexuality in the lower plants. *Symp. Soc. Exp. Biol.* 11, 143-165.
Raper, J. R. (1960). The control of sex in fungi. *Am. J. Bot.* 47, 794-808.
Shear, C. L., and Dodge, B.O. (1927). Life histories and heterothallism of the red bread mold fungi of the *Monilia* group. *J. Agric. Res.* 34, 1019-1042.
Shirin, R. A. (1974). Sex hormones in *Neurospora crassa.* M. S. Thesis, Univ. of Dacca, Dacca, Bangladesh.
Singleton, J. R. (1953). Chromosome morphology and the chromosome cycle in the ascus of *Neurospora crassa. Am. J. Bot.* 40, 124-144.
Stanford *Neurospora* Methods (1963). *Neurospora Newsl.* 4, 21-25.
Vigfusson, N. V., Walker, D. G., Islam, M. S., and Weijer, J. (1971). The genetics and biochemical characterizations of sterility mutants in *Neurospora crassa. Folia Microbiol. (Prague)* 16, 166-196.
Vogel, H. J. (1956). A convenient growth medium for *Neurospora. Microb. Genet. Bull.* 13, 42-43.
Westergaard, M., and Mitchell, H. K. (1947). *Neurospora.* V. A. synthetic medium favouring sexual reproduction. *Am. J. Bot.* 34, 573-577.
Wheeler, H. E. (1954). Genetics and evolution of heterothallism in *Glomerella. Phytopathology* 44, 342-345.

Wheeler, H. E., and McGahen, J. W. (1952). Genetics of *Glomerella*. X. Genes affecting sexual reproduction. *Am. J. Bot.* **39**, 110–119.

Yanagishima, N. (1969). Sexual hormones in yeast. *Planta* **87**, 110–118.

Yanagishima, N. (1971). Induced production of a sexual hormone in yeast. *Physiol. Plant.* **24**, 260–263.

Zickler, H. (1952). Zur Entwicklungsgeschichte des Askomyzeten, *Bombardia lunata*. *Arch. Protistenkd.* **98**, 1–70.

7

The Role of the Steroid Sex Pheromone Antheridiol in Controlling the Development of Male Sex Organs in the Water Mold, *Achlya*

PAUL A. HORGEN

I.	Introduction	155
	A. Steroids and Sexual Reproduction	156
	B. Culturing Methods	158
II.	Survey of the Literature	159
III.	Current Research	161
	A. Organization of DNA and Chromatin	161
	B. Steroid Pheromones and Transcription	162
	C. Qualitative Effects of Antheridiol on Macromolecular Synthesis	168
IV.	Perspectives	172
	References	174

I. INTRODUCTION

The eukaryotic microbe *Achlya* belongs to a group of fungi commonly referred to as water molds. The genus is in a class of fungi known as the Oomycetes. This group possesses several characteristics that make them phlogenetically unique in the kingdom Mycota. Differences in the chemistry of the cell wall (Bartjnicki-Garcia, 1970), in the organization of certain biosynthetic pathways (Klein and Cronquist, 1967; Hütter and DeMoss, 1967), and in the molecular weights of the ribosomal RNA's (Lovett and Haselby, 1971; Horgen et al., 1975a) have established the group's evolutionary uniqueness.

To the developmental biologist, and more recently to the molecular biologist, *Achlya,* as well as other oomycetes, possesses certain characteristics that makes it an ideal system for studying a wide number of interesting problems. The many attributes of *Achlya* as a model organism has thoroughly been discussed (see review in Horgen, 1977a) and will not be reiterated. Problems related to steroid hormone-induced differentiation have been investigated at the ultrastructural level (Nolan and Bal, 1974; Mullins and Ellis, 1974), at the level of *in vivo* macromolecular synthesis (see review in Horgen, 1977a), at the level of protein induction (Thomas and Mullins, 1969; Groner et al., 1976), at the level of hormone–cytoplasm interaction (Horgen, 1977b), and at the level of transcription (Horgen, 1977b; Sutherland and Horgen, 1977). In addition, studies on the biochemical control of asexual differentiation have provided a number of interesting reports dealing with a multitude of regulatory problems (Griffin, 1966; Griffin and Breuker, 1969; Timberlake et al., 1973; O'Day and Horgen, 1974; Horgen and O'Day, 1975; Sutherland et al., 1976; MacLeod and Horgen, 1979). More recently, *Achlya* has been utilized to study problems in molecular biology dealing with the structure of eukaryotic chromatin (see review in Horgen and Silver, 1978; Silver, 1979), with the interspersion of repeated and unique DNA sequences (Hudspeth et al., 1977), with the processing and diversity of messenger RNA (Timberlake et al., 1977; Rozek et al., 1978; Law et al., 1978), and with the role of unusually phosphorylated nucleotides in controlling various cellular processes (see review in LeJohn et al., 1977).

Sexual reproduction in two species, *Achlya bisexualis* and *Achlya ambisexualis,* can occur only if there is cooperation between partners of different mating type. In the late 1940's and early 1950's, John Raper demonstrated that sexual reproduction in these species was initiated and controlled by diffusible substances that he called hormones (more correctly termed pheromones) that are reciprocally secreted by the sexually reacting partners (Raper, 1950). We now know that these pheromones in *Achlya* are sterols (Barksdale, 1969; McMorris et al., 1975).

A. Steroids and Sexual Reproduction

Mating and sexual reproduction in *Achlya* involves the following series of events (Fig. 1):

7. Pheromones in *Achlya*

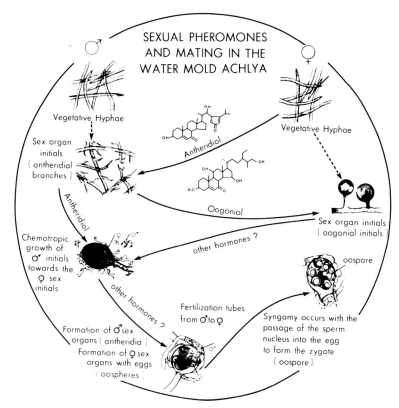

Fig. 1. Pheromones and sexual reproduction in the oomycete *Achlya*. Pen and ink drawings from microscopic observations. For an explanation of the diagram, see the text.

1. Females continually produce and secrete a sterol pheromone, antheridiol, which induces the formation of sex organ initials (antheridial branches) in male strains.

2. Male strains of *Achlya* exposed to antheridiol are stimulated to secrete a second pheromone, oogoniol, which acts upon the female mating partner by causing the formation of female sex organ initials (oogonial initials).

3. While the female sex organ develops, it continually secretes antheridiol which causes chemotropic growth of the male structures toward the female.

4. During the development of the female sex organ, a cross wall is formed, delimiting the organ (the oogonia) from the remainder of the hyphae. Nuclei within the oogonia undergo meiosis and the eggs (oospheres) are eventually formed. The sequence of events in the for-

mation of oogonia may involve only oogoniol (Barksdale, 1967) or perhaps other as yet uncharacterized hormones (Barksdale, 1969).

5. The male sex organs continue to grow toward the female and finally become physically attached to the oogonia. During the final stages of development of the antheridia, cross walls form which separate the sex organ from the remainder of the antheridial hypha. Cross wall formation is also controlled by antheridiol (Barksdale, 1967). Nuclei within the sex organs undergo meiosis (Barksdale, 1968; Bryant and Howard, 1969) to produce the sperm nuclei.

6. A fertilization tube forms between the male and female sex organs and the male gametic nuclei pass into the oogonium. The sperm nuclei fuse with the oospheres to produce the zygotes (oospores). The zygotes, under proper environmental conditions, will then germinate and produce new diploid individuals.

Achlya is the most primitive eukaryotic organism to produce and respond developmentally to well-characterized steroid sex pheromones. In higher eukaryotes one has to be concerned with distinguishing between cells that produce and secrete the hormone and cells that respond to the action of the hormone (target cells). In *Achlya* the pheromones are produced and released by opposing sexual partners and have an effect on the entire thallus of the partner. Therefore, the entire organism, either male or female, becomes the target tissue. The basic sterol structure of the *Achlya* pheromones is similar to mammalian steroid sex hormones (androgens and estrogens) and to ecdysone, an insect growth and molting hormone. *Achlya* therefore becomes an excellent system with which to investigate the basic action of steroid hormones.

B. Culturing Methods

Stock cultures of *Achlya ambisexualis* Raper (strain E87♂) are maintained on Emerson's YPSS (Difco) agar slants at $4° \pm 1°C$ and are transferred monthly. For maintaining day to day cultures the mycelium is transferred onto PYG (1.25 gm of yeast extract, 1.25 gm of peptone, 3.0 gm of glucose per liter of distilled H_2O) agar plates. Spore inocula are prepared according to the method of Griffin and Breuker (1969). Spore cysts are inoculated into PYG broth, grown for 24 hr at $24° \pm 1°C$ on a rotary shaker (130 rpm) until a finely suspended mycelia has formed. This was used immediately or stored at $4° \pm 1°C$ for no longer than 48 hr. Of this finely suspended mycelium, 30 ml was added per 1 liter of mating medium [2.4 gm glucose, 400 mg Edamin (hydrolyzed lactalbumin), 100 mg Tris, 80 mg calcium glycerophosphate, 0.5 ml of 2 M KCL, 1.0 ml of 0.5 M MgSO$_4 \cdot$7H$_2$O, 1.0 ml metal mix no. 4 (2 mg/ml; Barksdale, 1963) per liter of distilled water.] Cultures were usually grown for 48 hr at 24° ± 1°C. At this time cultures either received antheridiol at 5×10^{-11} gm/ml (106 pmoles) or were left to grow in the absence of the pheromone. It has been reported that *A. ambisexualis* (E87♂) will respond to as little as 6×10^{-12} gm/ml antheridiol and we have reasoned that a 10-fold higher concentration of the pheromone may be representative of the levels of the steroid the organism would encounter in nature. Other workers have chosen to work with considerably higher levels (1×10^{-9} gm/ml) of antheridiol (Timberlake, 1976; Michalski, 1978).

II. SURVEY OF THE LITERATURE

The early work on the biology of *Achlya* and the isolation and characterization of antheridiol has been adequately reviewed by Barksdale (1967, 1969). Work on the chemistry of the *Achlya* sterols has been recently reviewed by McMorris (1978). Detailed accounts of the biochemical events of morphogenesis of the male sex organ in *Achlya* has been reviewed by van den Ende (1976) and by Horgen (1977a).

The addition of antheridiol to undifferentiated cultures of *Achlya* elicits several changes in macromolecular synthesis during the ontogeny of antheridial branch formation. Studies with the *Achlya* system indicate that there is a concomitant enhancement of both RNA and protein synthesis (Silver and Horgen, 1974; Timberlake, 1976). The hormone appears to affect both gene activation and gene product accumulation. The first genes that appear to be activated by antheridiol are the ribosomal genes; the gene products that accumulate are the 26 S and 18 S ribosomal RNA's as new cellular ribosomes (Fig. 2) (Horgen et al., 1975b). Associated with this ribosomal RNA synthesis is a corresponding synthesis of ribosomal proteins (Horgen, 1977a; Michalski, 1978). The increase in rRNA occurs early (within 30 min) after the addition of antheridiol.

Immediately preceding the morphological appearance of the antheridial branches (at 3 hr) the pheromone also stimulates the synthesis of poly(A)$^+$ mRNA (Fig. 2). However, prior to this enhancement of mRNA synthesis, there is a modification of the chromosomal proteins of *Achlya*. This modification involves an acetylation of specific histones in the *Achlya* chromatin (Horgen and Ball, 1974). The addition of acetyl groups to histones is thought to make them less positive in charge,

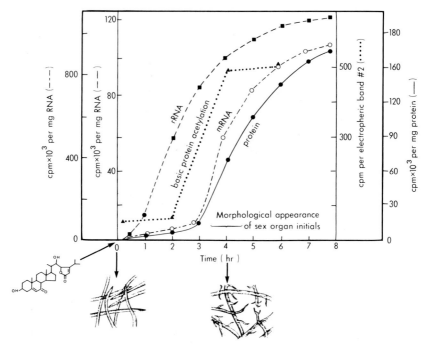

Fig. 2. Antheridiol and gene expression during male sex organ initial differentation. This figure is a summation of the data discussed by Horgen (1977a). The curves are all from hormone treated tissue and the increases shown represent differences compared to control (non-hormone-treated) cultures.

decreasing their affinity for the negatively charged DNA. It seemed, therefore, reasonable to suggest that this chromosomal protein modification could be involved in the activation of genes required for the morphogenesis of male sex organ initials. Indeed after the observed histone acetylation, there is an enhancement of mRNA synthesis and this is associated with a corresponding stimulation of protein synthesis (Silver and Horgen, 1974; Timberlake, 1976). This enhancement of total protein synthesis occurs immediately prior to the morphological appearance of male sex organ initials (Fig. 2). While these observations are suggestive of a set of new genes required for male sex organ initiation being activated by antheridiol, it should be clearly stated that the changes that have been measured (stimulation of rRNA, mRNA, and protein synthesis) to date are quantitative changes. With the exception of the reports of cellulase synthesis (Thomas and Mullins, 1969) and the reports of specific changes in protein synthesis (Groner et al., 1976; Michalski, 1978), very little evidence exists for specific qualitative changes induced by antheridol.

III. CURRENT RESEARCH

A. Organization of DNA and Chromatin

Two general patterns of DNA sequence organization arrangements have been found in the chromatin of complex plants and animals:

1. Short period interspersion pattern—genomes for which the major fraction of the DNA consists of repetitive sequences averaging 200-400 nucleotide pairs linked to single copy DNA at intervals of less than 2000 nucleotide pairs

2. Long period interspersion pattern—genomes lacking short repetitive sequences but containing long stretches (>4000 nucleotide pairs) of relatively nondivergent repeated DNA sequences, contiguous to single copy sequences at fairly long intervals (> 10,000 nucleotide pairs)

Short period patterns have been reported for a large number of animals and several flowering plants (Davidson et al., 1975; Zimmerman and Goldberg, 1977). Long period interspersion has been reported for several insect species (Crain et al., 1976) and for certain avian species (Arthur and Straus, 1978). Hudspeth et al. (1977) reported that repetitive and single copy sequences in *Achlya* are arranged in a long-period interspersion pattern. Estimates of the spacing intervals between repetitive and single copy DNA in *Achlya*, however, indicate that the interspersion pattern is longer than has been previously reported in other eukaryotes. Hudspeth et al. (1977) estimated that the average genomic lengths of repetitive and single copy DNA in *Achlya* were 2.7 $\times 10^4$ and 1.35×10^5 nucleotide pairs. These values were about five times longer than had been estimated for repetitive and single copy DNA in *Drosophila* (Hudspeth et al., 1977). It is interesting to note that recent reports indicate that certain other fungi also possess this unusual long period interspersion pattern or no interspersion at all (Timberlake, 1978; Hereford and Roshbash, 1977; Lauer et al., 1977).

Nucleosomes are the repeating nucleoprotein subunits of eukaryotic chromatin (Kornberg, 1977). In higher eukaryotes each nucleosome is composed of a core particle of 140 base pairs of DNA plus two molecules each of histones H2A, H2B, H3, and H4 and an intermediate region (linker) containing 40-60 base pairs of DNA associated with histone H1. The "repeat size" of the chromatin is the length of DNA found in the core plus linker region. Most higher eukaryotes have chromatin with a nucleosomal DNA repeat size of approximately 200 base pairs. Silver (1979) reported that the nucleosomal repeat size for *Achlya* chromatin was 159 base pairs. She also reported that the length of the DNA in the nucleosomal core of *Achlya* was 140 base pairs and thus was comparable in length to that of other examined eukaryotes (Silver, 1979). While the length of the DNA of the core particle seems to be consistent in all eukaryotic organisms that have been studied, it is interesting that repeat sizes which are 30–50 base pairs shorter than values reported for higher eukaryotes have been reproducibly found in a number of different eukaryotic microbes (see review, in Horgen and Silver, 1978).

B. Steroid Pheromones and Transcription

Tissues of animals that respond to steroid hormone contain binding macromolecules, or steroid receptors, with which the hormone interacts by a three stage mechanism (Jensen and DeSombere, 1973): (1) hormone enters the cells and binds to specific extranuclear receptor proteins characteristic of the target tissues; (2) the receptor–hormone complex is then translocated to the nucleus and en route encounters an alteration of the receptor protein, a phenomenon called receptor transformation; (3) the steroid–protein complex then enters the nucleus where it associates with an acceptor site on the chromatin and affects RNA synthesis both in a quantitative and qualitative manner (O'Malley and Means, 1974). Are these cellular control mechanisms of gene activation unique to highly evolved animal systems or do similar mechanisms exist in more primitive eukaryotes?

We examined the effects of cytosol–antheridiol complexes on the *in vitro* transcriptional capacity of chromatin isolated from non-antheridiol-induced cultures of *Achlya*. Table I shows the effect of preincubating *Achlya* chromatin with cytosol and/or pheromone under a series of defined conditions. When assayed by addition of *E. coli* RNA polymerase to chromatin, neither added antheridiol nor cytosol alone significantly affected *in vitro* transcription. Preincubating chromatin with both hormone and cytosol from E87♂, however, dramatically increases heterologous polymerase transcription of *Achlya* chromatin (Table I). Cytosol from a female strain 734 (which does not respond to antheridiol), preincubated with antheridiol and chromatin, does not stimulate RNA synthesis. These results demonstrate that there is "target tissue" specificity associated with cytosol–hormone interactions in *Achlya* and that preincubation of *Achlya* chromatin with hormone and target tissue cytosol dramatically modifies the chromatin template available for transcription. Similar results were obtained when cytosol–hormone mixtures were measured for their effect on

7. Pheromones in *Achlya*

TABLE I

The Effects of Cytosol, Antheridiol, and Cytosol–Antheridiol Mixtures on Transcription of *Achlya* Chromatin

Chromatin preincubated with[a]	Incubation temperature (°C)	Specific activity	Control (%)
(A) No additions (control)	30	111 ± 4	100
E87 cytosol	30	112 ± 6	101
Antheridiol	30	97 ± 2	86
E87 cytosol + antheridiol	30	415 ± 8	370
E87 cytosol + antheridiol preincubated at 0°C	0	27 ± 2	29
E87 cytosol + antheridiol preincubated at 25°C	0	76 ± 4	67
734 cytosol + antheridiol	30	103 ± 9	104
(B) No additions (control)	30	82 ± 2	100
E87 cytosol + antheridiol	30	271 ± 6	330
E87 cytosol	30	85 ± 1	104
E87 antheridiol	30	75 ± 4	92

[a] (A) Transcription with heterologous RNA polymerase. Fifty μg of chromatin DNA (strain E87) was added along with 10 μg of *E. coli* RNA polymerase to each reaction mixture. Either 1×10^{-9} gm/ml (2.1 nM) antheridiol or 1.25–1.50 mg of cytosol or a combination of both were added to each preincubation mixture. The final concentrations in the 1 ml reaction mixture were 50 mM Tris-HCl (pH 8.0), 10 mM dithiothreitol, 10 mM $MgCl_2$, 20% glycerol, 0.1 M KCl, 200 mM ammonium sulfate, and 0.4 mM K_2HPO_4. Unless otherwise indicated, these mixtures were preincubated for 15 min at 25°C. 0.4 mM UTP, 0.4 mM ATP, 0.4 mM CTP, 0.4 mM GTP, and 10 μCi [³H]UTP (specific activity 42 Ci/mmole) were added to begin the RNA synthetic reaction. Each reaction mixture was then incubated at the designated temperatures for 15 min. Reactions were terminated; products were collected and the radioactivity determined as described by Horgen and Key (23). (B) Transcription with endogenous *Achlya* RNA polymerases. Preincubation mixtures were similar to (A) except 200 μg chromatin DNA was added and no *E. coli* RNA polymerase was added. All data represent an average of four replications. The data are expressed as picomoles incorporated per minute per 50 μg chromatin DNA (specific activity).

stimulating RNA synthesis in isolated chromatin directed by the endogenous *Achlya* polymerases (Table I).

In mammalian systems a temperature dependent shift is required to convert the initial 4 S hormone–receptor complex to an active 5 S complex which then associates with the nuclear acceptor sites and activates transcription (Jensen and DeSombere, 1973). More RNA synthesis occurs in *Achlya* if the cytosol–hormone complex is preincubated for 15 min at 25°C, than when the cytosol–hormone complex is preincubated at 0°C (Table I). Therefore, temperature-dependent interactions are necessary before hormone–cytosol complexes can activate *in vitro* RNA synthesis in *Achlya* (Horgen, 1977b). The assays in these experiments were performed at 0°C to prevent any conversion of hormone–cytosol complexes during the incubation of the nucleoside triphosphates. In addition to the temperature-dependent incubations of the steroid–cytosol complexes, high ionic strength is required for optimal antheridiol–cytosol stimulation of *in vitro* transcription (Horgen, 1977b). These results are strongly suggestive that hormone–cytosol interactions which are similar to the hormone–receptor mechanisms operable in complex animal systems exist in the eukaryotic microbe *Achlya*.

In more complex eukaryotes, the induction of specific structural proteins is mediated by changes in transcription (O'Malley and Means, 1974; O'Malley *et al.*, 1969). In animal systems, estrogen stimulates the *de novo* synthesis of specific messenger RNA which codes for ovalbumin synthesis, while progesterone effects the synthesis of avidin messenger RNA (O'Malley and Means, 1974). Prior to the increase in steroid hormone-induced protein synthesis, dramatic changes in the levels of nuclear RNA polymerase activity, chromatin template activity, and chromatin function have been reported (Towle *et al.*, 1976). Evidence also exists which suggests that hormones may function by elevating chromatin template capacity (Breuer and Florini, 1966).

Methods have been developed which allow workers to measure the number of RNA synthetic initiation sites on eukaryotic chromatin isolated from different developmental stages. These methods utilize eukaryotic chromatin, *Escherichia coli* RNA polymerase, and the antibiotic rifampicin. Results from complex animal systems suggest that steroid hormones enhance transcription by making more areas of the genome free to bind RNA polymerase (Tsai *et al.*, 1975; Schwartz *et al.*, 1975).

We have investigated the effects of antheridiol on the transcriptional ability of chromatin isolated from *Achlya* at selected times during the early stages of male sex organ development. The characteristics of the chromatin isolated from *A. ambisexualis* were similar to those of chromatin isolated from several other lower eukaryotes. The ratio of RNA:DNA for *Achlya* chromatin was 0.7, whereas the ratio of total chromosomal protein to DNA was 4.5:1 (Sutherland and Horgen, 1977). The ratio of histone to DNA in *Achlya* was similar to the values reported for *Aspergillus, Saccharomyces,* and various other eukaryotes (Sutherland and Horgen, 1977). Furthermore, as has previously been discussed, *Achlya* possesses a typical nucleosome chromatin subunit structure (Silver, 1979).

7. Pheromones in Achlya

Chromatin from *Achlya* was capable of being transcribed by heterologous *E. coli* RNA polymerase. *In vitro* transcription was dependent on the presence of added chromatin DNA, the four nucleoside triphosphates, and was inhibited by the addition of actinomycin D. The *in vitro* assay followed normal kinetics of incorporation and the product synthesized was established as being RNA (Sutherland and Horgen, 1977).

Chromatin template activity was used as a means of assessing changes in transcription after the addition of antheridiol. The process of RNA chain initiation has been separated from other transcriptional events by titrating *Achlya* DNA or chromatin with increasing amounts of *E. coli* RNA polymerase in the presence of the drug rifampicin. Rifampicin specifically blocks initiation prior to the formation of stable initiation complexes. If *E. coli* RNA polymerase is preincubated with *Achlya* chromatin, stable initiation complexes will form. If one then initiates RNA synthesis by the addition of a labeled nucleoside triphosphate solution and rifampicin, RNA synthesis (chain elongation) will occur until the RNA polymerase recognizes the termination signal. At this point RNA synthesis ceases since reinitiation cannot occur in the presence of the antibiotic. Using this "rifampicin challenge assay," one can make measurements on the number of RNA polymerase binding and initiation sites on chromatin fractions. If chromatin is isolated from control (no pheromone) and treated (pheromone added at one or 4 hr) tissue, a number of interesting measurements can be made. In this regard, fixed concentrations (25 μg of chromatin DNA) of *Achlya* chromatin were titrated with increasing concentrations of *E. coli* RNA polymerase and preincubated together for 15 min at 37°C. RNA synthesis was initiated by the addition of the nucleoside triphosphates and rifampicin. The total reaction mixture was allowed to synthesize RNA. With increasing amounts of the bacterial polymerase, a transition point was reached. This transition point represents the number of RNA polymerase molecules needed to saturate high affinity initiation sites on the chromatin under conditions of drug inhibition of RNA chain reinitiation (Fig. 3). At 1 and 4 hr after the addition of antheridiol, a greater quantity of RNA polymerase was required to saturate the *Achlya* chromatin (Fig. 3). The amount of RNA polymerase needed to saturate chromatin is representative of the number of RNA polymerase binding sites available on the chromatin (Sutherland and Horgen, 1977). Within 1 hr after the administration of antheridiol to untreated cultures, the amount of RNA polymerase needed to saturate the chromatin is increased from 8 μg/25 μg of

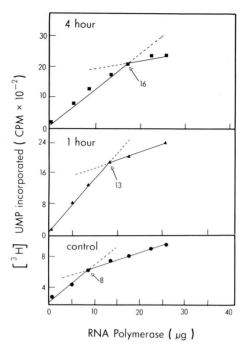

Fig. 3. Titration of *Achlya* chromatin with *E. coli* RNA polymerases. Increasing amounts of RNA polymerase were preincubated with (25 μg of chromatin DNA) *Achlya* chromatin isolated from control, 1 and 4 hr pheromone-treated *Achlya* cultures for 15 min at 37°C. Chromatin was isolated as described by Sutherland and Horgen (1977). RNA synthesis was started by the addition of the four nucleoside triphosphates and 10 μg of rifampicin. Samples were assayed for the incorporation of [^{3}H]UMP into trichloroacetic acid insoluble material as described by Sutherland and Horgen (1977).

chromatin DNA in control chromatin to 13 μg/25 μg chromatin DNA. At 4 hr a further increase (16 μg/25 μg chromatin DNA) in treated cultures was measured (Fig. 3).

The reciprocal experiment is shown in Fig. 4. If the RNA polymerase concentration is held constant and the chromatin DNA levels are varied, the results obtained again suggest that chromatin from pheromone-treated tissue has more RNA polymerase binding sites (Fig. 4).

From data generated in the kinds of experiments shown in Figs. 3 and 4, one can calculate the number of rifampicin-resistant initiation sites (Sutherland and Horgen, 1977). Chromatin from non-hormone-treated *Achlya* initiated 8000 RNA chains/pg of DNA, while chromatin from *Achlya* that had been exposed 4 hr to antheridiol initiated 22,000 chains/pg of DNA (Sutherland and Horgen, 1977). Nearly 50% of the DNA from unstimulated *Achlya* chromatin is available to bind RNA polymerase. This value increases to 64% when the chromatin was isolated from cultures that had been exposed to antheridiol for 4 hr.

The *in vitro* data generated from the experiments utilizing the rifampicin challenge assay indicate rather important differences between chromatin isolated from control and hormone-treated tissues. Whether the numerical data generated from this type of experiment reflects correctly the *in vivo* situation cannot at this time be determined. Nevertheless, utilizing the heterologous chromatin–RNA polymerase assay, it appears that antheridiol dramatically affects the transcription of *Achlya* chromatin.

The rationale for initially using a heterologous chromatin–polymerase system was because highly purified fungal RNA polymerases were extremely difficult to isolate (see review in Griffin *et al.*, 1975). Recently, techniques were applied to fungal systems which allow for the isolation and characterization of highly purified RNA polymerases (Vaisius and Horgen, 1979). We have utilized the polyethylenimine procedure to purify RNA polymerase II from *Achlya*.

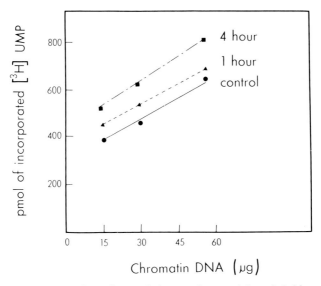

Fig. 4. Concentration dependence of the template activity of *Achlya* chromatin. *Achlya* chromatin from 0, 1 and 4 hr antheridiol-treated cultures was incubated with 22 µg of *E. coli* RNA polymerase. Nucleoside triphosphates were added and the reactions were assayed as described by Sutherland and Horgen (1977). No incorporation of labeled nucleoside into RNA was observed at zero concentration of chromatin DNA.

The purification procedure involves precipitation with polyethylenimine, selective elution of the polymerase from the polyethylenimine precipitate, ammonium sulfate fractionation, DEAE-cellulose chromatography, CM-cellulose chromatography, P-cellulose chromatography, and exclusion chromatography on Bio-Gel A-1.5M. A summary of the RNA polymerase II purification from *Achlya* is shown in Table II. The procedure is quite effective for purifying relatively large quantities of enzyme from the aquatic fungus. Table III shows a comparison of the subunit structure of the *Achlya* polymerase with several other eukaryotic enzymes. *Achlya* polymerase II is quite comparable to other eukaryotic polymerases II with regard to subunit composition (Roeder, 1976). Furthermore, the enzyme from *Achlya* was quite typical of other eukaryotic RNA polymerases with regard to template preference, salt optima, and divalent cation optima (Horgen, 1980). With purified *Achlya* enzyme, we are now in a position to examine the effects of antheridiol on the transcription process in a completely homologous system.

C. Qualitative Effects of Antheridiol on Macromolecular Synthesis

As previously mentioned, the changes that have been observed in *Achlya* for the most part have been quantitative changes. A typical response of some animal target tissues to steroid sex hormones is a

TABLE II

Summary of the Purification of RNA Polymerase II from *Achlya ambisexualis*[a]

Fraction	Volume (ml)	Protein (mg)	Total activity (units)	Specific activity (units/mg)	Purification factor	Yield (%)
Crude	2200	9150	25,992	2.8	1	100
0.25 M PEI	760	878	25,586	29.14	10.7	98
Ammonium sulfate precipitate	200	210	17,288	82.32	29.4	66.5
DEAE-cellulose peak	96	35	13,466	384.7	137.8	52
CM-Cellulose	79	24	13,166	548.6	196	51
P-Cellulose	20	16	13,054	815.9	291	50.5
Bio-Gel A-1.5M	8	10	12,987	1298.7	463	50

[a] 1 Unit RNA polymerase activity is defined as the incorporation of 1 nmole of labeled nucleotide into RNA per 15 min at 30°C. RNA polymerase II was isolated according to the procedure of Vaisius and Horgen (1979).

TABLE III

A Comparison of the Subunit Composition of *Achlya* RNA Polymerase II with Calf Thymus and the Mushroom, *Agaricus bisporus*[a]

Achlya ambisexualis II	Calf thymus (MW × 10⁻³)		Agaricus bisporus II
	IIa	IIb	
	214		
180		180	182
140	140	140	140
89			89
69			69
53			53
41			41
37.5	34	34	37
			31
29			29
25	25	25	25
20	20.5	20.5	19
16.5	16.5	16.5	16.5

[a] RNA polymerases II were purified by the polyethylenimine procedure as described by Vaisius and Horgen (1979). The purified enzymes were subjected to electrophoresis into sodium dodecyl sulfate polyacrylamide gels as described by Vaisius and Horgen (1979). The molecular weights of the polypeptide subunits were determined from running many different gels by use of the following marker proteins: myosin (200,000), *E. coli* RNA polymerase β (165,000), *E. coli* RNA polymerase β' (155,000), β-galactosidase (130,000), *E. coli* RNA polymerase σ (95,000), phosphorylase B (94,000), bovine serum albumin (68,000), ovalbumin (43,000), *E. coli* RNA polymerase α (39,000), carbonic anhydrase (30,000), soybean trypsin inhibitor (21,000), and lysozyme (14,300).

general stimulation of RNA and protein synthesis (Oka and Schimke, 1969; Tata and Barker, 1975). There are numerous examples in the literature of higher eukaryotes where steroid hormones alter the spectrum of mRNA's in the cell. Translation of hormone-specific mRNA's results in the production of proteins which are involved in the functional response of the cell. Numerous examples of steroid hormones causing the accumulation of one or more gene products include estrogen and progesterone-induced synthesis of egg white proteins in chicken (Oka and Schimke, 1969; O'Malley *et al.*, 1969), esterdiol-induced synthesis of egg yolk protein in *Xenopus* and chicken (Wallace and Jared, 1968; Tata, 1976), estrogen-induced accumulation of "induced protein" (IP) in rat uterus (Katzenellenbager and Gorski, 1972), dexamethasone-mediated induction of mouse mammary tumor virus (Ringold *et al.*, 1975), and the ecdysone-induced puffing in polytene chromosomes of *Drosophila* (Ashburner, 1973). Several circumstantial lines of evidence seem to suggest that in addition to having quantitative effects on macromolecular synthesis in *Achlya,* antheridiol may induce the selective transcription and accumulation of mRNA's. Sex organ initial production was dependent on continued RNA and protein synthesis (Kane *et al.,* 1973; Silver and Horgen, 1974; Timberlake, 1976) suggesting that selective gene transcription may mediate sexual morphogenesis. Cellulase was reported to be a key enzyme in *Achlya* males and increases in this enzyme were concomitant with sexual development (Thomas and Mullins, 1967; Mullins, 1973). Inhibitor studies showed that cellulase synthesis was dependent on continued RNA and protein synthesis. Groner *et al.* (1976), Horgen (1977a), and Michalski (1978) reported that antheridiol could stimulate both quantitative and specific qualitative changes in protein synthesis. It has been suggested that antheridiol may alter specific enzymes involved in steroid metabolism (Musgrave and Nieuwenhuis, 1975; McMorris and White, 1977).

Very recently, Rozek and Timberlake (1980) investigated messenger RNA populations isolated from control and from pheromone-treated cells to determine to what extent antheridiol causes the accumulation of new poly(A)$^+$ mRNA sequences or the loss of preexisting (vegetative) poly(A)$^+$ mRNA sequences. They employed a wide variety of experimental approaches having a broad range of sensitivities; some were potentially capable of detecting very small changes in mRNA populations. The techniques involved radioactive labeling and electrophoresis of the RNA, production of complimentary DNA from *Achlya* mRNA using reverse transcriptase followed by hybridization studies, hybridization of single copy DNA to excess amounts of RNA, and an approach called "differential plaque hybridization" which involves the screening of large numbers of recombinant DNA clones. The procedures that they employed failed to reveal any alterations in the spectrum of mRNA sequences transcribed and accumulated by *Achlya* males during sex organ initial formation. These authors conclude that any qualitative changes in gene activity induced by antheridiol during early male sex organ morphogenesis must be extremely subtle. Thus, they suggest that the changes occurring in *Achlya* may be comparable to the subtle changes occurring in rat uterus following estrogen administration (Notides and Gorski, 1966; Katzenellenbager and Gorski, 1972) or in rat seminal vesicle following testosterone administration (Higgens *et al.,* 1978).

While Rozek and Timberlake (1980) were interested in examining transcription products during antheridiol-induced sexual morphogenesis, Gwynne and Brandhorst (1980) were interested in carefully examining the products of translation. Proteins labeled with [^{35}S]methionine at different developmental stages were examined by both one-dimensional SDS gel electrophoresis and two-dimensional gel electrophoresis. All the proteins that were detectable during vegetative growth continued to be synthesized during sexual differentiation. A peptide with a molecular weight of 60,000 could be observed to increase dramatically 30 min after the addition of antheridiol. The high levels of this protein remained detectable for the next 60 min; the level dropping back to control 90 min after the addition of pheromone. Although the authors detected this peptide on one-dimensional gels, they were unable to resolve the change on two-dimensional gels. No other qualitative changes were observed by the authors during antheridial branch development. In addition, Gwynne and Brandhorst (1980) isolated poly(A)$^+$ mRNA at specific times during differentiation and translated this mRNA in a wheat germ cell free translation system. Again, analysis of the translation products on polyacrylamide gels showed no detectable qualitative differences during sex organ initiation. These authors concluded that antheridiol-induced differentiation is accompanied by minor qualitative transcriptional and translational changes.

With the exception of *Physarum,* which has been well studied, only minimal information is available on the non-histone proteins from eukaryotic microbes (Horgen and Silver, 1978). As a result, we have begun an examination of the non-histone proteins in *Achlya.* Heterogeneous acid-insoluble proteins ranging in molecular weight from ~15,000 to ~200,000 have been found. Polypeptides with molecular weights similar to mammalian actin and myosin have been identified. In animal systems, nonhistone proteins have been implicated as positive regulators of gene expression. Specifically, some of the nonhistone proteins have been implicated as being involved with the actions of steroid hormones with target cell nuclei (Spelsberg, 1974). Utilizing a technique called gel fluorography, we have compared the non-histone proteins which had been extracted from chromatin of control and 4 hr pheromone-treated tissue. Major differences were observed between control and treated tissues with bands with molecular weights of ~180,000 and ~75,000, respectively, being much more exposed in fluorographs of acidic proteins of hormone-treated tissues as compared to control (Fig. 5). In addition, less dramatic differences were observed between peptide bands at molecular weights of ~138,000 and ~85,000, in both cases the more intensely exposed bands were present in the 4 hr hormone-treated compared to control. Whether these differ-

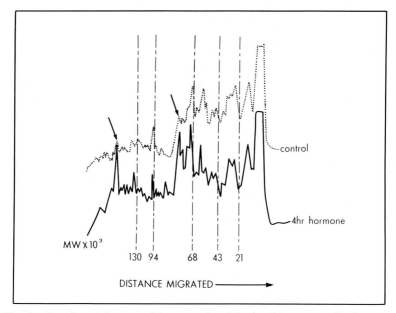

Fig. 5. Densitometric scans of fluorographs of the non-histone proteins from control and 4 hr antheridiol-treated cultures. A growing culture of *A. ambisexualis* (E87 ♂) was divided into two flasks. Each flask received 1 μCi per ml of [³H]leucine. To one culture 5 × 10⁻¹¹ gm per ml antheridiol was added. After 4 hr, chromatin was isolated from each of the cultures according to the procedure of Sutherland and Horgen (1977). Non-histone proteins were extracted according to the methods of LeStourgeon and Wray (1974) and electrophoresed into 10 × 1 SDS acrylamide gels (Hjérten, 1962). Fluorographs were prepared according to the procedures of Bonner and Laskey (1974) and then the developed, exposed X-ray film was scanned in a Gilford gel scanner. Known molecular weight standards were coelectrophoresed in the slab gels for comparative purposes (see Table III).

ences are qualitative or simply represent more quantitative differences between control and antheridiol-treated tissues cannot be ascertained from the preliminary experiments.

IV. PERSPECTIVES

During the last five years a considerable amount of information has been generated with regard to macromolecular events occurring during the ontogeny of male sex organ initials in the water mold *Achlya*. Addition of antheridiol to vegetatively growing cultures stimulates a quantitative increase in both RNA and protein synthesis. Recent evidence seems to suggest that qualitative differences are not easily detectable. The quantitative increases seem to be the result of increases in the synthesis of the same gene products that are present in vegetatively growing cultures. An important question to pose is how are the developmental changes that occur when antheridial branches form modulated? As has been suggested by Gwynne and Brandhorst (1980), perhaps posttranslational regulatory events may be required for sexual differentiation? Perhaps the changes are so subtle that the present approaches being used are not sensitive enough to detect them? Most of the biochemical investigations have been made during the early stages of sex organ initiation. It would seem to be important to also investigate biochemical events that occur later in development as the antheridium itself is formed.

What mechanisms are involved as the quantitative changes in RNA synthesis occur? Certainly our data with *in vitro* heterologous transcription indicate that more RNA polymerase binding and initiation sites are available on chromatin isolated from hormone treated tissues. Since we have purified the RNA polymerase from *Achlya* responsible for the synthesis of mRNA, we are in a position to examine in a completely homologous system the effects of antheridiol on the transcription process. These experiments will involve the *in vitro* transcription of chromatin, the effects of hormone–cytosol(receptor) complexes on this *in vitro* transcription and the quantitative determination of polymerase levels during the development of sex organ initials. Furthermore, comparison of the peptide subunits from polymerase isolated from control and pheromone-treated tissue will indicate if qualitative changes in RNA polymerases are involved in the morphogenesis of antheridial initials.

Another exciting area for future investigation in the *Achlya* system is the nature of the steroid receptors. One important criterion for carrying out effective studies that involve isolation of receptor molecules will be the ability to produce radioactive antheridiol with a relatively high specific activity. Once the labeled antheridiol is available, receptor studies comparable to those done in animal systems (Kuhn *et al.*, 1975; Jensen and DeSombere, 1973) can be accomplished.

Although the early events of steroid hormone action are reasonably well understood, the actual biochemical mechanisms of response remain unknown. What are the effects of the hormone binding to the receptor molecule? How does the hormone receptor complex move to the nuclear sites? What does the receptor complex do to alter the activity of the chromatin? These key questions have been most difficult to approach because of the many complexities associated with the highly developed, multicellular animal systems now being studied. *Achlya* possesses the most simplified steroidal system known. In addition, *Achlya* appears to be one of the simplest eukaryotes studied in terms of its genetic composition and the processes that occur during gene expression.

A growing number of laboratories are utilizing *Achlya* for its experimental advantages to probe at questions associated with the hormonal regulation of gene expression. Results as they are accumulated in *Achlya* are compared to what is known in more complex systems. In addition, the *Achyla* system is of interest from the point of view of the development and evolutionary origins of steroid regulation and the degree of conservation of this form of regulation throughout plant and animal phylogeny.

ACKNOWLEDGMENTS

Research done by the author was supported by grants from the National Research Council of Canada, the National Sciences and Engineering Council of Canada, and by Research Corporation (Brown-Hazen).

REFERENCES

Arthur, R., and Straus, N. (1978). DNA-sequence organization in the genome of the domestic chicken (*Gallus domesticus*). *Can. J. Biochem.* **56**, 257–263.

Ashburner, M. (1973). Sequential gene activation in polytene chromosomes of *Drosphila melanogaster* I. Dependence upon ecdysone concentration. *Develop. Biol.* **35**, 47–61.

Barksdale, A. W. (1963). The uptake of exogenous hormone A by certain strains of *Achlya*. *Mycologia* **55**, 164–171.

Barksdale, A. W. (1967). The sexual hormones of the fungus *Achlya*. *Ann. N.Y. Acad. Sci.* **144**, 313–319.

Barksdale, A. W. (1968). Meiosis in the antheridium of *Achlya ambisexualis* E87. *J. Elisha Mitchell Sci. Soc.* **84**, 187–194.

Barksdale, A. W. (1969). Sexual hormones of *Achlya* and other fungi. *Science* **166**, 831–837.

Bartjnicki-Garcia, S. (1970). Cell wall composition and other biochemical markers in fungal phylogeny. *In* "Phytochemical Phylogeny" (J. B. Harborne, ed.), pp. 81–103. Academic Press, New York.

Bonner, W. H., and Laskey, R. A. (1974). A film detection method for tritium-labelled proteins and nuclei acids in polyacrylamide gels. *Eur. J. Biochem.* **46**, 83–90.

Breuer, C. B., and Florini, J. R. (1966). Effects of ammonium sulfate, growth hormone, and testosterone propionate on ribonucleic acid polymerase and chromatin activities in rat skeletal muscle. *Biochemistry* **5**, 3857–3865.

Bryant, T. R., and Howard, K. L. (1969). Meiosis in the Oömycetes: I Microspectrophotometric analysis of nuclear deoxyribonucleic acid in *Saprolegnia terrestris*. *Am. J. Bot.* **56**, 1075–1083.

Crain, W. R., Davidson, E. H., and Britten, R. J. (1976). Contrasting patterns of DNA sequence arrangement in *Apis mellifera* (Honeybee) and *Musca domestica* (Housefly). *Chromosoma* 59, 1–12.

Davidson, E. H., Galau, G. A., Angerer, R. C., and Britten, R. J. (1975). Comparative aspects of DNA organization in *Metazoa Chromosoma* 51, 253–259.

Griffin, D. H. (1966). Effect of electrolytes on differentiation of *Achlya* sp. *Plant Physiol.* 41, 1254–1256.

Griffin, D. H., and Breuker, C. (1969). Ribonucleic acid synthesis during the differentiation of sporangia in the water mold *Achlya*. *J. Bacteriol.* 98, 689–696.

Griffin, D. H., Timberlake, W., Cheny, J., and Horgen, P. A. (1975). The RNA polymerases of fungi: Glutarimides and the regulation of RNA polymerase I. In "Isozymes I, Molecular Structure" (C. Markert, ed.), pp. 69–87. Academic Press, New York.

Groner, B., Hynes, N., Sippel, A. E.,and Schutz, G. (1976). Induction of specific proteins in hyphae of *Achlya ambisexualis* by the steroid hormone antheridiol. *Nature (London)* 261, 599–601.

Gwynne, D. I., and Brandhorst, B. P. (1980). Antheridiol induced differentiation of *Achlya* in the absence of detectable synthesis of new proteins. *Exp. Mycology* (in press).

Hereford, L. M., and Roshbash, M. (1977). Number and distribution of polyadenylated RNA sequences in yeast. *Cell* 10, 453–458.

Higgens, S. J., Burchell, J. M., Parker, M. G., and Herries, D. C. (1978). Effects of testosterone on sequence complexity of polyadenylated RNA from rat seminal vesicle. *Eur. J. Biochem.* 91, 327–339.

Hjerten, S. (1962). Molecular sieve chromatography on polyacrylamide gels, prepared according to a simplified method. *Arch. Biochem. Biophys. Suppl.* 1, 147–151.

Horgen, P. A. (1977a). Steroid induction of differentiation: *Achlya* as a model system. In "Eukaryotic Microbes as Model Developmental Systems" (D. H. O'Day and P. A. Horgen, eds.), pp. 272–294. Dekker, New York.

Horgen, P. A. (1977b). Cytosol-hormone stimulation of transcription in the aquatic fungus, *Achlya ambisexualis*. *Biochem. Biophys. Res. Commun.* 75, 1022–1028.

Horgen, P. A. (1981). Purification and characterization of RNA polymerase II from the aquatic fungus, *Achlya ambisexualis Exp. Mycol.* (submitted).

Horgen, P. A., and Ball, S. F. (1974). Nuclear protein acetylation during hormone-induced sexual differentiation in *Achlya ambisexualis*. *Cytobios* 10, 181–185.

Horgen, P. A., and O'Day, D. H. (1975). The developmental patterns of lysosomal enzyme activities during Ca^{2+}-induced sporangium formation in *Achlya bisexualis*. II. α-Mannosidase. *Arch. Microbiol.* 102, 9–12.

Horgen, P. A., and Silver, J. C. (1978). Chromatin in eukaryotic microbes. *Annu. Rev. Microbiol.* 32, 249–284.

Horgen, P. A., Smith, R. J., and Silver, J. C. (1975a). The biosynthesis of ribosomal RNA in the aquatic fungus, *Achlya ambisexualis*. *Cytobios* 13, 193–199.

Horgen, P. A., Smith, R., Silver, J. C., and Craig, G. (1975b). Hormonal stimulation of ribosomal RNA synthesis in *Achlya ambisexualis*. *Can. J. Biochem.* 53, 1341–1350.

Hudspeth, M., Timberlake, W. E., and Goldberg, R. B. (1977). DNA sequence organization in the water mold *Achlya*. *Proc. Natl. Acad. Sci. U.S.A.* 74, 4332–4336.

Hütter, R., and DeMoss, J. A. (1967). Organization of the trytophan pathway: A phylogenetic study of the fungi. *J. Bacteriol.* 94, 1896–1907.

Jensen, E. V., and DeSombre, E. R. (1973). Estrogen-receptor interaction. *Science* 182, 126–134.

Kane, B. E., Reiskind, J. B., and Mullins, J. T. (1973). Hormonal control of sexual morphogenesis in *Achlya:* Dependence on protein and ribonucleic acid synthesis. *Science* **180,** 1192-1193.

Katzenellenbager, B. S., and Gorski, J. (1972). Estrogen action *in vitro. J. Biol. Chem.* **247,** 1299-1305.

Klein, R. M., and Cronquist, A. (1967). A consideration of the evolutionary and taxonomic significance of some biochemical, micromorphological, and physiological characters in the thallophytes. *Q. Rev. Biol.* **42,** 105-296.

Kornberg, R. D. (1977). Structure of chromatin. *Annu. Rev. Biochem.* **46,** 931-954.

Kuhn, R. W., Schrader, W. T., Smith, R. G., and O'Malley, B. (1975). Progesterone binding components of chick oviduct. X. Purification by affinity chromatography. *J. Biol. Chem.* **250,** 4220-4228.

Lauer, G. D., Roberts, T. M., and Klotz, L. C. (1977). Determination of the nuclear DNA content of *Saccharomyces cervisiae* and implications for the organization of DNA in yeast chromosomes. *J. Mol. Biol.* **114,** 507-526.

Law, D. J., Rozek, C. E., and Timberlake, W. E. (1978). Polyadenylate metabolism in *Achlya* ambisexualis. *Exp. Mycol.* **2,** 198-210.

LeJohn, H. B., Klassen, G., McNaughton, D. R., Cameron, L. E., Goh, S. H., and Meuser, R. U. (1977). Unusual phosphorylated compounds and transcriptional control in *Achlya* and other aquatic molds. *In* "Eukaryotic Microbes as Model Developmental Systems" (D. H. O'Day and P. A. Horgen, eds.), pp. 69-96. Dekker, New York.

LeStourgeon, W. M., and Wray, W. W. (1974). Extraction and characterization of the phenol-soluble acidic nuclear proteins. *In* "Acidic Proteins of the Nucleus" (I. L. Cameron and J. R. Jeter, Jr., eds.), pp. 59-102. Academic Press, New York.

Lovett, J. S., and Haselby, J. A. (1971). Molecular weights of the ribonucleic acid of fungi. *Arch. Microbiol.* **80,** 191-204.

MacLeod, H., and Horgen, P. A. (1979). Germination of the asexual spores of the aquatic fungus, *Achlya bisexualis. Exp. Mycol.* **3,** 70-82.

McMorris, T. C. (1978). Antheridiol and the oögoniols, steroid hormones which control sexual reproduction in *Achlya. Philos. Trans. R. Soc. London, Ser. B* **284,** 459-470.

McMorris, T. C., and White, R. M. (1977). The biosynthesis of the oögoniols, steroidal sex hormones of *Achlya:* The role of fucosterol. *Phytochemistry* **16,** 359-362.

McMorris, T., Seshardri, R., Weiher, G., and Barksdale, A. (1975). Structures of oögoniol-1, -2 and -3, steroidal sex hormones in the water mold, *Achlya. J. Am. Chem. Soc.* **97,** 2544-2545.

Michalski, C. J. (1978). Protein synthesis during hormone stimulation in the aquatic fungus, *Achlya. Biochem. Biophys. Res. Commun.* **84,** 417-427.

Mullins, J. T. (1973). Lateral branch formation and cellulase production in the water molds. *Mycologia* **65,** 1007-1014.

Mullins, J. T., and Ellis, E. A. (1974). Sexual morphogenesis in *Achlya:* Ultrastructural basis for hormone induction of antheridial hyphae. *Proc. Natl. Acad. Sci. U.S.A.* **71,** 1347-1350.

Musgrave, A., and Nieuwenhuis, D. (1975). Metabolism of radioactive antheridiol by *Achlya* species. *Arch. Microbiol.* **105,** 313-317.

Nolan, R. A., and Bal, A. K. (1974). Cellulase localization in hyphae of *Achlya ambisexualis. J. Bacteriol.* **117,** 840-843.

Notides, A., and Gorski, J. (1966). Estrogen-induced synthesis of a specific uterine protein. *Proc. Natl. Acad. Sci. U.S.A.* **56,** 230-235.

O'Day, D. H., and Horgen, P. A. (1974). The developmental patterns of lysosomal enzyme activities during Ca^{2+}-induced sporangium formation in *Achlya bisexualis.* I. Acid phosphatase. *Dev. Biol.* **39,** 116-124.

Oka, T., and Schimke, R. T. (1969). Interaction of estrogen and progesterone in chick oviduct development, *J. Cell Biol.* 43, 123–137.

O'Malley, B. W., and Means, A. R. (1974). Female steroid hormones and target cell nuclei. *Science* 183, 610–620.

O'Malley, B. W., McGuire, W., Kohler, P., and Korenman, S. (1969). Studies on the mechanism of steroid hormone regulation of synthesis of specific proteins. *Recent Prog. Horm. Res.* 25, 105–160.

Raper, J. R. (1950). Sexual hormones in *Achlya* VII. The hormonal mechanism in homothallic species. *Bot. Gaz.* 112, 1–24.

Ringold, G. M., Yamamoto, K. A., Tompkins, G. M., Bishop, J. M., and Varmus, H. (1975). Dexamethasone mediated induction of mouse mammary tumor virus RNA: A system for studying glucocorticoid action. *Cell* 6, 299–305.

Roeder, R. (1976). Eukaryotic nuclear RNA polymerases. *In* "RNA Polymerase" (R. Losick and M. Chamberlin, eds.), pp. 285–329. Cold Spring Harbor Press, New York.

Rozek, C. E., and Timberlake, W. E. (1980). Absence of evidence for changes in messenger RNA populations during steroid hormone-induced cell differentiation in *Achlya*. *Exp. Mycol.* 4, 34–47.

Rozek, C. E., Orr, W. C., and Timberlake, W. E. (1978). Diversity and abundance of polyadenylated RNA from *Achlya* ambisexualis. *Biochemistry* 17, 716–722.

Schwartz, R. J., Tasi, M. J., Tsai, S. Y., and O'Malley, B. W. (1975). Effect of estrogen on gene expression in the chick oviduct. V. Changes in the number of RNA polymerase binding and initiation sites in chromatin. *J. Biol. Chem.* 250, 5175–5182.

Silver, J. C. (1979). Chromatin organization in the Oömycete *Achlya ambisexualis*. *Biochim. Biophys. Acta* 561, 261–264.

Silver, J. C., and Horgen, P. A. (1974). Hormonal regulation of presumptive mRNA in the fungus, *Achlya ambisexualis*. *Nature (London)* 249, 252–254.

Spelsberg, T. (1974). The role of nuclear acidic proteins in binding steroid hormones. *In* "Acidic Proteins of the Nucleus" (I. L. Cameron and J. R. Jeter, Jr. eds.), pp. 247–296. Academic Press, New York.

Sutherland, R. B., and Horgen, P. A. (1977). Effects of the steroid sex hormone, antheridiol, on the initiation of RNA synthesis in the simple eukaryote, *Achlya ambisexualis*. *J. Biol. Chem.* 252, 8812–8820.

Sutherland, R. B., Schuerch, B. M., Ball, S. F., and Horgen, P. A. (1976). The developmental patterns of lysosomal enzyme activities during Ca^{2+}-induced sporangium formation in *Achlya bisexualis* III. Ribonucleases. *Arch. Microbiol.* 109, 289–294.

Tata, J. R. (1976). The expression of the vitellogenin gene. *Cell* 9, 1–14.

Tata, J. R., and Barker, B. (1975). Differential subnuclear distribution of polyadenylate-rich ribonucleic acid during induction of egg-yolk protein synthesis in male *Xenopus* liver by Oestradiol-17β. *Biochem. J.* 150, 345–355.

Thomas, D. Des S., and Mullins, J. T. (1967). Role of enzymatic wall-softening in plant morphogenesis: Hormonal induction in *Achlya*. *Science* 156, 84–85.

Thomas, D. des S., and Mullins, J. T. (1969). Cellulase induction and wall extension in the water mold *Achlya ambisexualis*. *Physiol. Plant.* 22, 347–353.

Timberlake, W. (1976). Alterations in RNA and protein synthesis associated with steroid hormone-induced sexual morphogenesis in the water mold *Achlya*. *Dev. Biol.* 51, 202–214.

Timberlake, W. E. (1978). Low repetitive DNA content in *Aspergillus nidulans*. *Science* 202, 973–975.

Timberlake, W. E., McDowell, L., Cheney, J., and Griffin, D. H. (1973). Protein synthesis during the differentiation of sporangia in the water mold *Achlya*. *J. Bacteriol.* **116**, 67–73.

Timberlake, W. E., Shumard, D. S., and Goldberg, R. B. (1977). Relationship between nuclear and polysomal RNA populations of *Achlya:* A simple eukaryotic system. *Cell* **10**, 623–632.

Towle, H. C., Tsai, M. J., Hirose, M., Tsai, S. Y., Schwartz, R., Parker, M. G., and O'Malley, B. W. (1976). Regulation of the transcription of the Eucaryotic genome. *In* "Molecular Biology of Hormone Action" (J. Papaconstantino, ed.), pp. 107–136. Academic Press, New York.

Tsai, M. J., Schwartz, R. J., Tsai, S. Y., and O'Malley, B. W. (1975). Effects of estrogen on gene expression in the chick oviduct. IV. Initiation of RNA synthesis on DNA and chromatin. *J. Biol. Chem.* **250**, 5165–5174.

Vaisius, A. C., and Horgen, P. A. (1979). Purification and characterization of RNA polymerase II resistant to α-amanitin from the mushroom *Agaricus bisporus*. *Biochemistry* **18**, 795–803.

van den Ende, H. (1976). Sex hormones in the water mold *Achlya*. *In* "Sexual Interactions in Plants," pp. 35–52. Academic Press, New York.

Wallace, R. A., and Jared, D. W. (1968). Estrogen induces lipophosphoprotein in serum of male *Xenopus laeris*. *Science* **169**, 91–92.

Zimmerman, J. L., and Goldberg, R. L. (1977). DNA sequence organization in the genome of Nicotiana tabacum. *Chromosoma* **59**, 227–252.

8

Sex Pheromones in *Mucor*

B. E. JONES, I. P. WILLIAMSON, AND G. W. GOODAY

I.	Introduction	179
	A. Life Cycles of the Mucorales	179
	B. Chemistry of the Mating Pheromones	183
II.	Survey of the Literature and Current Research	188
	A. Physiological Effects of Trisporic Acid	188
	B. Enzymology of Trisporic Acid Biosynthesis	189
	C. Trisporic Acid and the Control of Terpenoid Biosynthesis	191
	D. Effect of Trisporate on Its Own Biosynthesis	193
III.	Conclusion	195
	References	195

I. INTRODUCTION

A. Life Cycles of the Mucorales

The Mucorales form a large group of fungi, with 54 genera being recognized by Zycha *et al.* (1969). They are of importance to man as producers of industrial enzymes, such as amylases and rennins, of secondary metabolites, and of fermented foods and single-cell protein; as industrial tools for the conversions of sterols and other organic molecules; as degraders of stored materials; and as causative agents of mucormycosis and a few plant diseases.

They are ubiquitous, being particularly common in soil. Most are mesophilic, but there are thermophilic species such as *Mucor pusillus* and psychrophilic species such as *Mucor strictus*.

Their sexuality is termed "simple" or "primitive," and as such *Mucor* and *Rhizopus* are used as the introductory examples of sexual reproduction in most elementary biology textbooks. An annotated life cycle of *Mucor mucedo* (Fig. 1) serves to illustrate our understanding of reproduction in this heterothallic species.

The growing vegetative mycelium of *Mucor mecedo* is chiefly coenocytic, forming only a few septa. Single strains show asexual reproduction, with the formation of sporangiophores occurring from older mycelium. These structures are positively phototropic and negatively geotropic, and form terminal sporangia (the "pinheads" of this "pin

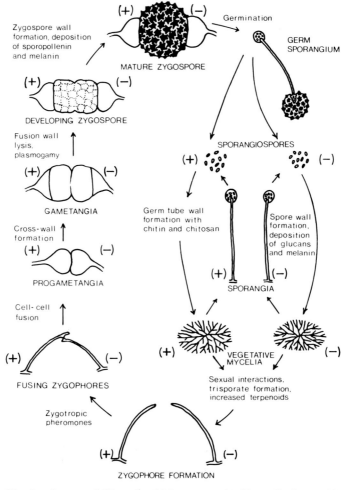

Fig. 1. Annotated life cycle of *Mucor mucedo*. (From Gooday, 1973.)

mold"). In nature, the sporangiospores are probably the major form of dispersal for this fungus.

When two mycelia of opposite mating type come close to one another, so that metabolites can diffuse through the substrate or air between them, sporangiophore formation does not occur. Instead, the distinctive sexual hyphae, the zygophores, are formed. These are not phototropic or geotropic, but are negatively autotropic (i.e., they grow out away from their subtending mycelium), and are strongly positively zygotropic—they grow toward zygophores of opposite mating type over a distance of several millimeters. This remarkable phenomenon results in their fusing in pairs, usually just behind their apices. After fusion, elongation ceases, and swelling occurs at the site of fusion, to give the two progametangia. A septum forms subapically across each progametangium to give a multinucleate gametangium from each mating type. The two gametangia have cytoplasmic continuity through plasmodesmata with the two suspensors subtending them (Hawker et al., 1966). The fusion walls break down to allow the intermingling of the cytoplasms of the two gametangia. The resultant compartment forms a thick, sculptured, multilayered wall, to give rise to the zygospore. New components in the wall of the zygospore are sporopollenin (a polymer of oxidised carotenoids) (Gooday et al., 1973) and the black melanin, both of which clearly have protective roles for the resting spore. Unfortunately, these components make study of the details of cytological events occurring after intermingling of the contents of the two gametangia very difficult. Probably only a few complementary nuclei fuse, the remainder presumably degenerating, and only one of the four products of meiosis survives, with the remainder degenerating. Nothing is known of the fate of other organelles, for example, whether the mitochondria of one mating type are destroyed by the cytoplasm of the other (in which case Blakeslee's long search for male and female in these fungi would be at an end). Eventually, after a dormant period, the zygospore germinates to give the germ sporangium, which in *Mucor hiemalis* usually produces germspores all of the same mating type (Gauger, 1965). Burnett (1965), in pointing out how erratic and poorly controlled zygospore formation and germination seem to be in nature, suggests that sexual reproduction plays only a small role in these fungi.

Other members of the Mucorales show variations of this basic pattern. Thus the zygophores of *Blakeslea trispora* and *Phycomyces blakesleeanus* (Sutter, 1977) differentiate as short ramified hyphae, usually below the surface of the substrate. In *Phycomyces blakesleeanus*, after fusion they elongate and swell, rising above the substrate surface, and resembling a pair of tongs. Branching thorns form on both suspensors to envelop the developing zygospore. By regenerating immature zygospores and their suspensor cells of *Rhizopus stolonifer,* Gauger (1977) has shown that meiosis occurs early during zygospore development, and, furthermore, that hyphae of opposite mating type can develop from a suspensor, indicating that nuclei can sometimes pass from the immature zygospore into either suspensor. In both *Rhizopus stolonifer* (Gauger, 1977) and *Phycomyces blakesleeanus* (Cerda-Olmedo, 1975), individual germ sporangia give rise to germ spores representing more than one meiotic product, and in *Phycomyces blakesleeanus* the number of different haploid nuclei undergoing meiosis is often more than one, but generally not more than two.

The regulation of sexual reproduction in the Mucorales was first elucidated by Blakeslee (1904). Previously zygospore formation had been a mystery, occurring sporadically and unpredictably in some isolates. Blakeslee showed that zygospores were formed by *Rhizopus nigricans* only at the line of meeting of two different isolates, which he designated plus and minus. All other isolates of *Rhizopus nigricans* could be assigned as plus or minus according to whether they formed zygospores with the minus or the plus, respectively. Blakeslee's observations over many years (e.g., Blakeslee and Cartledge, 1927) showed that by observing intraspecific and interspecific sexual reactions with tester strains, all his isolates of zygomycetes could be classified into four categories: heterothallic plus, heterothallic minus, homothallic, and sterile ("neutral"). None of the plus strains interact, or do any of the minus strains. A plus and a minus of the same species interact to give rise to zygospores. Plus and minus strains of different species never give true zygospores, but interact to varying degrees, showing formation of zygophores by one or both, or sometimes the formation of azygospores ("parthenospores," zygospore-like structures formed from one suspensor only).

The descriptions of these interspecific reactions give us our current tenet of belief—all mucorales share the same chemical control of their sexual differention, i.e., they share the same pheromone system. Although the term hormone persists in the literature on the Mucorales, the more precise term pheromone will be employed in this chapter.

The homothallic species are self-fertile, giving rise to zygospores on the same thallus. Some, such as *Rhizopus sexualis,* are isogamous (with the two zygophores showing indistinguishable behavior); others, such as *Zygorhynchus moelleri,* are anisogamous (with one zygophore developing differently to the other to give, for example, one suspensor larger than the other). The homothallic isolates also show interspecific reactions, some showing a plus tendency, i.e., interacting with minus tester strains, others showing a minus tendency, and others interacting with both plus and minus tester strains.

A direct approach to the unraveling of the genetic nature of "plusness" and "minusness" has been shown by Nielsen's (1978) experiments of mutagenesis of a plus strain of *Mucor pusillus*. Treatment with gamma rays yielded two stable mutants with minus phenotype (i.e., they formed normal zygospores with plus strains), and four stable homothallic mutants. Gauger (1975) described diploid strains of *Mucor hiemalis* obtained among single germ sporangiospore isolations. These diploids produced azygospores, acted as plus mating type in crosses with the parental strains, and were unstable in that they broke down, sometimes via bisexual heterokaryotic mycelia, to normal plus or minus strains.

Three species of fungi have been primarily used for nearly all work on sex pheromones.

Mucor mucedo (L.) Fresenius is very responsive to trisporic acid and its precursors, producing distinctive zygophores that are readily countable, and used for their bioassay (Gooday, 1978). Little is known of the genetic control of its sexuality, apart from the work of Köhler (1935) and Wurtz and Jockusch (1975). Its yield of trisporic acids is low.

Blakeslea trispora Thaxter produces very large amounts of trisporic acids, and so has been used for nearly all studies of the chemical steps leading to their formation (Bu'Lock *et al.*, 1976; van den Ende *et al.*, 1972; Sutter *et al.*, 1974). Nothing is known of its genetics. As its yield of trisporic acids is vastly in excess to that required for their pheromonal activity, it is probable that it has lost control of some regulatory step in their biosynthesis.

Phycomyces blakesleeanus Burgeff has recently been the subject of much detailed genetical research (Cerda-Olmedo, 1975; Eslava *et al.*, 1975). This work has yielded a number of well-defined mutants that have proved of use in elucidating the regulation of sexual development (Sutter, 1975). Its yield of trisporic acids is low, and it is poorly responsive to exogenous trisporic acids (Sutter, 1977).

B. Chemistry of the Mating Pheromones

Although zygospore production in Mucorales was long suspected to be a sexual process involving diffusable substances, the discovery of the controlling factor, trisporic acid, preempted the recognition of its pheromone function. Trisporic acid was isolated as a typical secondary metabolite from (*Choanephora*) *Blakeslea trispora* by Caglioti et al. (1964, 1967), and characterized as a group of related 18-carbon terpenoid acids (I, II; Fig. 2). They showed it to be the factor responsible for the dramatically enhanced β-carotene levels in mated fermentations. The profound physiological action of trisporic acid was not suspected until van den Ende (1968) and Gooday (1968) showed it to be identical with the zygophore-inducing substance from *Mucor mucedo*. The total synthesis of trisporic acid has been described, the synthetic product having the same biological activity as natural trisporic acid (Edwards et al., 1971). It is now believed that trisporic acid is the sex pheromone for all mucoraceous fungi because it will induce zygophore production in single strains, and zygophores can be mutually induced by interspecific matings. In addition, trisporic acid was instrumental in inducing zygophore formation in a previously sterile culture of the homothallic species *Syzygites megalocarpus* (Werkman and van den Ende, 1974).

The Mucorales are unusual in comparison with microorganisms possessing known sex pheromones in that both partners respond to a single substance. It requires, however, the equal participation of both strains to synthesize the hormone since mating of ^{14}C-labeled mycelium of *Blakeslea trispora* gives [^{14}C]trisporic acid irrespective of which strain is the labeled partner (van den Ende et al., 1972). Moreover, trisporic acid production requires enzyme synthesis *de novo* in both strains since inhibition of RNA synthesis by 5-fluorouracil of either partner prior to mating gives no trisporic acid (Bu'Lock and Winstanley, 1971; van den Ende et al., 1972).

The structural similarity between β-carotene and trisporic acid suggested a probable biosynthetic relationship. This view was strengthened by the evidence that the inhibition of β-carotene synthe-

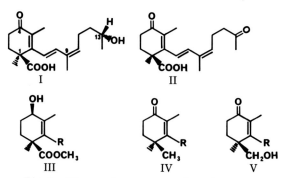

Fig. 2. The sex pheromones and propheromones.

sis by diphenylamine also inhibits trisporic acid production (Austin et al., 1969) and that *car* mutants (no carotene) of *Phycomyces blakesleeanus* are defective for the mating process (Bergman et al., 1969) and produce no trisporic acid (Sutter, 1975). Radioactive tracer studies proved that trisporic acid is biosynthesized from β-carotene as an extension of general isoprenoid biosynthesis from mevalonate (Austin et al., 1970). In addition, several postulated C_{18} and C_{20} intermediates including retinol were incorporated into trisporic acid in mated cultures of *B. trispora* (Bu'Lock et al., 1974a). The excellent incorporation of a β-C_{18} ketone and its 4-hydroxy derivative (Fig. 3) suggest that C-4 hydroxylation and C-11,12 hydrogenation are early steps in the pathway. The precise sequence is not known since none of these early metabolites has been isolated although retinal has been found in *Phycomyces blakesleeanus* (Bergman et al., 1969). However, trisporic acid synthesis by unmated strains of Mucorales is negligible, if it occurs at all, but the plus and minus strains produce small quantities of sex-specific metabolites which can only be converted to trisporic acid by the opposite mating partner (Sutter et al., 1973). Bu'Lock et al. (1973) recognized this to be a true mating type distinction, the plus strain producing compounds with a hydroxyl group at C-4 and methyl ester at C-1 (e.g., 4-hydroxymethyl trisporate, III, Fig. 2) while minus strains produced compounds having a 4-oxo group and *gem*-dimethyl or hydroxymethyl at C-1 (e.g., trisporins, IV; and trisporols, V; Fig. 2). It was further recognized that these mating type-specific single strain products although undoubtably related to each other and to trisporic acid could not possibly be intermediates in a common linear biosynthetic pathway. Therefore a "metabolic grid" was proposed to account for the differing abilities of plus and minus strains to carry out a complete oxidation at C-1 and C-4. Other experiments (Bu'Lock et al., 1972) had shown that both strains are capable of the necessary side-chain modifications and this, with the 9-cis/trans isomerization, is irrelevant to the concept of sex-linked strain differences, although the cis isomer and the B-series (13-oxo) products are more potent zygophore inducers in bioassay than the trans isomer and C-series (13-hydroxy) compounds (Bu'Lock et al., 1972).

The available evidence shows that the sex pheromone, trisporic acid, is produced only in response to the sexual process by a remarkable scheme of collaborative biosynthesis. The broad outlines of this biosynthesis from β-carotene via mating type-specific precursors ("propheromones," previously called "prohormones") now seems clear and a possible scheme is outlined in Figs. 3 and 4.

The first and probably rate-limiting step involves oxidative 15,15'

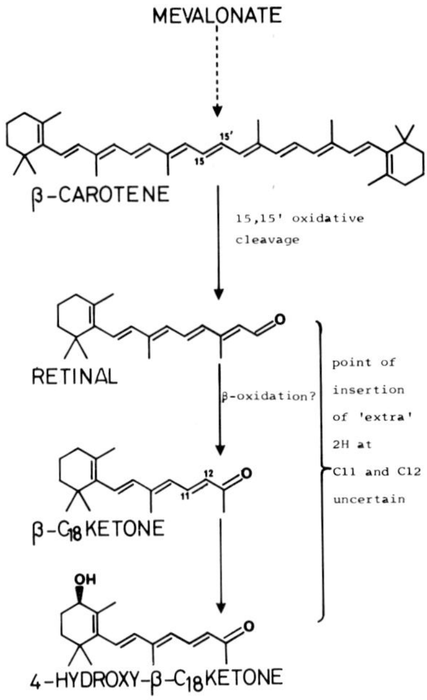

Fig. 3. Biosynthetic scheme for trisporic acid showing early reactions common to both mating types.

cleavage of β-carotene to retinal, although this remains unproved. There follows the loss of a 2-carbon unit, reduction of the 11,12 double bond and a series of oxidations some of which are mating type-specific. The precise order of these reactions is unclear and is probably unimportant. The proposed metabolic grid allows for a variation in pathways, although a preferred pathway (possibly species-variable) may operate during mating.

The nature of this metabolic grid explains why trisporic acid is only synthesized when the two thalli are in diffusion contact since only then can the "propheromones" diffuse from one cell to the other to complete the metabolic pathway. Hence, it is the biosynthesis and not the presence of trisporic acid which confers the mating type distinction.

On the basis of two alleles of a single gene (MT) a scheme for the control of the sexual process has been proposed (Bu'Lock, 1975). Since the difference between the two genotypes is small, it is envisaged that the allelic products of the mating type loci, MT^+ and MT^-, are slightly dissimilar repressor proteins, R^+ and R^-. In the unmated situation, in single strains these proteins repress to varying degrees the gene sites coding for some of the enzymes for the β-carotene to trisporate pathway. Thus each strain produces a low level of unique constitutive enzymes which continuously convert the available substrate into a low level mixture of characteristic compounds ("propheromones"). These are usually degraded to inactive compounds, such as trisporone (Cainelli et al., 1967) and other known cometabolites (Bu'Lock et al., 1973, 1976; Sutter, 1979). This may operate as an "overflow" mechanism which maintains a continuous low level throughput of material to maintain the pathways and keep a "constant head" of precursors in readiness for the interactions with a potential mate.

During mating the propheromones can diffuse to the opposite mating

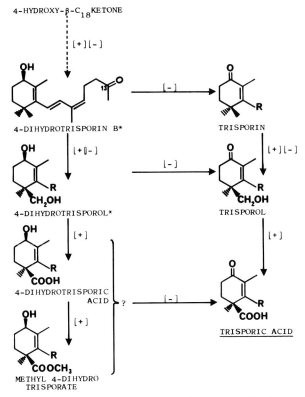

Fig. 4. Proposed "metabolic grid" scheme for the collaborative biosynthesis of trisporic acid showing mating type-specific reactions. For clarity the side chain has been omitted. The difference between the 13-oxo (B) series and 13-hydroxyl (C) series is not a point of mating type distinction: both are fully interconvertible by both strains and presumably a parallel sequence of reactions exists. Also omitted is the 9-cis/trans isomerization; its relevance is not clear and may be non-specific. The compounds marked * are postulated and have not been isolated.

partner and be converted to trisporic acid by the complementary constitutive enzymes. The trisporic acid so formed may then act as a derepressor of the pathway for its own synthesis. The system is now self-amplifying and a chain reaction of events ensues since trisporic acid has been shown to act by feedback induction on the overall derepressed pathway. Thus, the system changes from being enzyme limited to one of substrate limitation. Trisporic acid has been shown to induce the production of its immediate precursors (Werkman and van den Ende, 1973) and such events would serve to explain the small quantities of propheromone extracted from single strains: 1.6 mg of plus propheromone from 15 liters of plus *B. trispora* culture medium compared with nearly 600 mg of trisporic acid from a similar volume of a mated culture (Bu'Lock et al., 1972, 1974b).

II. SURVEY OF THE LITERATURE AND CURRENT RESEARCH

A. Physiological Effects of Trisporic Acid

The main problem remaining is that we have little information on the primary site of trisporic acid action or its effect at the molecular level. We could speculate by analogy with other pheromone or hormone systems on the existence of a trisporate receptor molecule. Since it is established that trisporic acid can diffuse away through the medium from its site of synthesis to affect distal mycelium (Gooday, 1978), we can assume that there is a mechanism for its uptake and reception by the cell. We have preliminary evidence that in minus *M. mucedo* trisporic acid is taken up more rapidly in the ionized form (Gooday et al., 1979) and the existence of a specific "permease" has been inferred from the work of Wurtz and Jockusch (1975) and Sutter (1977).

Cyclic AMP (cAMP) levels in the mycelium of mated *M. mucedo* have been shown to be significantly higher than those of vegetative hyphae and treatment of single strains with physiological concentrations of trisporic acid results in a transient but significant elevation of cAMP levels, maximal about 4 hr after the addition of pheromone and immediately prior to zygophore initiation (Bu'Lock et al., 1976). Whether the effects of trisporic acid are transduced by cAMP is still unclear. The time lag for this effect make this seem doubtful although it should be noted that the kinetics of the increase in cAMP levels are remarkably similar to those for the trisporic acid-induced increases in RNA synthesis in *B. trispora* (Bu'Lock et al., 1976). The morphological events following the treatment of *M. hiemalis* and *M. mucedo* with $N^6,O^{2'}$-dibutyryl adenosine 3',5'-cyclic monophosphate have been interpreted to represent at least in part an uncoordinated attempt at some of the changes associated with sexual development (Jones and Bu'Lock, 1977).

One event that is caused directly or indirectly by trisporic acid is the production of zygophores. These are the only cells in *Mucor* that undergo cell-cell fusion but only in pairs of opposite mating type. The opposite zygophores grow toward one another due to volatile zygotropic signals, perhaps of trisporate precursors (Mesland et al., 1974), possibly the propheromones themselves. The zygophores then fuse on contact. This rapid fusion infers a form of specific cell-cell recognition mediated by surface components. The surfaces of *M. mucedo* zygophores have been investigated using immunological techniques (Jones and Gooday, 1978) and although zygophore-specific antigens have been identified, no mating type-specific differences have been detected. A study of the surface properties of zygophores has been made using fluorescent lectins (which bind to specific polysaccharides). Concanavalin A (binds α-mannose), soya bean agglutinin (binds N-acetylgalactosamine) and a lectin from *Tetragonolobus purpureas* (binds fucose) had little or no affinity for vegetative or sexual cells of *M. mucedo*. However, zygophores in particular have a higher affinity for wheat germ agglutinin (binds N-acetylglucosamine) than either vegetative hyphae or sporangiophores (Jones and Gooday, 1977). Thus the zygophore has a distinctive surface chemistry but it is unclear how this relates to its function and how its formation could be regulated by trisporic acid.

B. Enzymology of Trisporic Acid Biosynthesis

The biosynthetic pathway leading to trisporic acid is shown in Figs. 3 and 4. For the purposes of study, it can be divided into three sections. First, the formation of β-carotene from mevalonate and its precursors occurs by the well-established terpenoid pathway. Detailed consideration of the general enzymology of this section (other than aspects of control which will be discussed later) is beyond the scope of this chapter. The area is well reviewed by Britton (1976) and Goodwin (1979). The second and third sections, namely, the formation of the β-C_{18} ketone and its subsequent conversion via the mating type-specific reactions of the metabolic grid to trisporic acid, are possibly unique to the Mucorales. The reactions involved are less well understood, and while the general features of the collaborative biosynthesis are not in doubt, few of the implied enzyme activities have been demonstrated directly.

The 15,15' cleavage of β-carotene to form retinal has not been demonstrated in fungi, but a well-characterized carotene 15,15'-oxygenase is found in mammalian intestinal mucosa and liver, where it is involved in vitamin A (retinol) formation from dietary β-carotene (Fidge et al., 1969).

Conversion of retinal to the C_{18} ketone requires, formally, the (oxidative?) removal of two carbon atoms. A prerequisite of such a shortening almost certainly involves oxidation to retinoic acid. Again, this reaction has not been demonstrated in fungi, but enzymes catalyzing the reaction are present in rat intestine and liver. Both enzymes would appear to be dehydrogenases rather than oxidases, but the nature of the accepting coenzyme is unclear (Moffa et al., 1970).

A novel reaction sequence can be proposed to convert retinoate to β-C_{18} ketone. After the probable formation of retinoyl-CoA, the trans double bond β to the carboxyl might be hydrated by an enoyl-CoA hydratase. The resulting β-hydroxyacyl-CoA might undergo aldol cleavage losing two-carbon fragment (acetyl-CoA) and yielding the β-C_{18} ketone. An analogous reaction is the cleavage of 3-hydroxy-3-methylglutaryl-CoA by a specific lyase during leucine catabolism or mammalian ketone body synthesis.

The reactions involved in the conversion of the β-C_{18} ketone to the 4-dihydrotrisporins have been ascribed, by precursor studies, to both mating strains, but are otherwise uncharacterized.

The C-4 hydroxylation is likely to be catalyzed by a monooxygenase system (Gunsalus et al., 1975) which may be related to those which bring about the C-3,3' hydroxylation of carotene to form the xanthophylls or the C-4,4' hydroxylations of algal "secondary carotenoid" synthesis (Goodwin, 1971).

Saturation of the C-11 to C-2 double bond of the isoprenoid side chain to form that characteristic of the trisporates could involve either flavoprotein or pyridine nucleotide-linked hydrogen donors.

The oxidation of C_4-OH to C_4-oxo has been stated by Bu'Lock et al. (1976) to be that activity which specifically characterizes the minus mating type. An enzyme catalyzing the oxidation of methyl-4-dihydrotrisporate to methyl trisporate has been located in the soluble fraction of extracts of Mucor mucedo by Werkman (1976). It has a pH optimum of 7.6, a K_m for substrate of 1 mM, and requires NADP as coenzyme. It remains to be established whether this enzyme is responsible for all the horizontal 4-dehydrogenations of Fig. 4, i.e., whether it has broad specificity toward substrates of differing substitution at C-1.

Other than their restriction to mating-type, nothing is known of the enzymes catalyzing 1-methyl hydroxylation (distribution between plus and minus appears to be variable and species specific) and 1-CH_2OH

oxidation. This latter is the activity characteristic of plus mating type (Bu'Lock *et al.*, 1976).

The reaction sequence (4-dihydro)trisporin → trisporol → (trisporal) → trisporic acid is similar to that which occurs at the C_4-methyl of lanosterol during its converstion to cholesterol (Bechtold *et al.*, 1972) and could well involve analogous methyl-group hydroxylation by a cytochrome P-450 system. Subsequent dehydrogenations of the alcohol and aldehyde might be expected to be pyridine nucleotide linked.

The esterification of the 4-carboxylate to yield the methyl trisporates is a plus strain activity but is otherwise uncharacterized. Werkman (1976) has demonstrated the presence of methyl esterase in extracts of *Mucor mucedo* (minus).

C. Trisporic Acid and the Control of Terpenoid Biosynthesis

The effects of mating and of trisporic acid treatment on the content of carotenoids and other terpenoids in both *Blakeslea trispora* and *Mucor mucedo* have been documented by Gooday (1978) and will not be detailed here. The increased carotene content resulting from the action of trisporic acid can be considered as part of a positive feedback mechanism whereby trisporic acid stimulates its own biosynthesis. In addition, a proportion of the carotene becomes incorporated as sporopollenin into the protective outer layer of the zygospores (Gooday *et al.*, 1973). The increases in content of steroids, polyprenols, and ubiquinones may be adventitious, reflecting only the early point of action of trisporic acid in polyisoprenoid biosynthesis. However, Bu'Lock and Osagie (1973) have suggested that the formation of zygophores and later stages in sexual differentiation may have requirements for membranes, surface polymers, and respiratory chain components.

Available evidence indicates that trisporic acid exerts its effects on the carotenogenic pathway by derepression of one or more enzymes. Thomas *et al.* (1967) showed that the increase in carotene content of trisporic acid-treated *B. trispora* (minus) could be inhibited by cycloheximide. Eslava *et al.* (1974) find a similar cycloheximide inhibition of the vitamin A and β-ionone stimulation of carotenogenesis in *Phycomyces blakesleeanus*. It is likely that these two compounds are acting as structural analogues of trisporic acid, mimicking its effects, albeit at much greater concentration.

Most of our knowledge of the control of terpenoid synthesis has come from studies of cholesterol synthesis in mammals. In most systems studied, the point at which control is effected is at 3-hydroxy-3-

methylglutaryl-CoA reductase (HMG-CoA reductase) (Brown and Rodwell, 1979). The activity of this enzyme is modulated by a variety of processes (not all believed to operate simultaneously), including changes in rate of synthesis and breakdown and phosphorylation/dephosphorylation. Another point where some control might be effected has been identified as mevalonate kinase. Preparations of the enzyme from both animal (Dorsey and Porter, 1968) and plant sources (Gray and Kekwick, 1972) are strongly inhibited by prenyl pyrophosphates but it is not known whether such feedback regulation operates *in vivo*.

Recent work has been directed toward the location of the point of action of trisporic acid. The approach taken has been that of direct enzyme assay. A rate-limiting enzyme might be expected to show significant increases in activity on trisporic acid treatment. Preliminary experiments with the mevalonate kinase of *M. mucedo* suggested that this was probably not a limiting enzyme for terpenoid biosynthesis (Gooday, 1978). However, the activity of mevalonate kinase in cultures of *B. trispora* was reported by Desai and Modi (1977) to double on mating or on trisporic acid treatment of single strains. A more extensive study of *Mucor* mevalonate kinase has been made by H. A. Sani and I. P. Williamson (unpublished). The enzyme has been partially purified from the soluble fraction of mycellial extracts, and is similar in general properties to that isolated from *Neurospora crassa* (Imblum and Rodwell, 1974). It has apparent K_m values for mevalonate and ATP of 0.23 mM and 1.20 mM, respectively, and its molecular size as determined by gel filtration and sucrose density gradient centrifugation is 100,000. Its specific activity is unchanged on treatment of minus *M. mucedo* mycelium with concentrations of trisporic acid sufficient to induce maximum rates of carotenogenesis. It therefore seems unlikely that, in *Mucor*, mevalonate kinase plays a primary role in the regulation of terpenoid synthesis.

Throughout this study it was found necessary to adopt stringent precautions to minimize inactivation of the enzyme by the endogenous fungal proteases (Pringle, 1975; Lumsden and Coggins, 1977) if reproducible activity determinations were to be made and enzyme preparations obtained with half-lives measured in hours rather than minutes. It may be that *Mucor* mevalonate kinase is particularly susceptible to proteolytic attack but similar inactivation of other enzymes in mycelial extracts might be expected.

Similar studies have been made on the HMG-CoA reductase of *M. mucedo* by I. P. Williamson and V. W. Rodwell (unpublished). The enzyme is similar in properties to that of *Neurospora crassa* (Imblum and Rodwell, 1974). When appropriate precautions to minimise proteolysis are taken, the enzyme may be assayed readily in a "microsomal" fraction. It uses NADPH as reductant. Its activity has been determined in cultures of *M. mucedo* (*minus*) grown in the presence or absence of trisporic acid (Fig. 5). Although massive and immediate increases in β-carotene content of mycelia were elicited by pheromone treatment, the specific activity of reductase did not differ significantly from untreated controls.

It is concluded that HMG-CoA reductase is unlikely to be the rate-limiting enzyme in terpenoid biosynthesis in *Mucor*. The point of action of trisporic acid is therefore still an unanswered question.

While the controlling role of HMG-CoA reductase in mammalian sterol synthesis is an established fact, teleological arguments suggest that other enzymes might be regulatory in other systems or act in addition to the primary regulator in mammals. In cytoplasmic sterol synthesis the reaction catalyzed by HMG-CoA reductase is not that which commits acetate carbon to a polyisoprenoid fate. That function is performed by HMG-CoA synthase. (The equilibrium of the thiolase reaction lies so far in the direction of cleavage of acetoacetyl-CoA as to rule out consideration of that enzyme.)

In recent years, convincing evidence has been obtained for the modulation of HMG-CoA synthase in the regulation of cholesterol synthesis in liver (Clinkenbeard *et al.*, 1975) and in rat adrenal gland (Balasubramanian *et al.*, 1977). In these mammalian tissues the regulatory role of HMG-CoA synthase is secondary to regulation by HMG-CoA reductase, and the synthase must be considered as a contender for the controlling enzyme of carotenogenesis in *Mucor*.

The arguments proposed above for the identification of the rate-limiting enzyme have been based on the assumption that the carotenogenic carbon flux in the Mucorales is through acetyl-CoA. This assumption is not proved; even if so, it is possible that the increased flux evident on hormone treatment has other origins. Friend *et al.* (1955) showed that the amino acids leucine and valine could be strongly carotenogenic in *Phycomyces blakesleanus* under certain growth conditions. Stimulation of pathways that normally are minor contributors of carbon to the overall synthetic pathway must be considered a possibility.

D. Effect of Trisporate on Its Own Biosynthesis

The qualitative aspects of the production of trisporic acid by the collaboration of mated cultures are explained earlier in this chapter.

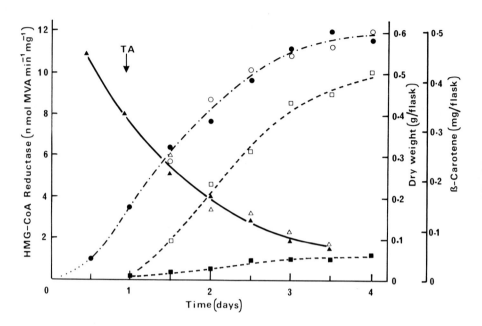

Fig. 5. The effect of trisporic acid on growth, β-carotene content and activity of HMG-CoA reductase of *Mucor mucedo* (minus). Two sets of cultures were grown on 4.3% malt extract, 0.7% mycological peptone in a gyrotary shaker in the dark at 23°C. After 1 day's growth, 0.5 mg trisporic acid were added to each of one set of flasks. At 12 hr intervals, 1 flask was harvested from each set and dry weight (-·-·-), β-carotene (- - - -), and HMG-CoA reductase activity (—) determined. Solid symbols, untreated cultures. Open symbols, trisporic acid treated cultures. (Results of I. P. Williamson and V. W. Rodwell, unpublished.)

The mating type-specific precursors are detected only in very small quantities in unmated cultures, leaving a problem as to how much larger amounts of trisporic acid can be made. Trisporic acid itself has been shown to stimulate the production of these precursors in unmated cultures of *B. trispora* and *M. mucedo* (Werkman and van den Ende, 1973, 1974; Bu'Lock et al., 1976). The effect was inhibited by 5-fluorouracil and so presumably involves protein synthesis. The specific activities of the previously mentioned 4-OH dehydrogenase and methyl esterase (Werkman, 1976) were doubled by treatment of the mycelium with trisporic acid; this effect was inhibited by cycloheximide.

III. CONCLUSION

The "trisporic acid story" has many remarkable facets, and has provided us with much intellectual stimulation during its unraveling. It certainly holds surprises for us yet. The remarkable collaborative biosynthesis of trisporic acid appears unique in nature, but may be a particularly clear-cut example of a widespread biochemical phenomenon.

REFERENCES

Austin, D. J., Bu'Lock, J. D., and Winstanley, D. J. (1969). Trisporic acid biosynthesis and carotenogenesis in *Blakeslea trispora. Biochem. J.* 133, 34P.

Austin, D. J., Bu'Lock, J. D., and Drake, D. (1970). The biosynthesis of trisporic acids from β-carotene via retinal and trisporol. *Experientia* 26, 348–349.

Balasubramaniam, S., Goldstein, J. L., and Brown, M. S. (1977). Regulation of cholesterol synthesis in rat adrenal gland through the coordinate control of 3-hydroxy-3-methyl-glutaryl coenzyme A synthase and reductase activities. *Proc. Natl. Acad. Sci. U.S.A.* 74, 1421–1425.

Bechtold, M. M., Delwiche, C. V., Cornai, K., and Gaylor, G. L. (1972). Investigation of the component reactions of oxidative sterol demethylation. *J. Biol. Chem.* 247, 7650–7656.

Bergman, K., Burke, P. V., Cerda-Olmedo, E., David, C. N., Delbruck, M., Foster, K. W., Goodill, E. W., Heisenberg, M., Meissner, G., Zalokar, M., Dennison, D. S., and Shropshire, W. (1969). Phycomyces. *Bacteriol. Rev.* 33, 99–157.

Blakeslee, A. F. (1904). Sexual reproduction in the Mucorineae. *Proc. Am. Acad. Arts Sci.* 40, 205–319.

Blakeslee, A. F., and Cartledge, J. L. (1927). Sexual dimorphism in Mucorales. II Interspecific reactions. *Bot. Gaz. (Chicago)* 84, 51–57.

Britton, G. (1976). Biosynthesis of carotenoids *In* "Chemistry and Biochemistry of Plant Pigments" (T. W. Goodwin, ed.), 2nd ed., Vol. 1, pp. 262–327. Academic Press, New York.

Brown, W. E., and Rodwell, V. W. (1979). Hydroxymethyl-glutaryl CoA reductase. *In* "Dehydrogenases Requiring Nicotinamide Coenzymes" (J. Jeffery, ed.), pp. 232–272. Birkheuser, Basel.

Bu'Lock, J. D. (1975). Cascade expression of the mating-type-locus in Mucorales. *Proc. Int. Symp. Genet. Ind. Micro-organisms, 2nd, 1974* pp. 497–509.

Bu'Lock, J. D., and Osagie, A. U. (1973). Prenols and ubiquinones in single strain and mated cultures of *Blakeslea trispora. J. Gen. Microbiol.* 76, 77–83.

Bu'Lock, J. D., and Winstanley, D. J. (1971). Carotenoid metabolism and sexuality in Mucorales. *J. Gen. Microbiol.* 68, xvi–xvii.

Bu'Lock, J. D., Drake, D., and Winstanley, D. J. (1972). Specificity and transformations of the trisporic acid series of fungal sex hormones. *Phytochemistry* 11, 2011–2018.

Bu'Lock, J. D., Jones, B. E., Quarrie, S. A., and Winskill, N. (1973). The biochemical basis of sexuality in Mucorales. *Naturwissenschaften* 60, 550–551.

Bu'Lock, J. D., Jones, B. E., Taylor, D., Winskill, N., and Quarrie, S. A. (1974a). Sex hormones in Mucorales: The incorporation of C_{20} and C_{18} precursors into trisporic acids. *J. Gen. Microbiol.* **80,** 301–306.

Bu'Lock, J. D., Winskill, N., and Jones, B. E. (1974b). Structures of the mating-type-specific prohormones of Mucorales. *Chem. Commun.* pp. 708–711.

Bu'Lock, J. D., Jones, B. E., and Winskill, N. (1976). The apocarotenoid system of sex hormones and prohormones in Mucorales. *Pure Appl. Chem.* **47,** 191–202.

Burnett, J. H. (1965). The natural history of recombination systems. *In* "Incompatibility in Fungi" (K. Esser and J. R. Raper, eds.), pp. 98–113. Springer-Verlag, Berlin and New York.

Caglioti, L., Cainelli, G., Camerino, B., Mondelli, R., Prieto, A., Quilico, A., Salvatori, T., and Selva, A. (1964). Sulla constituzione degli acidi trisporici. *Chim. Ind. (Milan)* **46,** 961–966.

Caglioti, L., Cainelli, G., Camerino, B., Mondelli, R., Prieto, A., Quilico, A., Salvatori, T., and Selva, A. (1967). The structure of trisporic acid C. *Tetrahedron, Suppl.* **7,** 175–187.

Cainelli, G., Camerino, B., Grasselli, P., Mondelli, R., Morrochi, S., Prieto, A., Quilico, A., and Selva, A. (1967). Structtura del trisporone e dell'anidotrisporone. *Chim. Ind. (Milan)* **49,** 748–751.

Cerda-Olmedo, E. (1975). The genetics of *Phycomyces blakesleeanus*. *Genet. Res.* **25,** 285–296.

Clinkenbeard, K. D., Sugiyama, T., Reed, W. D., and Lane, M. D. (1975). Cytoplasmic 3-hydroxy-3-methyl-glutaryl coenzyme A synthase from liver. *J. Biol. Chem.* **250,** 3124–3135.

Desai, H. G., and Modi, V. V. (1977). Physiology of mated cultures of *Blakeslea trispora* and mechanism of trisporic acid action. *Indian J. Exp. Biol.* **15,** 609–612.

Dorsey, J. K., and Porter, J. W. (1968). The inhibition of mevalonic kinase by geranyl and farnesyl pyrophosphates. *J. Biol. Chem.* **243,** 4667–4670.

Edwards, J. A., Schwarz, V., Fajkos, J., Maddox, M. L., and Fried, J. H. (1971). Fungal sex hormones. The synthesis of $(\pm)-7(t)$, $9(t)$-trisporic acid B methyl ester. The stereochemistry at C-9 of the trisporic acids. *Chem. Commun.* pp. 292–293.

Eslava, A. P., Alvarez, M. J., and Cerda-Olmedo, E. (1974). Regulation of carotene biosynthesis in *Phycomyces* by vitamin A and β-ionone. *Eur. J. Biochem.* **48,** 617–623.

Eslava, A. P., Alvarez, M. I., Burke, P. V., and Delbrück, M. (1975). Genetic recombination in sexual crosses of *Phycomyces*. *Genetics* **80,** 445–462.

Fidge, N. H., Smith, E. R., and Goodman, DeW. S. (1969). The enzymic conversion of β carotene into retinal in hog intestinal mucosa. *Biochem. J.* **114,** 689–694.

Friend, J., Goodwin, T. W., and Griffiths, L. A. (1955). Studies in Carotenogenesis 15. The role of carboxylic acids in the biosynthesis of β carotene by *Phycomyces blakesleanus*. *Biochem. J.* **60,** 649–655.

Gauger, W. L. (1965). The germination of zygospores of *Mucor hiemalis*. *Mycologia* **57,** 634–641.

Gauger, W. L. (1975). Further studies on sexuality in azygosporic strains of *Mucor hiemalis*. *Trans. Br. Mycol. Soc.* **64,** 113–118.

Gauger, W. L. (1977). Meiotic gene segregation in *Rhizopus stolonifer*. *J. Gen. Microbiol.* **101,** 211–217.

Gooday, G. W. (1968). Hormonal control of sexual reproduction in *Mucor mucedo*. *New Phytol.* **67,** 815–821.

Gooday, G. W. (1973). Differentiation in the Mucorales. *Symp. Soc. Gen. Microbiol.* **23,** 269–293.

Gooday, G. W. (1978). Functions of trisporic acid. *Philos. Trans. R. Soc. London, B Ser.* **284,** 509–520.

Gooday, G. W., Fawcett, P., Green, D., and Shaw, G. (1973). The formation of fungal sporopollenin in the zygospore wall of *Mucor mucedo:* A role for the sexual carotenogenesis in the Mucorales. *J. Gen. Microbiol.* 74, 233–239.

Gooday, G. W., Jones, B. E., and Leith, W. H. (1979). Trisporic acid and the control of sexual differentiation in the Mucorales. *In* "Regulation of Secondary Product and Plant Hormone Metabolism" (M. Luckner and K. Schreiber, eds.), pp. 221–230. Pergamon, Oxford.

Goodwin, T. W. (1971). Algal carotenoids. *In* "Aspects of Terpenoid Chemistry and Biochemistry" (T. W. Goodwin, ed.), pp. 315–356. Academic Press, New York.

Goodwin, T. W. (1979). Biosynthesis of terpenoids. *Annu. Rev. Plant Physiol.* 30, 369–404.

Gray, J. C., and Kekwick, R. G. O. (1972). The inhibition of plant mevalonate kinase preparations by prenyl pyrophosphates. *Biochim. Biophys. Acta* 279, 290–296.

Gunsalus, I. C., Pederson, T. C., and Sligar, S. G. (1975). Oxygenase-catalysed biological hydroxylations. *Annu. Rev. Biochem.* 44, 377–407.

Hawker, L. E., Gooday, M. A., and Bracker, C. E. (1966). Plasmodesmata in fungal cell walls. *Nature (London)* 212, 635.

Imblum, R. L., and Rodwell, V. W. (1974). The 3-hydroxy-3-methylglutaryl-CoA reductase and mevalonate kinase of *Neurospora crassa. J. Lipid Res.* 15, 211–222.

Jones, B. E., and Bu'Lock, J. D. (1977). The effect of $N^6,O^{2'}$-dibutyryl adenosine $3',5'$-cyclic monophosphate on morphogenesis in Mucorales. *J. Gen. Microbiol.* 103, 29–36.

Jones, B. E., and Gooday, G. W. (1977). Lectin binding to sexual cells in fungi. *Biochem. Soc. Trans.* 5, 719–721.

Jones, B. E., and Gooday, G. W. (1978). An immunofluorescent investigation of the zygophore surface of Mucorales. *FEMS Microbiol. Lett.* 4, 181–184.

Köhler, F. (1935). Genetische Studien an *Mucor mucedo* Brefeld. *Z. Indukt. Abstamm.-Vererbungsl.* 70, 1–54.

Lumsden, J., and Coggins, J. R. (1977). The subunit structure of the *arom* multienzyme complex of *Neurospora crassa. Biochem. J.* 161, 599–607.

Mesland, D. A. M., Huisman, J. G., and van den Ende, H. (1974). Volatile sex hormones in *Mucor mucedo. J. Gen. Microbiol.* 80, 111–117.

Moffa, D. J., Lotspeich, F. J., and Krause, R. F. (1970). Preparation and properties of retinal oxidizing enzyme from rat intestinal mucosa. *J. Biol. Chem.* 245, 439–447.

Nielsen, R. I. (1978). Sexual mutants of a heterothallic *Mucor* species, *Mucor pusillus. Exp. Mycol.* 2, 193–197.

Pringle, J. R. (1975). Methods for avoiding proteolytic artefacts in studies of enzymes and other proteins from yeasts. *Methods Cell Biol.* 12, 149–181.

Sutter, R. P. (1975). Mutations affecting sexual development in *Phycomyces blakesleeanus. Proc. Natl. Acad. Sci. U.S.A.* 72, 127–130.

Sutter, R. P. (1977). Regulation of the first stage of sexual development in *Phycomyces blakesleeanus* and in other mucoraceous fungi. *In* "Eukaryotic Microbes as Model Developmental Systems" (D. H. O'Day and P. A. Horgen, eds.), pp. 251–272. Dekker, New York.

Sutter, R. P. (1979). Precursors of trisporic acids in *Phycomyces:* methyl 4-dihydrotrisporate and methyl trisporate. *Int. Congr. Biochem., 11th, Toronto* Abstr. 07-8-H85.

Sutter, R. P., Capage, D. A., Harrison, T. L., and Keen, W. A. (1973). Trisporic acid biosynthesis in separate *plus* and *minus* cultures of *Blakeslea trispora:* Identification by *Mucor* bioassay of two mating-type-specific components. *J. Bacteriol.* 114, 1074–1082.

Sutter, R. P., Harrison, T. L., and Galasko, G. (1974). Trisporic acid biosynthesis in *Blakeslea trispora* via mating type-specific precursors. *J. Biol. Chem.* **249,** 2282–2284.

Thomas, D. M., Harris, R. C., Kirk, J. T. O., and Goodwin, T. W. (1967). Studies on carotenogenesis in *Blakeslea trispora*. II. The mode of action of trisporic acid. *Phytochemistry* **6,** 361–366.

van den Ende, H. (1968). Relationship between sexuality and carotene synthesis in *Blakeslea trispora*. *J. Bacteriol.* **96,** 1298–1303.

van den Ende, H., Werkman, B. A., and van den Briel, M. L. (1972). Trisporic acid synthesis in mated cultures of the fungus *Blakeslea trispora*. *Arch. Mikrobiol.* **86,** 175–184.

Werkman, B. A., and van den Ende, H. (1973). Trisporic acid synthesis in *Blakeslea trispora*. Interaction between *plus* and *minus* mating types. *Arch. Mikrobiol.* **90,** 365–374.

Werkman, B. A., and van den Ende, H. (1974). Trisporic acid synthesis in homothallic and heterothallic Mucorales. *J. Gen. Microbiol.* **82,** 273–278.

Werkman, T. A. (1976). Localization and partial characterization of a sex-specific enzyme in homothallic and heterothallic Mucorales. *Arch. Microbiol.* **109,** 209–213.

Wurtz, T., and Jockusch, H. (1975). Sexual differentiation in *Mucor:* Trisporic acid response mutants and mutants blocked in zygospore development. *Dev. Biol.* **43,** 213–220.

Zycha, H., Siepmann, R., and Linneman, G. (1969). "Mucorales." Cramer, Weinheim.

9

Pheromonal Interactions during Mating in *Dictyostelium*

DANTON H. O'DAY AND KEITH E. LEWIS

I. Introduction	199
A. The Cellular Slime Molds	199
B. Macrocyst Development	201
II. Survey of the Literature	204
A. The Macrocyst as a Sexual Structure	204
B. Mating Types Communicate via Pheromones	204
III. Current Research	206
A. Pheromonal Interactions in *Dictyostelium giganteum*	206
B. The Sex Pheromone Is Not the Chemoattractant for Aggregation	208
C. Zygote Formation Occurs before Aggregation	211
D. Ca^{2+} and Cell Fusion	213
E. Regulation of the Pheromonal System	216
F. Regulation of Giant Cell Formation	217
IV. Perspectives	218
References	220

I. INTRODUCTION

A. The Cellular Slime Molds

The cellular slime molds (Acrasiomycetes) are soil microorganisms. They are found in many environments but are most abundant in the decaying litter of forests or stands of deciduous trees where they feed on bacteria. They likely play a major role in detritus turnover in such environments. In their feeding phase, the cellular slime molds exist as individual amebas. When the local food supply is exhausted they may embark on one of three developmental pathways, the chosen pathway generally being determined by environmental conditions. The alternative developmental pathways that are open to the cellular slime molds are summarized in Fig. 1.

Fruiting body development is an asexual process involving the aggregation of up to 10^5 amebas which subsequently undergo a beautiful morphogenetic sequence that culminates in the production of a sorocarp comprised of a long, slender stalk supporting a ball of spores (Fig. 1). The details of this process are described in two excellent books

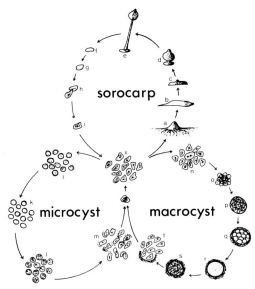

Fig. 1. Alternative developmental pathways in the cellular slime molds. After growth, depending on environmental conditions, the amebas (x) embark on one of three developmental pathways. On a solid substratum in the presence of light, as many at 10^5 amebas undergo a chemotactic aggregation (a) to form a multicellular pseudoplasmodium (b). This tissue-like structure, which shows pattern and polarity, may undergo a migration phase before undergoing morphogenesis (c, d) to produce the final sorocarp (e). The sorocarp consists of a long, slender stalk made up of dead, vacuolated stalk cells supporting a ball of thousands of single spore cells. Under the appropriate conditions each spore (f) swells (g) and splits (h) to liberate a single ameba (i). The second asexual developmental pathway is that of microcyst development. Under high osmotic conditions or in the presence of certain agents, amebas (x) round up and secrete a bilaminar cell wall around themselves (j) to become differentiated microcysts (k). When the microcysts are subsequently suspended in buffer or distilled water they swell (l) and enzymatically remove the microcyst wall, each liberating a single ameba (m). Macrocyst development is detailed in Fig. 2 and in the remainder of the text of this chapter.

(Bonner, 1967; Loomis, 1975). It is this asexual cycle that has established the slime molds, in general, and *Dictyostelium discoideum,* in particular, as one of the most important model systems for studies in cell and developmental biology. An alternative asexual process is that of microcyst development. In this process, single cells round up to form unicellular microcysts. Each microcyst germinates to yield a single ameba. The processes of encystment and germination so far have not been extensively studied (see the review in O'Day, 1977). The third pathway of development is the multicellular process of macrocyst development which is detailed in the remainder of this chapter.

All well-studied species are available from the American Type Culture Collection and, in addition, almost all slime mold workers are benevolent when it comes to supplying eager colleagues with these special beasts.

B. Macrocyst Development

Although the asexual (sorocarp) developmental cycle of the cellular slime molds has been under active study since the 1940's, detailed work on the macrocyst pathway has just recently begun. Blaskovics and Raper (1957) first described the structure of macrocysts and revealed aspects of their development but their significance as sexual structures eluded researchers until the 1970's. The morphological events that occur during the development of macrocysts have been described for several species by a number of workers (Blaskovics and Raper, 1957; Nickerson and Raper, 1973a; Erdos *et al.,* 1972; Filosa and Dengler, 1972; O'Day, 1979) and the basic events of this process appear to be essentially the same in all of the species that have been studied.

Figure 2 depicts the various stages of macrocyst development in *D. discoideum.* The first major morphological event in the developmental process is the aggregation of hundreds of amebas (Fig. 2A) to form tight, roughly spherical cell aggregates (Fig. 2B). Each aggregate is transformed into a precyst (Fig. 2C) by the secretion of a thin sheath (1° wall) around the cells. This 1° wall may be equivalent to the slime sheath that surrounds the asexual pseudoplasmodium (Blaskovics and Raper, 1957; Erdos *et al.,* 1972; Filosa and Dengler, 1972). Harrington and Raper (1968) have shown that the 1° wall contains cellulose and recent cytochemical and enzyme digestion analyses by Fukui (1976) have verified this. The presence of other carbohydrates in the 1° wall is likely (Fukui, 1976). Inside the precyst a large cell becomes evident which begins engulfing the smaller amebas (peripheral cells) that sur-

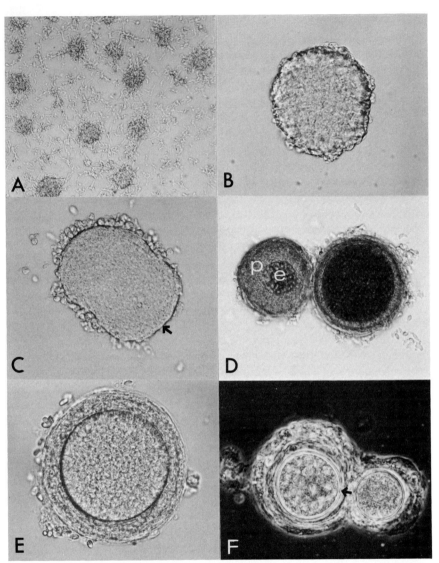

Fig. 2. Macrocyst development in mated cultures of *Dictyostelium discoideum*. Equal numbers of heat-shocked (45°C for 30 min) spores of the opposite mating types (NC4 and V12) to give a final spore concentration of 2×10^5 cells/ml were mixed with a dilute suspension of *E. coli* and 5.0-ml aliquots of this were plated on 0.1% lactose–proteose peptone plates (LPP) and stored in the dark according to a modified method of Nickerson and Raper (1973a; see O'Day, 1979). Under these conditions aggregation (A) begins after about 40 hr, resulting in the formation of tight aggregates (B) which contain hundreds of amebas. (C) A primary wall (arrow) forms, delineating the precyst. (D) Within the precyst the appearance of endocytes (e) reflects the phagocytic engulfment of the peripheral cells (p) by the zygote giant cell. (E) Phagocytosis continues until all of the peripheral cells (p) are engulfed. (D–E) During this time a loose 2° wall has been forming. (F) Finally, as maturation ensues, evidence of the endocytes disappears and a 3° cell wall (arrow) is synthesized.

round it (Fig. 2C and D). Because of its behavior and size, this cell has been termed a cytophagic giant cell (Filosa and Dengler, 1972; Fukui, 1976). The engulfed amebas (endocytes) are round and refractile under phase microscopy. Electron microscopic analysis has revealed that each endocyte is surrounded by a vacuolar membrane and that its nuclei and cytoplasm become electron dense. The phagocytic activity of the giant cell continues until all of the peripheral cells have been engulfed (Fig. 2E).

In all species a trilaminate macrocyst wall (3° wall; Fig. 2F) forms around the giant cell. This wall resembles the basic structure of the asexual spore wall but is much thicker (Erdos et al., 1972; Filosa and Dengler, 1972). In some species a 2° wall (Fig. 2D), which may be similar to the asexual microcyst wall, forms before the inner, trilayered 3° macrocyst wall is laid down (Erdos et al., 1972; O'Day, 1979). All the walls are rich in cellulose (Harrington and Raper, 1968) but other aspects of their biochemical constitution have not yet been clearly defined.

The final events of macrocyst maturation involve a shrinkage of the giant cell while it acquires dark pigmentation. The cytoplasm of the giant cell gradually becomes more uniform under light and phase microscopy (Fig. 2F) as the endocytes are digested. The endocytes probably break down as a result of the action of lysosomal enzymes but, as yet, no work has been done on this problem. During this maturation period the single, large nucleus of the giant cell becomes replaced by a large number of smaller nuclei. After a final period of dormancy the macrocyst will germinate, liberating hundreds of amebas (Erdos et al., 1973; Nickerson and Raper, 1973b).

Several physical factors have been shown to affect macrocyst development; moisture, pH, light, temperature, and phosphate ions (Nickerson and Raper, 1973a; Erdos et al., 1976; Blaskovics and Raper, 1957). Darkness is essential during the early growth phase of macrocyst cultures (0–24 hr; Erdos et al., 1976) but after this period light has no effect on macrocyst development. A similar temporal effect for temperature has also been noted (Erdos et al., 1976). Macrocysts form maximally at the upper temperatures of the critical 20°–25°C range. Wet conditions are essential for macrocyst development and they form best between pH 5.5 and 7.0 (Blaskovics and Raper, 1957). Although phosphate ions are inhibitory to sexual development (Erdos et al., 1973), the temporal nature of this sensitivity has not been defined. Furthermore, the mode of action has not been defined for any of these physical factors.

II. SURVEY OF THE LITERATURE

A. The Macrocyst as a Sexual Structure

An ultrastructural analysis of macrocyst development carried out with *Polysphondylium violaceum* by Erdos *et al.* (1973) revealed that a giant cell that appears in the center of macrocyst aggregates is at first binucleated but by the time its cytophagic behavior develops it becomes uninucleate. In addition, the morphological demonstration of a putative synaptonemal complex in the nucleus at the time of endocyte breakdown strongly suggested that the giant cell was a zygote that resulted from the fusion of two amebas. However, the first genetic evidence that meiosis was occurring in macrocysts was provided by MacInnes and Francis (1974) who used a homothallic strain of *D. mucoroides* (Dm7). The progeny that were generated from the various combinations of genetic markers used in this work were consistent with the occurrence of meiosis. Furthermore their data revealed that only one zygote nucleus existed per macrocyst. A subsequent genetic analysis has been carried out using the *D. giganteum* macrocyst cycle and this work also was supportive of a meiotic event in macrocyst development (Erdos *et al.*, 1975). A recent study with *D. discoideum* was less conclusive because of the poor germination that occurs in this species (Wallace and Raper, 1979). So far, the macrocyst system has not been used to any significant degree in the genetic analysis of cellular slime molds. The inability to obtain a high percentage of germinating macrocysts in *D. discoideum* has so far prevented the use of this system for the fine structural genetic analysis that is not possible with the currently used, parasexual genetic system.

B. Mating Types Communicate via Pheromones

Early work revealed that macrocysts occur extensively in certain species of cellular slime molds. For example, in *D. mucoroides* approximately 50% of the strains which were isolated from the soil formed macrocysts when cultured alone (Filosa and Chan, 1972). On the other hand, several species never seemed to form macrocysts. An exhaustive study by Clark *et al.* (1973) using five species of cellular slime molds revealed that this was due to the existence of mating types in many species and that when the proper pairings of soil isolates were made in four of the species studied, macrocyst development was invoked. The existence of mating types in *D. discoideum* was simultaneously reported by Erdos *et al.* (1973). Thus both homothallic and heterothallic strains of cellular slime molds exist. Mating type interactions in cellular slime molds and their use in the genetic analysis of development have previously been reviewed in detail by Francis et al. (1977).

In an attempt to learn something about the way mating-type strains communicate their matability we began an investigation into the presence of factors that regulate sexual development in *D. discoideum*. In these studies we used the classical strains of this species (NC4 and V12) which are of opposite mating type. Since these strains are heterothallic, cells of each strain will not make macrocysts when cultured alone. In these experiments the cell-free conditioned medium (CFCM) from a single strain of culture was added to cells of the opposite mating type (O'Day and Lewis, 1975). Controls consisted of adding CFCM of one strain back to cells of the same strain and of adding distilled water in lieu of CFCM. When these experiments were done, CFCM from NC4 cultures was shown to induce macrocyst development in V12 cells, but V12 CFCM could not induce NC4 cells to form macrocysts (O'Day and Lewis, 1975). Furthermore, the interaction was time dependent, with CFCM from 24-hr cultures being most effective on cells of the same age. This would explain why an early attempt to define the presence of sex pheromones in *D. discoideum* was unsuccessful since the previous workers used 2-day-old cultures as their source of CFCM (Erdos et al., 1973). Thus, mating interactions in *D. discoideum* were interpreted as operating via a secretor–responder system, with NC4 acting as the secretor of pheromone while V12 responded to it. Additional studies with mating types separated by a dialysis membrane supported this one-way interaction and further showed that the pheromones were of low molecular weight since they passed through such membranes. This one-way interaction was subsequently verified by Machac and Bonner (1975).

Through similar studies with *Dictyostelium purpureum*, identical conclusions about the mode of pheromonal communication were reached (Lewis and O'Day, 1976). Using three heterothallic strains (Dp2, Dp6, and Dp7) in which strains Dp6 and Dp7 will mate with strain Dp2 but not with each other, we carried out CFCM studies in a similar manner to those described above. The results of these experiments showed that macrocysts formed in cultures of Dp2 only when either Dp6 of Dp7 CFCM was added. Furthermore, this interaction was maximal (i.e., the largest number of macrocysts formed) when 24-hr-old cultures were used as the source of both CFCM and cells. Thus, Dp6 and Dp7 acts as secretors of pheromone while Dp2 is a responder strain. Heterospecific combinations revealed that the sex pheromones were species-specific since CFCM from a secretor of one species could not induce responder cells of the opposite species to produce macrocysts.

In an attempt to learn more about the nature of the pheromonal interactions, a simple diffusion culture chamber was designed which would permit cells to communicate only via the gas phase (Lewis and O'Day, 1977). Weinkauff and Filosa (1965) had previously presented some preliminary data which suggested that volatile molecules were important for macrocyst development in homothallic strains of *D. mucoroides*. We had also noticed that mated cultures of *D. discoideum* yielded more macrocysts when the petri dishes had been sealed with tape. The diffusion chamber setup we utilized for these experiments consisted of three small, open petri dishes (30 × 10 mm) in which the cultures were placed, inside a larger (100 × 15 mm), sealed dish which provided a common air space for communication (Lewis and O'Day, 1977). Complete macrocyst development occurred when the diffusion chamber contained two NC4 cultures and one V12 culture. In this setup macrocysts formed in the V12 culture, revealing that the sex pheromone secreted by NC4 was volatile. In keeping with an earlier result macrocysts never formed in the NC4 cultures. However, these experiments divulged another interesting point about the pheromonal control of macrocyst development. When only one NC4 culture was present, macrocyst development was begun by the V12 cells but they did not develop past the precyst stage. This suggested that development up to the precyst stage was controlled by a low level of pheromone while continued development (i.e., the development of phagocytic competence) required a higher level of the pheromone or possibly a second pheromone. Filosa (1979 and personal communication) has since indicated that two pheromones may be involved in regulating macrocyst development in *D. mucoroides* (Dm7) as well.

III. CURRENT RESEARCH

A. Pheromonal Interactions in *Dictyostelium giganteum*

As a result of our work with *D. discoideum* and *D. purpureum* we asked ourselves a simple question: Do all mating interactions in the cellular slime molds operate via pheromonal secretor–responder interactions? An analysis of the data of Erdos *et al.* (1975) suggested that the secretor–responder pheromonal interactions that appear to be operative during sexual development in *D. discoideum* and *D. purpureum* could not be the system that functions during mating in *D. giganteum*. In this species, the four strains that had been isolated and studied in detail all intermated successfully. This suggested to us that

9. Pheromonal Interactions in *Dictyostelium*

each strain might secrete its own pheromone. To test this, CFCM and cells were prepared for each strain and intermixed in all heterotypic combinations. As before, controls consisted of adding distilled water to cells or of adding CFCM back to cells of its own strain (a homotypic combination). The results showed that the CFCM of each of the four *D. giganteum* strains could induce all of the other strains to form macrocysts (Fig. 3) (Lewis and O'Day, 1979). Thus, as suspected, each of the strains produces its own pheromone. Furthermore, when each strain was compared on the basis of its ability to induce each of the other strains, it was clear that a mating hierarchy could be set up (Fig. 3). Thus on the basis of pheromone activity: WS589 > WS606 > WS607 > WS588; while on the basis of ability to respond to CFCM from the other strains: WS588 > WS607 > WS606 > WS589. Clearly, these patterns are the reverse of each other.

Mating hierarchies have been seen in other microorganisms such as *Achlya* (see Chapter 7 by Horgen) but no other system is known which is comparable to that operating in the cellular slime mold *D. giganteum*. The existence of this hierarchy must also affect our interpretation of the mating interactions in *D. discoideum* and *D. purpureum*. Rather than defining these interactions as true secretor–responder systems we must accept the possibility that we may be dealing with strains which represent the extreme ends of mating hierarchies. Thus, for example, V12 may secrete a pheromone but NC4 may be too far

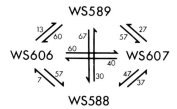

Fig. 3. Pheromonal interactions in *Dictyostelium giganteum*. The four strains were cultured separately on 0.1% lactose–bactopeptone agar plates (LP) under the conditions of Nickerson and Raper (1973a) as described in the legend to Fig. 2. A cell-free conditioned medium (CFCM) was prepared from 24-hr cultures of each strain by passing the 3000 *g* supernatant through a sterile, 0.45 μg Millipore filter (Millipore Corp.). Cells from an opposite mating-type strain were pelleted by low-speed centrifugation, resuspended in the CFCM, replated on fresh 0.1% lactose–proteose peptone plates, and stored at 23 ± 1°C. Controls consisted of cells which were resuspended in distilled water or their own CFCM. After 2 weeks the cultures were scored for the percent of the cultures that contained macrocysts. The numbers on the arrows represent this percent induction of the CFCM of one strain on cells of the other. For example, CFCM from WS589 cells induced 67% of WS588 cultures from macrocysts. Macrocysts never formed in control cultures. The methods are detailed in Lewis and O'Day (1979).

down the responder hierarchy to "understand" the message. A further, more detailed analysis of mating types in *D. discoideum* and *D. purpureum* and in other cellular slime molds might clarify this situation.

B. The Sex Pheromone Is Not the Chemoattractant for Aggregation

In *D. discoideum* a chemotactic process appears to direct the aggregation of individual amebas to form multicellular precysts since amebas show oriented movement toward late aggregates (O'Day and Durston, 1979). In the previously discussed diffusion chamber experiments, a small source of pheromone (i.e., a low number of NC4 cells) resulted in a stoppage of development at the precyst (postaggregation) stage while a larger source of pheromone (a high number of NC4 cells) resulted in complete development (Lewis and O'Day, 1977). As discussed earlier, this was interpreted as indicating that development to the precyst required a low level of volatile pheromone while further development required a higher level of the pheromone or possibly a second, volatile pheromone. Thus, one function of the pheromone might be to direct the aggregation of amebas to form the multicellular precyst within which zygote formation might be favored in a similar manner to that which occurs in yeasts (see Chapter 10 by Calleja *et al.* and Chapter 11 by Yanagishima and Yoshida) and in certain ciliates (see Chapter 14 by Hiwatashi).

In order to learn more about the aggregation process in *D. discoideum*, the early events of macrocyst development were studied by static photomicrography and microcinematography. Early movies of the aggregation process showed that many of the amebas in the culture demonstrated oriented movements toward macrocyst aggregates and this oriented movement of cells continued throughout development. When tracings of the movements of individual cells were made it became clear that almost all the amebas showed very directed movements toward the aggregates (Fig. 4A), suggesting that a chemotactic process was occurring (O'Day and Durston, 1979). When cultures were made in which the nutrients were increased, high cell densities resulted and the aggregating cells moved in streams rather than as individuals (O'Day, 1979). As can be seen in Fig. 4B, the amebas in these streams have the same morphological appearance as amebas from sorocarp aggregates (i.e., the amebas are elongate with distinct head–tail orientations) (Bonner, 1947, 1967). Fukui (1976) has previously shown that amebas liberated from precysts have the same end-to-end contacts as those which occur in cells during the aggregation which results in asexual pseudoplasmodium formation. Movie

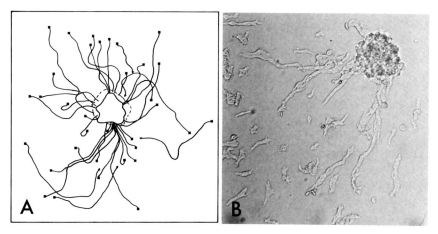

Fig. 4. Aggregation during macrocyst development in *D. discoideum*. In mated cultures where low cell densities are generated (i.e., on LP agar) the aggregation movements of individual amebas can be traced from time-lapse films (A), revealing that the oriented movement of amebas toward aggregation centers is occurring. Under conditions where high cell numbers are formed (i.e., LPP agar) the amebas move into the aggregate in streams (B). The amebas in these streams show a head-tail polarity. See the legend to Fig. 2 for details about the culture methods.

analysis of the aggregation process in dense cell cultures has revealed that pulsatile streaming occurs but its periodicity has not been defined (O'Day, 1979, and unpublished). All these data suggest a strong similarity between sorocarp and macrocyst aggregation.

Cyclic AMP (cAMP) is the chemoattractant for sorocarp aggregation (Konijn et al., 1967, 1968). If macrocyst aggregation involves the same mechanism as sorocarp aggregation then at least two additional criteria have to be met: (1) macrocyst aggregates must be sources of cyclic AMP and (2) macrocyst amoebae from aggregation-phase cultures must be competent to respond chemotactically to cAMP. A bioassay for cAMP exists in which amebas from sorocarp cultures can be tested for their ability to chemotactically orient toward putative cAMP sources (Konijn, 1970). Using this idea, macrocyst aggregates were plated on a lawn of cAMP-sensitive amebas. When movies were made of these experiments, the amebas showed directed movements toward the macrocyst aggregates, forming large aggregates around them (O'Day, 1979). Thus, macrocyst aggregates are sources of cyclic AMP. Since the amebas of the original macrocyst aggregates continued to develop into macrocysts while the newly arrived cells developed into fruiting bodies, it is clear that the plating procedure did not alter the developmental competences of the cells.

To test the ability of macrocyst amebas to chemotactically orient toward a source of cAMP, a modified Konijn (1970) test and Bonner *et al.* (1966) cellophane square test bioassay technique was used (O'Day, 1979). When chemotactically competent amebas are plated as a drop on hydrophobic water agar they usually remain within the perimeter of that drop as in Fig. 5A. However, when cyclic AMP is contained in the agar, aggregation phase cells from sorocarp cultures which are chemotactically responsive to cAMP and which are producing an extracellular phosphodiesterase (ePDE) show an oriented movement out of the drop area which is detectable after 1 hr and very evident after 3 hr (Fig. 5B). This behavioral response is due to the action of the PDE which hydrolyzes local cAMP thus setting up a cAMP gradient which the cells can orient to chemotactically. When amebas from aggregation-phase cultures were plated on these cyclic AMP agar plates they showed the behavior identical to that of aggregation-competent cells from sorocarp cultures. Furthermore, drops of these macrocyst cells on plain water agar plates continued development to mature macrocysts while the sorocarp amebas formed fruiting bodies, revealing that the plated cells had not lost their developmental compe-

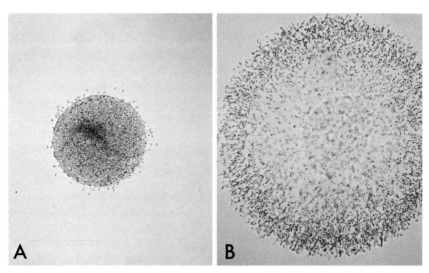

Fig. 5. The chemotactic response of amebas plated on dilute agar containing cyclic AMP. When drops of aggregation-phase amebas from sorocarp or macrocyst cultures are plated on 0.4% purified agar (Difco Laboratories), the amebas tend to remain within the area defined by the drop perimeter (A). Drops of these aggregation-phase amebas placed on 0.4% agar containing cyclic AMP (10^{-5}–10^{-7} M final conc.) show an outward migration beyond the drop perimeter (B), revealing their chemotactic sensitivity to this chemical.

tence. Thus, aggregation phase cells from macrocyst cultures orient chemotactically to cAMP and also must produce ePDE.

In total, the data reveal that aggregation during asexual and sexual development in *D. discoideum* is identical. The same kinds of experiments have recently been carried out with amebas of *D. purpureum*, which also uses cAMP as its chemoattractant during sorocarp development (G. Clarke and D. H. O'Day, unpublished). The results were the same as those with *D. discoideum,* suggesting that possibly all species of cellular slime molds use the same chemoattractant for both sexual and asexual aggregation.

The aggregation of amebas occurs in both mated and unmated cultures. However, the aggregates in single-strain cultures do not form a primary wall; they remain loose and progress no further in development (O'Day, 1979; Filosa and Dengler, 1972). This, plus evidence that the pheromone functions before aggregation, puts further emphasis on the argument that the sex pheromone is not involved in the aggregation process. Since the volatile sex pheromone of *D. discoideum* is not the chemotactic agent and it functions prior to aggregation, we feel that it is directly involved in zygote formation. This concept is supported by recent work on the time course of zygote formation.

C. Zygote Formation Occurs before Aggregation

Early ultrastructural studies on homothallic strains indicated that the zygote formed after aggregation (Erdos *et al.*, 1972; Filosa and Dengler, 1972). In fact, Filosa and Dengler (1972) proposed that the environmental conditions at the center of the precyst were responsible for inducing giant cell formation in a manner that would be analogous to that occurring during sexual interactions in other eukaryotic microbes as mentioned earlier. However, a careful analysis of the early events of macrocyst development has revealed that in *D. discoideum* giant cells first appear before aggregation begins (Erdos, personal communication; O'Day, 1979).

Using the plate culture technique as detailed by Erdos *et al.* (1976), zygote giant cells appear about 24 hr after the initiation of the mixed mating-type culture (Erdos, personal communication; O'Day, 1979). Zygote giant cells are easily distinguished from unfused amebas on the basis of their size, shape (Fig. 6), and refractility under phase microscopy. These young precytophagic giant cells have been observed to contain two nuclei when the cells were stained with fluorescent nuclear stains (O'Day, unpublished). Shortly after their formation the giant cells begin their cytophagic activity while also acting as centers

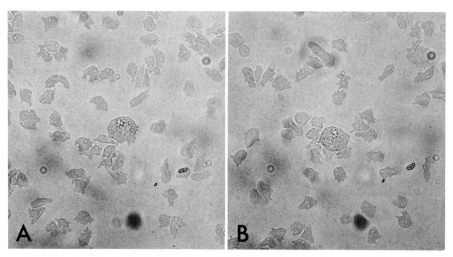

Fig. 6. Zygote giant cells formed in mated plate (LP agar) cultures of *D. discoideum*. When NC4 and V12 mating type strains are mixed, zygote giant cells form prior to aggregation (A) and subsequently act as centers for the aggregation process (B). The photographs in A and B were taken 10 min apart and tracings of the movements of the cells (not shown) have revealed that the amebas show oriented movements toward giant cells, suggesting that the zygote is chemically attracting them. See the legend to Fig. 2 for culture details.

for aggregation (O'Day, 1979), presumably through the secretion of cAMP. Thus they act in a way which might be compared to the founder cells of aggregation seen in some slime mold species (Shaffer, 1962).

Preliminary ultrastructural studies (J. A. Percy and D. H. O'Day, unpublished) and work with fluorescent nuclear stains (O'Day, unpublished) has revealed some information about nuclear events during cell fusion for zygote formation. The nuclei of unfused amebas are roughly spherical. When two amebas fuse, it seems that their nuclei migrate to the center of the giant cell where nuclear fusion ensues. The resulting zygote nucleus is larger than those of unfused amebas and is irregular in shape. When cells are treated with Hoechst 33258 (Filion et al., 1976), a fluorescent nuclear stain, the zygote nuclei appear larger and less bright than those of the unfused amebas. Nuclear fusion generally occurs before endocytosis begins but the events do not appear to be linked. since many giant cells begin endocytosis before nuclear fusion has occurred. These results correlate well with aspects of the ultrastructural work done on *Polysphondylium violaceum* (Erdos et al., 1972).

D. Ca^{2+} and Cell Fusion

The development of a liquid culture system has greatly facilitated an analysis of the factors involved in zygote giant cell formation and has allowed us to pursue the timing of the events of this differentiation process. Previously we had been discussing results based on work done with lactose–bactopeptone (LP) or lactose–proteose peptone (LPP) agar plate macrocyst cultures. In general, agar cultures have several drawbacks. For example, the agar absorbs materials, it is difficult to add exogenous chemicals quantitatively, the cell yield is low, and cell harvesting is difficult. As a result the potential to carry out certain types of experimental approaches with agar cultures is restricted or impossible. To overcome these problems we developed a liquid culture system which is based on the enriched lactose–proteose peptone (LPP) agar culture (O'Day, 1979; Chagla et al., 1980; Table I). This culture system has enabled us to begin quantitative studies on the effects of various chemicals on sexual development. In addition, it is now possible to grow cells in flasks of any size or in volumes as small as 50–80 μl in Disposo-Trays (Linbro Scientific, Inc., Hamden, Connecticut).

TABLE I

The Effect of Various Chemicals on the Number of Giant Cells Produced in Mated Liquid Cultures of *Dictyostelium discoideum*[a]

Chemical added	Molarity in culture	Giant cells at 28 hr (%)
None	—	0.25
$CaCl_2$	1 mM	14.21
$CaCl_2$	20 μM	1.50
$MgCl_2$	1 mM	1.95
NaCl	1 mM	1.91
EDTA	10 mM	0
EGTA	10 mM	0
A23187 + $CaCl_2$	1 μM + 1 mM	14.0
A23187 + $CaCl_2$	1 μM + 20 μM	1.3

[a] Spores of NC4 and V12 were mixed in equal numbers, heat-shocked at 45 ± 1°C for 30 min to promote synchronous germination, mixed with a dilute suspension of *E. coli*, then added to a liquid medium consisting of 0.1% lactose and 0.1% proteose peptone to give a final spore density of 2 × 10^6 spores/ml. The liquid medium (100 ml) was contained in 250-ml Erlenmeyer flasks which had been completely covered in black vinyl electrical tape to provide a dark environment for macrocyst development. $CaCl_2$, $MgCl_2$, and NaCl were all added at the start of the cultures while EGTA, EDTA, and the ionophore (A23187) were added at 18 hr.

Many reports for a variety of membrane fusion events (e.g., sperm–egg, secretory granule–plasma membrane, myoblasts, macrophage) have shown that Ca^{2+} is essential and, based on extensive information, Papahadjopoulos et al. (1979) have proposed a new model for membrane fusion in which Ca^{2+} plays key roles. Ca^{2+}, in fact, is believed to play two roles in this process. First, it allows the adjacent membranes to come into close apposition by decreasing electrostatic repulsion and, second, they suggest that it induces membrane destabilization by causing the formation of regions at which membrane fusion can occur. These fusion areas are proposed to be sites of acid phospholipids which have been forced into rigid crystalline domains by the Ca^{2+}. Fusion occurs when the domains of adjacent membranes coincide. Most of the work which has resulted in this proposal has come from research on fusion of artificial lipid vesicles. As a result, a question as to its applicability to living systems requires that related work with natural fusion systems is essential.

Another aspect of artificial systems is that they lack the specificity of fusion that occurs in natural ones. In this regard, cell recognition and cell-specific adhesion (or "docking") play a critical preliminary function in biological membrane fusion events. Ca^{2+} is known to function in cell–cell adhesion, thus adding a third role for this ion in the fusion of natural membranes. Because of the availability of excellent methods for membrane isolation in *D. discoideum* and a rapidly growing literature on the subject, we felt that by finding means to increase the amount of cell fusion in *D. discoideum* we might be able to exploit this system for the analysis of the sequence of events in the fusion of biomembranes.

In keeping with the results of others, we have found that Ca^{2+} is essential for cell fusion in *D. discoideum* (Chagla et al., 1980). The chelation of Ca^{2+} with 10 mM EDTA or EGTA prevents zygote giant cell formation in mated cultures of *D. discoideum* while augmentation of the level of Ca^{2+} in the liquid culture medium with $CaCl_2$ results in an increase in the total number of zygote giant cells formed by 28 hr (Table I; Fig. 7A). The requirement is specific for Ca^{2+} since other divalent cations (e.g., Mg^{2+}) and various other chloride salts do not give the enhancement of fusion that Ca does (Table I). Furthermore, the Ca^{2+}-ionophore A23187 does not enhance zygote giant cell formation, suggesting that the function of Ca^{2+} is mediated solely at the cell surface. This is reasonable since all the events of the fusion sequence (cell–cell adhesion and the actual events of lipid merging) occur at the outer surfaces of the cell membranes.

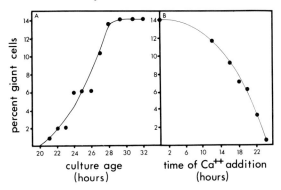

Fig. 7. The time course of the appearance of zygote giant cells in mated liquid cultures of *D. discoideum* treated with 1 mM Ca^{2+}. (A) When $CaCl_2$ is added when the mated culture was started (0 hr), zygote giant cells appear between 20 and 21 hr, then increase in number to a maximum of about 14% by 28 hr. (B) The number of giant cells produced is related to the time of addition of the $CaCl_2$. The later the salt is added, the fewer is the number of giant cells produced. See Table I for culture details.

The role Ca^{2+} may be playing in cell fusion in *D. discoideum* is confused by the temporal pattern of its effect. The later the $CaCl_2$ is added to macrocyst cultures the lower is the number of zygotes that form (Fig. 7B). This indicates that part of the Ca^{2+} effect occurs early in the developmental cycle. On the other hand, chelation of the Ca^{2+} at 18 hr completely inhibits giant cell formation, indicating that Ca^{2+} also functions later in the developmental process.

Recent detailed studies with the fluorescent nuclear stain Hoechst 33258 have clarified the timing of the nuclear events of sexual development (A. H. Chagla, S. P. Szabo, and D. H. O'Day, unpublished results). The actual fusion of amoebas begins at approximately 10 hr after the start of the culture and continues until about 25 hr. The appearance of morphologically distinct zygote giant cells begins around 20 hours, indicating that after fusion there is a differentiative phase wherein the fusion products are transformed into zygote giant cells. This suggests that Ca^{2+} plays a role first in the fusion process and second in the morphological differentiation of zygote giant cells.

The significance of our results with Ca^{2+} is that we can increase the number of cell fusions approximately 57-fold to yield over 14% zygotes (Chagla *et al.*, 1980; A. Chagla *et al.*, unpublished). We feel that the precise timing of this fusion sequence and the large number of fusions that occur will make *D. discoideum* an appealing system for the analysis of the events of biomembrane fusion in eukaryotes.

E. Regulation of the Pheromonal System

Results obtained from experiments involving the crossing of cells and CFCM of the same age were interpreted as indicating that a discrete period of pheromone secretion occurred during mating in both *D. discoideum* and *D. purpureum* (Lewis and O'Day, 1976; O'Day and Lewis, 1975). However, this conclusion was premature since CFCM and cells of different ages had not been mixed. Our continued experiments have indicated the importance of realizing that the communicative interaction between two mating types is composed of at least two separate events: pheromone production and pheromone reception. When a series of combinations of CFCM and cells of *D. discoideum* mating type of various ages was made, it became clear that the control of the mating event lies more with the responder than with the secretor.

The pheromone of *D. discoideum* is apparently produced constitutively by NC4 but must increase to adequate levels in the culture medium before the CFCM is able to induce sexual development in V12. The pheromone reaches this level about 18–20 hr after the initiation of an LPP liquid culture and CFCM prepared from cultures any time after this period will induce V12 cells. On the other hand, V12 cells only remain competent to respond to the hormone for a limited time. The developmental window for the response of V12 is between 18 and 28 hr for cells grown in the dark on the LPP agar medium.

Using LP agar plate cultures, early workers (Erdos *et al.*, 1976; Nickerson and Raper, 1973a) had shown that light inhibited macrocyst formation. However, the specific event that was light-sensitive was not explored. Since the period of light sensitivity (0–24 hr) correlated with the period of maturation of V12 cells and the onset of giant cell formation, we decided to investigate the effects of light on the pheromonal system. In preliminary experiments mixed mating type agar cultures were kept in the dark for various periods during the light-sensitive period (Erdos *et al.*, 1976) and then exposed to the light. Macrocyst development was scored after 48 hr. Cultures exposed to the light after 24 hr showed normal development while those exposed between 18 and 22 hr revealed a marked decrease in macrocyst number. Macrocyst development in cultures exposed to light before 18 hr was completely inhibited. Subsequently, we cultured NC4 and V12 separately for 24 hr under both light and dark conditions, mixed the cells in the various combinations, and then placed the newly mated cultures in the light. The presence of macrocysts in these cultures was determined after 72 hr. When both NC4 and V12 were grown in the dark and mixed after 24 hr, they gave identical results to those cultures which were mated

9. Pheromonal Interactions in *Dictyostelium* 217

at the start of the culturing (0 hr). Similarly, light-grown NC4 mixed with dark-grown V12 possessed the same number of macrocysts as control cultures. However, no macrocysts formed when only the V12 cells were grown in the light. These results suggest that the maturation of V12 is light-sensitive but pheromone production by NC4 is not.

F. Regulation of Giant Cell Formation

Ca^{2+} greatly enhances giant cell formation but still there is an upper limit to the total number of zygotes that are formed (approx. 14%) and the time during which they appear. It thus seems that something must control the number and timing of the fusion events. In an attempt to learn more about this, CFCM experiments were carried out using mixed mating type cultures. In these experiments CFCM and cells were isolated from cultures before cell fusion had begun (20 hr) and after the fusion process had essentially ended (28 hr) then intermixed homo- and heterotypically (S. P. Szabo *et al.*, unpublished). The number of giant cells was then determined. Figure 8 reveals the results of some of these experiments. It can be seen that when 28 CFCM is added to 20-hr

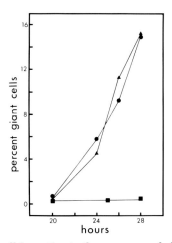

Fig. 8. Zygote grant cell formation in the presence and absence of CFCM from 28-hr mixed mating-type cultures. Liquid mixed mating-type cultures were made as detailed in Table I. As controls, one set of cultures was allowed to develop undisturbed (●) while the cells from another set were harvested at 20 hr, washed several times using centrifugation to remove the bacterial food source, and resuspended in 1 mM Tris-HCl (pH 6.5) buffer (▲). In the experimental cultures, cells harvested at 20 hr were resuspended in CFCM which had been prepared from a 28-hr culture (■). The percent giant cells was determined at specific times as detailed earlier.

cells zygote giant cell formation is almost completely prevented. This was not due to the removal of any required factors since washing and resuspending 20-hr cells in buffer did not prevent the normal pattern of zygote giant cell formation. Thus, an inhibitor of giant cell formation appears to be present in 28-hr cultures. The removal of this inhibitor from 28-hr cultures and replacement with CFCM from 20-hr cells resulted in a slight augmentation of the number of giant cells that are formed (data not shown). Initial work has indicated that the inhibitor has a low molecular weight. We are currently investigating the details of this regulatory event.

IV. PERSPECTIVES

Several features of sexual (macrocyst) development in *Dictyostelium discoideum* make it a valuable system for further study. The mating process of this species is regulated by volatile sexual pheromones which appear to be produced constitutively by only one of the mating type strains. The first observable event of macrocyst development is the fusion of pairs of amoebae to form zygote giant cells. Our work suggests that the primary role of the sex pheromone is to induce changes in responsive amebas such that they become fusion competent. The developmental appearance of the competence of amebas to respond to the pheromone is a photosensitive event whereas the secretion of the pheromone is not. After their formation, the zygote giant cells gain a unique phagocytic ability while simultaneously acting as centers for the chemotactic aggregation of unfused amebas, which results in the formation of multicellular aggregates. It is likely that this aggregation process is directed by cyclic AMP in a manner which is identical to the asexual, sorocarp aggregation process. The phagocytic behavior of the giant cells develops some time between their formation and the onset of aggregation and continues throughout the developmental sequence until all the aggregated amebas have been engulfed and the endocyte-filled giant cell resides alone within the macrocyst walls.

In summary, at least three major events occur during macrocyst development which are worthy of further analysis: (1) membrane fusion (zygote formation); (2) phagocytosis (endocyte formation); and (3) membrane synthesis (phagosome and zygote cell membrane).

Membrane fusion is a basic cellular event in all eukaryotic cells (e.g., pinocytosis, phagocytosis, exocytosis, intracellular vacuole fusion, etc.) and, in addition, is a critical event for specific developmental processes (e.g., sperm–egg and myoblast fusion). In *D. discoideum,* using Ca^{2+} we have been able to enhance the fusion process for zygote giant cell formation to such a level that it now rivals other natural systems in terms of its synchrony and the length of the fusion period. With the availability of a liquid culture method, we feel zygote giant cell formation presents an exciting system for the analysis of the molecular events of biomembrane fusion. Similarly, a vast amount of phagocytosis occurs during macrocyst development, presenting a potentially useful system for the analysis of the membrane changes that are associated with this process. What molecular changes in the zygote cause it to engulf amebas of the same genotypes from which it was derived? Related to the phagocytosis of several hundred amebas is the vast increase in size the giant cell must undergo. Calculations indicate that the surface area of the giant cell increases about 1000-fold during the developmental sequence. Thus, the zygote giant cell must synthesize enough membrane to surround a few hundred amebas with phagocytic membranes while simultaneously undergoing a massive increase in volume. Since only slightly more than 24 hr elapses from the time of formation of the zygote to the full endocyte stage, the synthesis of membrane components during this time must be phenomenal. Because these postaggregative processes are occurring essentially in a closed system (i.e., no nutrients from outside), this presents a potentially valuable system for the analysis of the events of membrane synthesis in a eukaryotic organism.

Some more subtle but equally appealing biological problems also exist. Regarding the pheromonal system that initiates all these processes, we feel that there are some especially interesting aspects to pursue. The purification and characterization of the pheromone of *D. discoideum* is of special interest. The availability of pure preparations of this molecule would provide another valuable probe for getting at the molecular regulatory events of this important model system. The photolability of the development of pheromone responsiveness might be a benefit in the detection and subsequent isolation of the pheromone receptor.

In conclusion, sexual development in *D. discoideum* presents many interesting problems for future study. Although the analysis of macrocyst development is still in its infancy, it has behind it the vast amount of knowledge generated from the study of other pathways of development in this organism. As a result many techniques and methods of approaching problems have been defined and need only be applied to this alternative developmental pathway.

REFERENCES

Blaskovics, J. C., and Raper, K. B. (1957). Encystment stages of *Dictyostelium. Biol. Bull. (Woods Hole, Mass.)* 113, 58-88.

Bonner, J. T. (1947). Evidence for the formation of cell aggregates by chemotaxies in the development of the slime mold *Dictyostelium discoideum. J. Exp. Zool.* 106, 1-26.

Bonner, J. T. (1967). "The Cellular Slime Molds." Princeton Univ. Press, Princeton, New Jersey.

Bonner, J. T., Kelso, A. P., and Gilmour, R. G. (1966). A new approach to the problem of aggregation in the cellular slime molds. *Biol. Bull. (Woods Hole, Mass.)* 130, 28-42.

Chagla, A., Lewis, K. E., and O'Day, D. H. (1980). Ca^{++} and cell fusion during sexual development in liquid cultures of *Dictyostelium discoideum. Exp. Cell Res.* 126, 501-505.

Clark, M. A., Francis, D., and Eisenberg, R. (1973). Mating types in cellular slime molds. *Biochem. Biophys. Res. Commun.* 52, 672-678.

Erdos, G. W., Nickerson, A. W., and Raper, K. B. (1972). Fine structure of macrocysts in *Polysphondylium violaceum. Cytobiologie* 6, 351-366.

Erdos, G. W., Raper, K. B., and Vogen, L. K. (1973). Mating types and macrocyst formation in *Dictyostelium discoideum. Proc. Natl. Acad. Sci. U.S.A.* 70, 1828-1830.

Erdos, G. W., Raper, K. B., and Vogen, L. K. (1975). Sexuality in the cellular slime mold *Dictyostelium giganteum. Proc. Natl. Acad. Sci. U.S.A.* 72, 970-973.

Erdos, G. W., Raper, K. B., and Vogen, L. K. (1976). Effects of light and temperature on macrocyst formation in paired mating types of *Dictyostelium discoideum. J. Bacteriol.* 128, 495-497.

Filion, W. G., MacPherson, P., Blakey, D., Yen, S., and Culpeper, A. (1976). Enhanced Hoechst 33258 fluorescence in plants. *Exp. Cell Res.* 99, 204-206.

Filosa, M. F. (1979). Macrocyst formation in the cellular slime mold *Dictyostelium mucoroides:* involvement of light and volatile morphogenetic substance(s). *J. Exp. Zool.* 207, 491-495.

Filosa, M. F., and Chan, M. (1972). The isolation from soil of macrocyst-forming strains of the cellular slime mould *Dictyostelium discoideum. J. Gen. Microbiol.* 71, 413-414.

Filosa, M. F., and Dengler, R. E. (1972). Ultrastructure of macrocyst formation in the cellular slime mold, *Dictyostelium mucoroides:* extensive phagocytosis of amoebae by a specialized cell. *Dev. Biol.* 29, 1-16.

Francis, D. W., Eisenberg, R. M., and MacInnes, M. A. (1977). Genetics and development in *Dictyostelium* and *Polysphondylium. In* "Eucaryotic Microbes as Model Developmental Systems" (D. H. O'Day and P. A. Horgen, eds.), pp. 155-177. Dekker, New York.

Fukui, Y. (1976). Enzymatic dissociation of nascent macrocysts and partition of the liberated cytophagic giant cells in *Dictyostelium mucoroides. Dev., Growth Differ.* 18, 145-155.

Harrington, B. J., and Raper, K. B. (1968). Use of a fluorescent brightener to demonstrate cellulose in the cellular slime molds. *Appl. Microbiol.* 16, 106-113.

Konijn, T. M. (1970). Microbiological assay of cyclic 3',5'-AMP. *Experientia* 26, 367-369.

Konijn, T. M., Van de Meene, J. G. C., Bonner, J. T., and Barkley, D. S. (1967). The acrasin activity of adenosine 3'5'-cyclic phosphate. *Proc. Natl. Acad. Sci. U.S.A.* 58, 1152-1154.

Konijn, T. M., Barkley, D. S., Chang, Y. Y., and Bonner, J. T. (1968). Cyclic AMP: a naturally occurring acrasin in the cellular slime molds. *Am. J. Bot.* 102, 225-233.

Lewis, K. E., and O'Day, D. H. (1976). Sexual hormone in the cellular slime mould, *Dictyostelium purpureum*. *Can. J. Microbiol.* **22**, 1269-1273.

Lewis, K. E., and O'Day, D. H. (1977). Sex hormone of *Dictyostelium discoideum* is volatile. *Nature (London)* **268**, 730-731.

Lewis, K. E., and O'Day, D. H. (1979). Evidence for a hierarchical mating system operating via pheromones in *Dictyostelium giganteum*. *J. Bacteriol.* **138**, 251-253.

Loomis, W. F. (1975). "*Dictyostelium discoideum*. A Model Developmental System." Academic Press, New York.

Machac, M. A., and Bonner, J. T. (1975). Evidence for a sex hormone in *Dictyostelium discoideum*. *J. Bacteriol.* **124**, 1624-1625.

MacInnes, M. A., and Francis, D. W. (1974). Meiosis in *Dictyostelium mucoroides*. *Nature (London)* **251**, 321-324.

Nickerson, A. W., and Raper, K. B. (1973a). Macrocysts in the life cycle of the Dictyosteliaceae. I. Formation of the macrocyst. *Am. J. Bot.* **60**, 190-197.

Nickerson, A. W., and Raper, K. B. (1973b). Macrocysts in the life cycle of the Dictyosteliaceae. II. Germination of the macrocysts. *Am. J. Bot.* **60**, 247-254.

O'Day, D. H. (1977). Microcyst germination in the cellular slime mold Polysphondylium pallidium requirements for macromolecular synthesis and specific enzyme accumulation. *In* "Eucaryotic Microbes as Model Developmental Systems" (D. H. O'Day and P. A. Horgen, eds.), pp. 353-372. Dekker, New York.

O'Day, D. H. (1979). Aggregation during sexual development in *Dictyostelium discoideum*. *Can. J. Microbiol.* **25**, 1416-1426.

O'Day, D. H., and Durston, A. J. (1979). Evidence for chemotaxis during mating in *Dictyostelium discoideum*. *Can. J. Microbiol.* **25**, 542-544.

O'Day, D. H., and Lewis, K. E. (1975). Diffusible mating-type factors induce macrocyst development in *Dictyostelium discoideum*. *Nature (London)* **254**, 431-432.

Papahadjopoulos, D., Poste, G., and Vail, W. J. (1979). Studies on membrane fusion with natural and model membranes. *Methods Membr. Biol.* **10**, 1-121.

Shaffer, B. M. (1962). The Acrasina. *Adv. Morphog.* **2**, 109-182.

Wallace, M. A., and Raper, K. B. (1979). Genetic exchanges in the macrocysts of *Dictyostelium discoideum*. *J. Gen. Microbiol.* **113**, 327-337.

Weinkauff, A. M., and Filosa, M. F. (1965). Factors involved in the formation of macrocysts by the cellular slime mold *Dictyostelium mucoroides*. *Can. J. Microbiol.* **11**, 385-387.

Part III
CELL SURFACE INTERACTIONS

10

The Cell Wall as Sex Organelle in Fission Yeast

G. B. CALLEJA, BYRON F. JOHNSON, AND B. Y. YOO

I. Introduction 225
   A. Sex and the Cell Surface 225
   B. The Cell Wall as Sex Organelle 226
   C. The Fission Yeast *Schizosaccharomyces pombe* 227
II. Survey of the Literature 230
   A. Gross Phenomenology 230
   B. Quantitation of Flocculation 232
   C. The Nature of the Forces 232
   D. Flocculation: A Prerequisite to Sex 234
   E. Induction to Competence 235
   F. Fine Structure of the Interactions 237
   G. Macromolecular Changes during the Sequence 239
   H. Conjugation-Induced Lysis 239
   I. Thermosensitivity of the Developmental Program 240
   J. Sex Interconversion in a Homothallic Strain 241
III. Current Research 242
   A. Effect of 2-Deoxyglucose on Sex Cells 242
   B. The Ratio of P Cells to d Cells in a Floc 247
   C. Sexual Preferences, Positions, and Orientations 248
IV. Perspectives 253
   A. Summing Up 253
   B. Directions of Things to Come 254
   References 255

I. INTRODUCTION

A. Sex and the Cell Surface

Sex in unicellular organisms may be described as the transfer of genetic material from one cell to another. It need not involve the physical union of cells nor the transfer of a whole genetic complement. It can be consummated directly by intercellular contact or indirectly at a distance. When sex is consummated at a distance, the genetic material may be introduced by way of a virus, as in transduction, or as a fragment of naked DNA, as in transformation. Directly introduced, the genetic material is passed on from one cell to another by way of either a transient union or an irreversible fusion of cells. Transient union is exemplified by conjugation in *Escherichia coli* and results in a merodiplont. Irreversible fusion is exemplified by conjugation in a yeast and results in a dikaryon. The formation of a zygote, the resultant genetic recombination, and the subsequent expression of the transferred and combined information are the consequences of sex, not part of it. In microorganisms, sex is separable from, and not essential to, reproduction. It is more recreational than procreational in function.

The cell surface—membrane, wall, and appendages—has multifarious functions. Some of these functions derive from its being a surface, others from its being a bag, still others from special cellular adaptation. Of prime importance to our subject, the cell surface is the site of cell–cell interaction. Signals from other cells are first received and processed by the cell surface through specific receptors. When two cells approach each other, it is their cell surfaces that are in contact first. Sex, however it is consummated, is largely a phenomenon of the cell surface. It is then through the study of the cell surface that a number of problems of cell biology may be profitably approached. And in cell–cell interaction and sex, the cell surface not only cannot be avoided, but must indeed be the focus of attention.

The cell surface of yeasts includes the plasma membrane, the periplasmic layer, the cell wall, and appendages on the cell wall. The plasma membrane, which surrounds the cytoplasm and is directly in contact with it, is made up chiefly of lipids and proteins. The cell wall, which covers the plasma membrane, is made up chiefly of carbohydrates. In between them is a periplasmic layer, wherein enzymes are presumed to be located. Embedded in the wall and in the membrane, and protruding inward and outward, are proteins with various functions. All these constitute the cell surface, to which the cell wall contributes the most in bulk.

B. The Cell Wall as Sex Organelle

What makes the cell wall a sex organelle rather than just an erogenous zone? At the level of the cell, an erogenous zone would be very difficult to define. When all things are considered, the entire cell may be thought of as constituting an erogenous zone. Virtually every large enough component of the cell partakes in the sexual act. A sex organelle, in the present context, must serve as a channel through which the genetic material must pass from one cell to another. Unless the transfer occurs indirectly at a distance, the sex organelles of two cells must be in intimate material contact when the transfer is made. Moreover, the site of initial contact must be specific and complementary. In *Schizosaccharomyces pombe*, the nucleus, the mitochondrion, the cell membrane, all partake in the sexual act but are not sex organelles. Only the cell wall appears to qualify by our definition.

In a strict sense, the irreversible fusion of two cells during conjugation in fission yeast does not lead to a transfer of genes from one cell to another. At the moment that the crosswall separating the conjugants from each other breaks down, the two separate cytoplasms become one and the erstwhile pair of conjugants become, morphologically and functionally, one cell. Neither one of the two original cells can be considered as the donor or the recipient. But both can be considered conceptually as simultaneous donors. In this sense, there is a unidirectional transfer of genetic material—from two cells to a dikaryon.

C. The Fission Yeast *Schizosaccharomyces pombe*

Schizosaccharomyces pombe belongs to the group of ascomycetous yeasts (Slooff, 1970). The group comprises 22 genera, among them *Hansenula* and *Saccharomyces*. First described by Lindner (1893), *S. pombe* is the type species of the genus *Schizosaccharomyces* Lindner, which includes four species and two varieties. It was originally isolated in East Africa from the Bantu beer which is called *pombe,* hence its specific name. Since then, it has been isolated from cider and grape must in temperate regions of the world and from various sugar-containing products, such as cane molasses and palm wine, in the tropics. Unlike the more popular budding yeasts, it multiplies by fission, as the generic name denotes. Initially through the efforts of Leupold (1950, 1970) in genetics and of Mitchison (1957, 1970) in physiology, it has become the model organism for many a biological problem. Its genetics has been fairly well studied. Three chromosomes have been cytologically identified (Robinow, 1977) and recent chromosome maps include 118 genetic markers (Kohli *et al.,* 1977).

Like all yeasts, *S. pombe* is a unicellular organism. Cells are characteristically cylindrical (Fig. 1). Grown in malt extract broth, cells in logarithmic phase measure $3-5 \times 7-20$ μm. Prior to stationary phase, cells divide for the last time in a batch culture and become much shorter so that some cells are almost as wide as they are long. Ul-

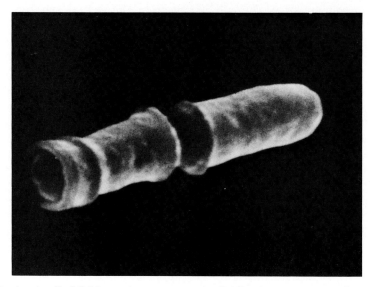

Fig. 1. A cell of *Schizosaccharomyces pombe* dividing into two as seen through the scanning electron microscope. (From Prescott, 1976.)

trathin sections examined by electron microscopy reveal cytomorphological features common to yeasts: a nucleus delimited by a double membrane, mitochondria, Golgi bodies, endoplasmic reticulum, vacuoles, a plasma membrane, periplasmic space, and a cell wall (Fig. 2). Encompassing the whole cell is a cell wall which is around 0.15 μm thick. It contains fission scars, which are telltale relics of past fission events, and fuscannels, which are fission-associated electron-dense rings found close to the fission scars (Johnson *et al.*, 1973).

The life cycle of a homothallic strain of *S. pombe* is shown in Fig. 3 (Calleja *et al.*, 1977b). The stably haploid organism grows profusely as a still culture in malt extract broth at 32°C. The culture attains a stationary population density of 5×10^7 cells ml^{-1}. Under these conditions, the generation time is about 120 min. When the culture is aerated during the early portions of its stationary phase, cells begin to form grossly visible flocs within 1 hr after the start of aeration. The size and the number of flocs increase in the next 5 to 6 hr. During this aeration procedure, cells are induced to become competent to form flocs. Random collision due to agitation and thermal motion allows the sorting-out of cells. The competent cells stick together to form flocs, while the incompetent ones remain free. Flocculation, essentially the aggregation of cells by hydrogen bonding, is followed by copulation, the

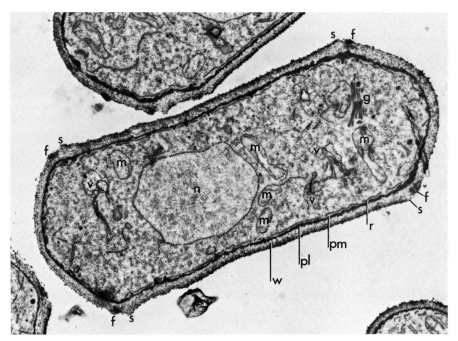

Fig. 2. An ultrathin section of a cell of *Schizosaccharomyces pombe* as seen through the transmission electron microscope. Symbols: f, fuscannel; g, Golgi apparatus; m, mitochondrion; n, nucleus; pl, periplasmic layer; pm, plasma membrane; r, endoplasmic reticulum; s, fission scar; v, vacuole; and w, cell wall.

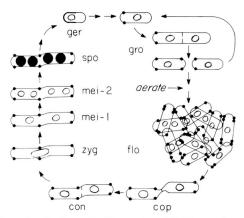

Fig. 3. The life cycle of a homothallic strain of *Schizosaccharomyces pombe*. Abbreviations: gro, growth and cell division in the mitotic cycle; flo, flocculation; cop, copulation; con, conjugation; zyg, zygote formation and nuclear fusion; mei-1, first meiotic division; mei-2, second meiotic division; spo, sporulation; and ger, germination of ascospore. Black dots are fuscannels.

covalent union between cells in flocs. Soon after copulation, conjugation takes place. Conjugation is synonymous with cytoplasmic fusion or dikaryon formation. It is indicated by the elimination of the crosswall that separates the pair of cells. A zygote is formed when the nuclei in the dikaryon fuse. Meiosis I occurs, which is followed closely by meiosis II. The four nuclear products of the sequential meiotic events are then packaged individually into four spores within an ascus derived from the original copulating pair. The first asci in the population may be observed within 8 hr after the start of aeration. Further incubation of the culture may result in the liberation of spores from their asci. When transferred to fresh medium, these spores, liberated or still confined within their respective asci, germinate to become dividing vegetative cells once more, thus completing the cycle. The sequence from the start of induction to conjugation is what concerns us in this chapter.

There are variations to the cycle (Gutz et al., 1974; Crandall et al., 1977). If the culture is aerated from the time of inoculation, the cells become induced as soon as the medium is exhausted. Heterothallic cultures remain nonflocculent until the complementary mating types are mixed together, whereupon they flocculate, but only after an extended period of apparently mutual stimulation (Egel, 1971). In some strains under certain conditions, meiosis may be dissociated from its usual position in the cycle. The zygote once formed may simply remain a diplont and replicate as such (Egel, 1973). The diploid cell may then either undergo meiosis and sporulate or reduce its ploidy and convert itself into two haploid cells. In other species of fission yeast, such as *S. octosporus* or *S. japonicus*, the second meiotic division is immediately followed by mitotic replication of the meiotic products before they are packaged into eight spores (Conti and Naylor, 1960; Slooff, 1970; Tsuboi et al., 1978).

II. SURVEY OF THE LITERATURE

A. Gross Phenomenology

Our preoccupation with the cell wall as sex organelle began as a side observation a few years ago (Calleja, 1970). Cultures of *S. pombe* NCYC 132 (ATCC 26192), obtained from Prof. J. M. Mitchison and initially studied in our laboratory solely as a model of cell division and growth, were frequently observed to form clumps—an observation that was a source of annoyance to us, for this required disposing of the errant culture and starting a new one. Similar observations have been made in other laboratories (Mitchison, 1970; Slooff, 1970).

The clumping phenomenon is best demonstrated in aerated cultures grown to stationary phase in malt extract broth (Calleja, 1970). Actively dividing cells and cells in nonaerated cultures do not form clumps. We call the phenomenon flocculation, in an attempt to find similarities with yeast flocculation in the brewery (Geilenkotten and Nyns, 1971; Stewart et al., 1976). In an aerated culture, the first flocs, microscopic in size, appear after the end of logarithmic growth. Then, flocs increase in size and number until maximum flocculation is attained within a period of 5 hr. As much as 70% of the cell population may be found in flocs. The biggest flocs comprise up to 10^5 cells each.

Flocs are not the consequence of sibs failing to separate after fission. Cells in flocs do not divide—unless they are transferred to fresh medium. An hour or so before the appearance of flocs, no aggregates of greater than 10 cells each are observed. Furthermore, ultrasonic treatment of stationary-phase cultures does not prevent flocculation. We have defined *cell aggregation* as *the gathering together of cells to form fairly stable, contiguous, multicellular associations under physiological conditions* (Calleja and Johnson, 1980). Clearly, the flocculation we observe among fission yeast cells is a form of cell aggregation.

Flocs are fairly stable during dilution in distilled water, ruling out the presence of a flocculating agent in the culture medium. They can be separated from free cells by gravity sedimentation. We allow a flocculated culture to stand undisturbed in a tapered tube for 5 min. The flocs settle to the bottom of the tube while the free cells remain in the supernatant fluid, which may then be decanted and replaced with distilled water. The purification of flocs consists of repeated sedimentation in distilled water or in a buffer solution. Addition of fresh medium to purified flocs does not cause deflocculation until the cells begin to grow once more. This observation excludes the possibility that there is in the medium a deflocculating agent that confers nonflocculence to growing cells.

Purified flocs are stable to temperature up to 50°C (Calleja, 1970). Higher temperatures cause deflocculation, which is complete at about 80°C. T_m, the temperature at which one-half of the cells in flocs are dispersed to free cells, is 61°C in distilled water. In spite of its lethal effect, deflocculation by high temperatures is reversible. When free cells derived from thermally dispersed flocs, and heat-killed in the process, are left to stand at ambient temperature, they reflocculate. Enhanced by gentle agitation, reflocculation of heat-killed cells indicates that flocculation itself, as distinguished from induction to competence to form flocs, is not a vital process. Heat-killed cells not derived from flocs do not flocculate. Isolated cell walls of competent cells do flocculate among themselves, those from incompetent cells do not. It is obvious that the difference between free, incompetent cells and competent cells in flocs lies in their cell walls. A change in the material composition of the cell wall may then be envisaged as having occurred during the transition of a cell from incompetence to competence.

B. Quantitation of Flocculation

A floc is defined as an aggregate of more than 10 cells (Calleja and Johnson, 1977). Any cell in a group of 10 or less is considered free. The cutoff number, seemingly arbitrary, is based both on convenience and on the distribution of aggregation numbers in a population of uninduced, nonflocculating cells. A microfloc is an aggregate of no more than 100 cells. Obviously, it is identifiable only with the aid of a microscope. An aggregate of more than 100 cells but still not visible to the unaided eye is called a minifloc. Gross flocs are visible to the unaided eye.

We routinely use three methods for monitoring the development of flocculation in a culture or for estimating the purity of a floc suspension: hemocytometric, turbidimetric, or visual (Calleja and Johnson, 1977). The definitions above are important in the hemocytometric method, which requires counting in a hemocytometer the number of free cells in a culture before and after complete deflocculation. The difference between the two counts is equivalent to the number of cells in flocs. The free cells before deflocculation are best scored when separated from the flocs by differential sedimentation. With the turbidimetric method, a tube of flocs is read in a photoelectric turbidimeter for turbidity of the suspension well above the sedimentation line and then left undisturbed for 5 min, until flocs are settled to the bottom of the tube. Once again the turbidity is read. The difference between readings is equivalent to the turbidity contribution of flocs. Almost instantaneous estimation and reasonably quantitative comparison of many cultures may be made without instrumentation by the unaided eye.

C. The Nature of the Forces

An aqueous suspension of purified flocs may be deflocculated by heat, reflocculated on cooling, and deflocculated again (Calleja, 1974a). The sequence can be repeated on the same suspension without significantly altering the ability of competent cells to reflocculate. The presence of NaCl or $CaCl_2$ increases the stability of flocs as reflected in an increase in T_m. Like gentle agitation, it also increases the rate of reflocculation. The nonspecific role of salt in increasing the rate of reflocculation may be superseded by forceful contact in a centrifugal field. Heat-dispersed competent cells, packed to a pellet, do flocculate when resuspended, while incompetent cells do not. Agitation simply increases the probability of collision among cell particles by turbulent motion; salt does likewise. By serving as counterions to the negatively charged cell surface, the cations lower the energy barrier between two cells approaching each other.

Reversible deflocculation may be effected by other means. Although stable over a wide range of hydrogen ion concentrations, flocs may be reversibly deflocculated at extremes of pH, provided they are not left too long at those extremes. Mechanical shear, such as blender treatment, causes partial and reversible deflocculation. Also reversible but complete is deflocculation caused by ultrasonic treatment. Among many chemicals tested at various concentrations, urea (5 M) and guanidinium chloride (3 M) are rapid and most effective deflocculants. Upon removal of either reagent, reflocculation takes place. In contrast, deflocculation by sodium dodecyl sulfate is slow but also reversible.

Of the many commercial enzymes tested, only proteinases cause deflocculation. Among these are Pronase, trypsin, papain, and clostripain. Carbohydrases, lipases, nucleases, serum albumin, and heat-denatured proteinases are ineffective. Proteolytic deflocculation is complete and irreversible in that the deflocculating activity cannot be reversed by repeated washing of the dispersed cells. High concentrations of dithiothreitol or mercaptoethanol also cause irreversible deflocculation. Mono- and disaccharides reported to cause deflocculation in brewer's yeast (Eddy, 1955; Mill, 1964) are not effective deflocculants even at concentrations as high as 50% (w/v). High concentrations (up to 5 M) of NaCl or of $CaCl_2$ cause only marginal instability of flocs at ambient temperature.

Reflocculation of cells reversibly dispersed, in whatever manner, from purified flocs is usually incomplete. Reflocculation values range from 80 to 95%. The hysteretic loss is partly due to incompetent cells found in the original flocs and now unable to compete favorably for places in the reconstituted flocs. Incompetent cells, although unable to form flocs among themselves, are able to bind weakly to competent cells. When found at the periphery of a floc, they are easily dislodged by thermal motion. Surrounded by competent cells, they may find themselves ensconsed in a floc. Addition of incompetent cells to purified flocs causes an increase in the rate of deflocculation and a decrease in the rate of subsequent reflocculation.

Cell–cell interaction during flocculation is likely not ionic in character, for common salts are ineffective deflocculants. Reversible deflocculation by mechanical shear, ultrasonic treatment, heat, urea, or guanidinium chloride strongly suggests the participation of hydrogen bonds. Slow, reversible deflocculation by sodium dodecyl sulfate points to a minor contribution by hydrophobic interaction. Reversible deflocculation, however effected, excludes covalent linkages between cell surfaces. It likewise excludes bifunctional ligands, which should be washed away during removal of reagents causing reversible deflocculation. A protein-mediated interaction is indicated by the irreversible deflocculation by proteinases and by the lack of effect of nonproteolytic enzymes. It is suggested also by irreversible deflocculation by disulfide-reducing reagents, such as dithiothreitol. The protein receptors are most likely covalently bonded to the cell wall.

D. Flocculation: A Prerequisite to Sex

Because the strain we started with was not known to sporulate, it had always been considered as sterile. A floc cannot be easily examined with the microscope for internal details because of the compact arrangement of the many thousands of cells it contains. But the component cells can be examined when the floc is dispersed into free cells. Sampling a flocculated culture for free cells does not readily reveal copulant pairs, conjugant pairs, asci, or spores (Calleja and Johnson, 1971). As deflocculation progresses, the number of these bizarre forms (as we called them before we learned of their sexuality) increases in the subpopulation of free cells. Under physiological conditions that prevent flocculation or the development of competence to form flocs, no spores are observed. Conjugant pairs and asci are found only in flocs and never among free cells in the culture. They appear only after flocculation. Furthermore, nonflocculent mutants do not form flocs nor spores.

It is now clear that in aerated broth cultures, flocculation is a prerequisite to sex. Hence, we call it sex-directed flocculation. Moreover, there is no doubt that NCYC 132 is a homothallic strain. Sex-directed flocculation is analogous to sexual agglutination in other yeasts, such as species of *Hansenula, Saccharomyces,* and *Citeromyces* (Wickerham, 1969; Crandall, 1978; Yanagishima, 1978). It now seems categorically different from the apparently nonsexual yeast flocculation in the brewery.

The time course of flocculation and of the subsequent events scorable with the light microscope is shown in Fig. 4. The simplified sequence consists of flocculation, copulation, conjugation, and then sporulation. Copulation is defined as the covalent union of cells. Unlike flocculation, such a union is no longer sensitive to disruption by heat, urea, sodium dodecyl sulfate, or proteolytic enzymes. It stops short of conjugation, whose immediate product is a dikaryon. Note that once flocculation attains a maximum, it remains high throughout the sequence. Slow, irreversible deflocculation, after sporulation, occurs over an extended period of several days. We call it postdevelopmental deflocculation (Calleja and Johnson, 1979).

E. Induction to Competence

Cells grown anaerobically do not flocculate, but can be made to do so when aerated in stationary phase. Actively dividing cells, aerated or nonaerated, do not flocculate. Neither do cells grown in a defined medium consisting of Wickerham's yeast nitrogen base (Difco) plus carbon source, nor those grown in yeast extract plus glucose. The deficiency of the defined medium can be corrected by the addition of peptones or vitamin-free casamino acids (Difco). Obviously, because it occurs only under fairly circumscribed conditions, sex-directed flocculation is inducible.

Aeration, rather than mere agitation, is required for induction to occur, because cells grown unshaken in thin layers of broth on shallow plates flocculate, conjugate, and sporulate (Calleja and Johnson, 1971). Also, sporulation is observed on agar plates, but only after the cells have become competent to form flocs. Competence is confirmed by suspending the agar-grown cells in water.

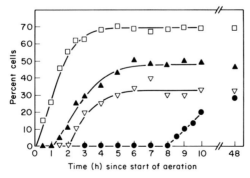

Fig. 4. The time course of flocculation (□), copulation (▲), conjugation (▽), and sporulation (●). Still cultures were grown for 24 hr to stationary phase and then aerated at 0 hr. At each of the times indicated, a whole culture was sacrificed for cell counts in a hemocytometer. (From Calleja and Johnson, 1979.)

We routinely induce cells to competence in the following manner. Cells are grown to stationary phase as a still culture in 10 ml of malt extract broth (2% w/v, Oxoid) in a tightly capped 20-ml universal bottle. Incubated at 32°C, the culture grows for 6 to 7 generations from an inoculum of 5×10^6 cells taken from a 24-hr culture grown the same way. It is transferred, 24 hr later, to a 125-ml Erlenmeyer flask and shaken at 150 rpm on a rotary shaker at 32°C. Induction is thus achieved by aeration in stationary phase of cells grown as a still culture.

Incubated as still cultures for various lengths of time and then aerated, cultures show different induction lags and different capacities for induction (Calleja, 1973). The capacity for induction decays as a culture ages in stationary phase. Heat-killed cells, like cells in stationary phase for more than 2 weeks, are not inducible, but do flocculate if induced before they are killed. Thus, whereas flocculation per se is a mechanical gathering and sticking together, the induction to competence to form flocs is a vital process. However, massive ultraviolet irradiation, which results in the failure of cells to form colonies when plated on agar, does not stop induction, thus ruling out the need for

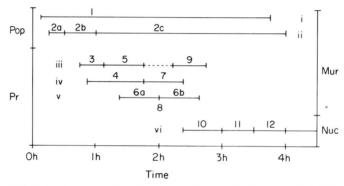

Fig. 5. Schedule of mural and nuclear events from induction to meiosis I. Segments represent scorable events from initiation to completion: (1) induction to competence to form flocs; (2) flocculation (a, microflocs; b, miniflocs; c, gross flocs); (3) copulation or covalent-bond formation; (4) conjugation tube formation; (5) crosswall formation; (6) crosswall erosion (a, thinning and softening of wall; b, widening of perforation); (7) conjugation tube expansion; (8) cytoplasmic fusion or dikaryon formation; (9) dedifferentiation of site of union; (10) nuclear migration; (11) karyogamy or nuclear fusion; and (12) meiosis I. Activity groups are designated by Roman numerals i to vi. Events in one group are similar or associated activities or parts of a continuum. Pop, population; Pr, pair; Mur, mural; Nuc, nuclear. Groups i and ii represent many individual events occurring in the population. Groups iii to vi represent events occurring in a pair of cells. The lengths of the segments are approximate. Broken line between segments 5 and 9 indicates temporal discontinuity. Event 8, by definition, is a point in time. (From Calleja et al., 1977b.)

DNA replication; but events after flocculation are severely inhibited. Added at the start of aeration, metabolic inhibitors, such as cyanide, azide, and dinitrophenol, inhibit induction, as do membrane-specific drugs, such as nystatin and polymyxin, as well as inhibitors of cytoplasmic protein synthesis, such as cycloheximide. Chloramphenicol, an inhibitor of mitochondrial protein synthesis, has no effect on induction when added at the start of aeration, but, if added to aerobically growing cells long before the last generation, it inhibits floc formation. Taken together, these results indicate that induction to competence to form flocs requires energy, mitochondrial function, and cytoplasmic protein synthesis.

Later experiments show the inhibiting effect of glucose on the induction process (Calleja, 1974b; Calleja et al., 1980b). High concentrations of glucose (1% and above) cause catabolite repression, which is not relieved by cyclic nucleotides. Experiments with actinomycin D are somewhat perplexing (Calleja, 1972b). Added at the start of aeration, the drug inhibits flocculation induction. Added at the time of inoculation, it does not inhibit cell division or growth, but nonetheless completely prevents flocculation from taking place.

F. Fine Structure of the Interactions

We have described in detail the mural and nuclear events from induction, the initial event in the sexual process, to meiosis I, the start of the sporulation process (Calleja et al., 1977b). These mural and nuclear events are arranged in a graphic schedule to indicate their temporal position in the course of development (Fig. 5).

Induction to competence to form flocs does not render an induced cell morphologically distinguishable from an uninduced cell, except perhaps by the presence of sex hairs (Calleja et al., 1977b). Analogous to fimbriae in bacteria (Ottow, 1975; Sokatch, 1978) and in fungal species (Poon and Day, 1975; Day and Poon, 1975; Day et al., 1975; see also the chapter by Day and Cummins in this volume), sex hairs presumably serve as grappling hooks by hydrogen-bonding cells together during flocculation. Because sex hairs are distributed all over the surface, there seems no special structure on the cell wall to indicate a specific site for copulatory activity. Hydrogen bonds during flocculation are supplemented and then replaced by covalent bonds between two walls during copulation. Approximately coincident with copulation is the initiation of conjugation tubes, scored as deformed poles of cells in contact. The mural regions in contact now become a two-layer crosswall via fusion. Next follows the enzymatic erosion of the crosswall. It usually begins in the central region of both sides of the crosswall even before conjugation tube formation is finished. It continues long after dikaryon formation and conjugation tube expansion. When the hole appears, a dikaryon is formed. The conjugation tube, now a singular element derived from paired elements, is enlarged as the hole is widened. The crimp at the site of union is ironed out and the remnants of the suture are repaired in such a manner that the site of union can no longer be identified. Nuclear migration toward the conjugation tube, now the middle portion of the new cell that is the dikaryon, begins before crosswall erosion is complete. Karyogamy then takes place, followed by meiosis I, when the fused nucleus divides and the resultant nuclei separate poleward (Yoo et al., 1973). The two nuclei each undergo another division (meiosis II) to form four nuclei, which then develop separately into four spores.

The various mural and nuclear events are conveniently sorted into six groups (Fig. 5). The sole event in group i, induction to competence to form flocs, is presumed to be an event that mainly involves the synthesis of sex hairs on the cell wall and of those enzymes required for the subsequent morphogenic events. Group ii consists of the mechanical construction of microflocs, miniflocs, and gross flocs. Both groups are indicated here as events in the population, although only flocculation is strictly so, induction being an event that happens in the individual cell—perhaps, independently of other cells. The reason for our position is that it is not the insular cell which is scored for induction but the population.

The ligating activity between a pair of walls is found in group iii. Crosswall formation is indicated as a consequence and extension of copulation. Subsequent ligating activity occurs during mural dedifferentiation at the site of union. Conjugation tube formation and conjugation tube expansion are combined in group iv. Both glucaneogenic and glucanolytic activities are overtly implicated in this group, but the final outcome is a net synthesis of more wall material. In contrast, group v, which includes crosswall erosion, perforation, and perforation expansion, is primarily the result of net glucanolytic activity.

Bunched together in group vi are the nuclear events. Save for these, all the significant morphological events are mural. By the time meiosis I is completed, mural functions and activities connected with sex are finished. The morphogenetic activity of the wall ceases for a while, until the wall is once more reactivated when the developmental program nears completion. After sporulation, mural activity, directed now toward self-destruction, is chiefly glucanolytic. But by this final act of immolation, spores are liberated for a new generation.

G. Macromolecular Changes during the Sequence

The developmental sequence has been monitored for respiratory activity, macromolecular changes, catabolite repression, and commitment to sporulation (Calleja et al., 1980b). Respiratory activity, which is required for induction, increases fivefold prior to maximum flocculation and remains at that level up to the end of the sequence. Total protein and total RNA increase just before conjugation and gradually decrease shortly thereafter.

A round of premeiotic DNA synthesis occurs after copulation, presumably during conjugation. Earlier work by the Egels has shown that premeiotic DNA synthesis occurs before karyogamy (Egel and Egel-Mitani, 1974). Prior to this synthesis, the DNA content per cell remains unchanged. The DNA content of a sex cell is equivalent to that of a spore. These observations indicate that cells before induction are in G_1, a condition that appears required for conjugation in fission yeast (Egel and Egel-Mitani, 1974; Streiblová, 1976) as well as in *Saccharomyces cerevisiae* (Hartwell, 1973; Bücking-Throm et al., 1973; see also Chapter 2 by Manney et al.).

The developmental sequence being subject to catabolite repression, commitment to sporulation may be monitored by adding glucose at various times and then noting the time at which development becomes refractory to glucose addition. Cells not committed are repressed by glucose and revert to their mitotic cycles. In contrast, committed cells proceed to sporulate in the presence of exogenous glucose. Commitment to sporulation appears to occur soon after premeiotic DNA synthesis.

H. Conjugation-Induced Lysis

Erosion of the crosswall during conjugation is fraught with danger to the cells. About 15% of conjugating pairs (copulating pairs and asci included in the population) lyse spontaneously during the conjugation process (Calleja et al., 1977a; Johnson et al., 1977). Up to 40% may be caused to lyse by dilution in distilled water. Lysis occurs at the site of union. The time at which the cells are most susceptible to lysis is about 2 hr after the start of the induction procedure. It corresponds to the beginning of conjugation.

The lytic phenomenon indicates a failure of controls over the glucanolytic activity of the cell wall. Fusion and erosion of cell walls during conjugation are presumed to be the concerted work of glucanases and glucan synthases (Johnson, 1968a). Lysis can occur if the lytic component becomes decidedly more active than the synthetic. A hyperactive glucanase with spillover activity might cause thinning of the wall region close to the site of union, bringing about an unscheduled rupture. Another possibility is premature dissolution of the two layers of crosswall between conjugating cells. If the enzymatic activity that catalyzes the formation of covalent bonds between walls is somehow defective such as to bring about fusion that is not leak-free, then erosion of the crosswall may bring about a lytic event. At any rate, the cytological evidence points to faulty fusion or badly coordinated glucanolytic activity during crosswall removal. Increased hydrolytic activity during yeast conjugation has been reported by others (Shimoda and Yanagishima, 1972; Kröning and Egel, 1974). Conjugation-induced lysis, which seems analogous to lethal zygosis in bacteria (Alfoldi et al., 1957; Skurray and Reeves, 1973), is evidence that the sexual process is subject to miscoordination.

I. Thermosensitivity of the Developmental Program

We routinely grow and induce cultures at 32°C. Cells grown to stationary phase at 32°C and then aerated at 37°C do not form spores (Calleja and Johnson, 1979). Grown to stationary phase at 37°C, cells are not immediately inducible when aerated later at 32°C. Growth for at least one generation at 32°C is necessary before they can become induced. To identify which events in the developmental program are thermosensitive, we grow and induce cultures at the permissive temperature and then, at different times, transfer them to the restrictive temperature. The following events are thermosensitive: development of respiratory sufficiency, readiness (defined as inducibility of a culture within 1 hr of aeration), induction to competence to form flocs, copulation, conjugation, and early sporulation (including meiosis). Cell division, respiration, flocculation, and spore maturation are thermoresistant. At the restrictive temperature, conjugation-induced lysis and postdevelopmental deflocculation are markedly stimulated, suggesting an enhancement of glucanolytic activity (presumed to cause lysis) and of extracellular proteolytic activity (presumed to cause irreversible deflocculation). Neither lysis nor deflocculation, however, can account for the thermosensitivity of the other events. Thermosensitivity of development is known in other fungal systems (Bowman, 1946; Hirsch, 1954; Hawker, 1966; Fowell, 1969; Lu, 1974; Doi and Yoshimura, 1977, 1978).

10. Cell Wall as Sex Organelle 241

J. Sex Interconversion in a Homothallic Strain

Exposure to iodine vapors gives agar colonies of *S. pombe* (and other fission yeasts) characteristic colors indicative of sporulation. Classical homothallic colonies are black; heterothallic colonies are yellowish. Furthermore, a black line, the iodine junction reaction, marks the boundary mutually shared by contiguous heterothallic colonies of complementary mating types (Leupold, 1950, 1970; Gutz *et al.*, 1974). The iodine reaction marks the presence of spores in asci, which are derived in a haploid culture from the zygotic union of cells. Thus, it is diagnostic of sexual interaction between complementary cells found in a colony or along the common boundary of contiguous colonies. However, although obviously homothallic, the strain we have been working with all these years does not yield homogeneously black colonies that are characteristic of the homothallic h^{90} strain of Leupold. Neither does it yield homogeneously yellowish colonies characteristic of the heterothallic h^+/h^- system. Instead, it yields colonies intermediate in appearance, mosaics of black streaks on a yellowish backdrop (Calleja, 1972a).

Stained with iodine, colonies of 360-2, a direct derivative of NCYC 132, are of two classes: P, with numerous black radial streaks, and d, with scarcely any (Calleja *et al.*, 1979). Neither class can be isolated as a pure culture. When replated on agar, a P colony generates a P plate, which comprises mostly P but also d colonies; a d colony generates a d plate, which comprises mostly d but also P colonies. The P/d colony ratio of a fresh isolate, if isolated as a P colony, is very high or, if isolated as a d colony, very low. On subsequent replatings of the same isolate, the ratio falls, if initially high, or rises, if initially low. An isolate that has been maintained for hundreds of generations attains an equilibrium P/d colony ratio of about 0.5. Contiguous P and d colonies, but not contiguous P nor contiguous d colonies, give the iodine junction reaction. P appears to correspond to the mating type of the tester strain 975 h^+ and d to the mating type of the tester strain 972 h^-. Tetrad analysis shows a 2:2 segregation ratio of the classes, indicating heterozygosity at a single locus. From these observations, we deduce that a homothallic clone is a community of two cell types: P cells, which generate P colonies, and d cells, which generate d colonies. The cell types are sexually complementary and interconvertible. The P→d interconversion is about twice as frequent as the d→P interconversion.

P and d colonies may be distinguished from each other by criteria other than the intracolonial iodine reaction, the iodine junction reaction between classes and with the mating-type testers, and the resultant plate on replating. Colonies may be identified by the frequency of depression lines on the colony surface (much more numerous on a P colony), by the coloration on malt extract broth agar (a P colony is a shade darker), and by the extent of sporulation (a P colony has more ascospores). But more important to the subject at hand, a P colony, inoculated into a flask of malt extract broth and allowed to shake overnight, gives rise to a culture more flocculent than a culture arisen from a d colony.

It is instructive to compare mating systems in fission yeast on the basis of sex-interconversion rates. The putative complementary mating types found in the homothallic h^{90} system cannot be isolated nor are they easily identified, because their interconversion rates are extremely high (Egel, 1976, 1977a,b). On the other hand, the two mating types that make up the heterothallic h^+/h^- system interconvert so infrequently ($<10^{-4}$ per cell division) that they fail to give the intracolonial iodine reaction (Gutz and Doe, 1973; Meade and Gutz, 1976). Isolatable as pure and fairly stable cultures they can be identified by the iodine junction reaction. The P/d system lies between the two extremes. Its complementary components cannot be isolated as pure cultures, but they are differentiated from each other by their intracolonial iodine reaction and the iodine junction reaction.

The P/d system, as well as the work of Egel (1977b), supports the thesis that there is no homothallic fission yeast cell, only a homothallic clone. A homothallic clone is made up of two cell types that are sexually complementary and interconvertible. The sexual event is an interaction between a P cell and a d cell, not between a P and a P, or between a d and a d. We assume that complementarity lies in the molecular arrangement of the complementary cell walls.

III. CURRENT RESEARCH

A. Effect of 2-Deoxyglucose on Sex Cells

The compound 2-deoxyglucose (2DG) is well known to cause lysis of growing yeast cells (Megnet, 1965). It has been used to localize mural extension and the activity *in vivo* of endoglucanases implicated in cell growth and septum formation (Berliner and Reca, 1970; Biely *et al.*, 1971, 1973; Johnson, 1968a,b; Johnson and Rupert, 1967; Poole and Lloyd, 1973; Svoboda and Smith, 1972). Its effect on conjugation of fission yeast has been examined by Kröning and Egel (1974). They find

10. Cell Wall as Sex Organelle

that 2DG causes swelling of the conjugation tubes, indicating the *in vivo* action of lytic enzymes.

We have reexamined this effect of 2DG on sex cells. To 24-hr old stationary, still cultures, we added various concentrations of 2DG (2-deoxy-D-*arabino*-hexose, Calbiochem) at the start of the flocculation-induction procedure. Flocculation was scored by eye, +10 being highest, +1 the slightest sign of flocculation, and 0, no gross flocs observed. Conjugation, sporulation, and cell lysis were scored in a hemocytometer.

Flocculation (including flocculation induction), conjugation, and sporulation, all were inhibited by 2DG (Fig. 6). Conjugation and sporulation were completely inhibited at concentrations as low as 10^{-4} M, whereas flocculation was completely inhibited only at much higher concentrations ($>3 \times 10^{-3}$ M). So long as the dosage of 2DG was low enough that flocculation was allowed to take place, however limited, the onset of flocculation induction was not greatly affected. The difference in induction lags between the control and the highest concentration at which flocculation was observed was no more than 1 hr. However, even doses as low as 10^{-4} M were sufficient to elicit an effect on flocculation when observed hours after the maximum. This effect was premature deflocculation. Postdevelopmental deflocculation at 32°C normally takes a week (Calleja and Johnson, 1979). In the presence of 2DG, deflocculation was accelerated so that it occurred within a day (Table I).

Because flocculation is a prerequisite to conjugation, it is conceivable that conjugation (and sporulation, subsequently) was inhibited by 2DG-induced deflocculation. However, deflocculation took at least 10 hr for completion, whereas uninhibited conjugation was complete within 5 to 6 hr. Another possible mechanism for conjugation inhibition could be cell lysis. But lysis occurred only within a limited range of

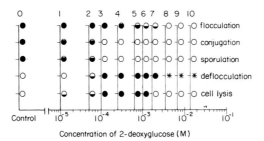

Fig. 6. Effects of 2-deoxyglucose (added at the start of the induction procedure) on flocculation, conjugation, and sporulation. Cultures were observed up to 96 hr since the start of aeration and addition of the inhibitor. Yes, ●; some, ◐; no, ○; not applicable, *.

TABLE I

Effect of Age on Sensitivity of Conjugating Cells to 2-Deoxyglucose Added at Start of Induction Procedure

Concentration of 2DG (M)	Maximum flocculation	Deflocculation	Cell lysis[b]	Conjugation	Sporulation
24-hr old[a]					
Control	+10	No	No	+4	+4
1×10^{-5}	+10	No	10% at 72 hr	+3	+3
1×10^{-4}	+10	Total at 22 hr	95% at 5 hr	0	0
72-hr old[a]					
Control	+9	No	No	+3	+3
1×10^{-5}	+9	No	No	+3	+3
1×10^{-4}	+9	No	No	+3	+2

[a] Time reckoned from the time of inoculation to the start of the induction procedure.
[b] No cell lysis means less than 5% at 72 hr.

concentrations. At a concentration that completely inhibited conjugation (10^{-4} M), close to 95% of the cells lysed within 5 hr (Table I). At concentrations above 10^{-3} M, lysis was suppressed; nevertheless, conjugation inhibition was not relieved (Fig. 6). The suppression of lytic activity of 2DG when used at high concentrations is most likely due to the inhibition of mural synthesis.

Microscopic examination revealed that at low concentrations of 2DG ($<10^{-4}$ M), lysis occurred mainly at the juncture region of a conjugating pair, whereas at high concentrations ($>10^{-4}$ M), lysis occurred at the corresponding end of a copulating half. In both cases, the mural site of response was the same, but the timing of lysis differed. At the low concentrations, a limited extent of conjugation was permitted to occur. But somehow, after the fusion of cells was effected, lysis took place at the site of fusion. At the high concentrations, lysis occurred much earlier, during the formation of conjugation tubes and before fusion was achieved. Also observed at low concentrations of 2DG was an enlargement of the copulating ends, making the cells look like drumsticks—an observation reported earlier (Kröning and Egel, 1974). When allowed to incubate for several days, most of these drumstick cells lysed or became enlarged to form bloated cells, up to 3 to 4 normal cell diameters, suggesting a net synthesis of wall material and that glucan synthesis was not abolished, although obviously deranged.

Added not at the time of induction but at the time of inoculation, 2DG at concentrations that allowed an increase in cell population inhibited flocculation, conjugation, and sporulation (Fig. 7). This time, however, the dose–response range was markedly different. A concentration less than that allowing almost normal flocculation in the previous experiments (column 4 of Fig. 6) completely prevented flocculation induction. This concentration allowed population growth, albeit slower than normal, and also resulted in the lysis of almost all cells during stationary phase. Upon further incubation, cells that escaped lysis were converted into monster cells of various shapes, up to 10 cell diameters in size. At concentrations of 10^{-3} M and higher, population growth was inhibited, but mural extension was allowed and led to rapid lysis. At even higher concentrations ($>10^{-2}$ M), lysis was suppressed, indicating that cellular metabolism had been overwhelmed by the inhibitor. Note also that the concentration at which lysis could take place was much higher when 2DG was added at the time of inoculation than at the start of the induction procedure (compare columns 4 and 5 of Fig. 7 with columns 6 and 7 of Fig. 6). Also, lysis during the growth phase of a culture was much faster than lysis during sexual development. This observation simply confirms the fairly obvious, that conjugation tube formation and allied activities require less cell wall synthesis than does cell division or mural extension.

It is well known that 2DG causes lysis of growing cells only (Megnet, 1965; Johnson, 1968a; Johnson and Rupert, 1967). The more mural extension activity there is, the more sensitive a cell is to cell lysis. We have observed that as cells grow older in stationary phase, the less their mural activity. They become less efficient in copulation, conjugation, and sporulation. Their conjugation tubes are less prominent. The longer they stay in stationary phase, the longer it takes to induce them (Calleja, 1973). Thus, it was of interest to determine whether age had any effect on the sensitivity of a cell to 2DG.

Two cultures of different ages are compared in Table I. A concentration of 2DG (10^{-4} M) added at the start of the induction procedure to

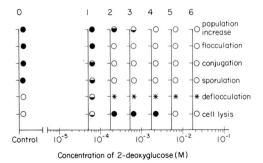

Fig. 7. Effects of 2-deoxyglucose (added at the time of inoculation) on population increase, flocculation, conjugation, and sporulation. Cultures were observed up to 96 hr of shaking since inoculation and addition of the inhibitor. Symbols as in Fig. 6.

the 24-hr-old culture led to complete inhibition of conjugation and sporulation, to lysis (about 95% of the cells within 5 hr), and subsequently to deflocculation of the lysed remnants (complete within 22 hr). The same concentration did allow conjugation and sporulation of the septuagenarians, which showed less conjugation activity to start with. More important, deflocculation was not observed and lysis among these old cells was minimal.

Although the proteinases of lysed cells can cause deflocculation, the 2DG-induced deflocculation described here cannot logically be ascribed to the lysis that happens earlier. Premature deflocculation was observed to occur even in the absence of lysis (column 7 of Fig. 6). Indeed, high concentrations which did not allow lysis caused even faster deflocculation (data not shown). High concentrations of 2DG suppress cell lysis but do not suppress premature deflocculation. Added at 1.5 hr, after flocculation induction but before conjugation, 2DG at a high concentration (1.2×10^{-2} M) completely inhibited conjugation (and sporulation, consequently) without causing cell lysis. But although deflocculation took place (complete within 10 hr), it was nonetheless too late to explain conjugation inhibition, which was apparently instantaneous.

Because low concentrations of 2DG (e.g., 10^{-4} M, column 3 of Fig. 6) brought about premature deflocculation without seemingly affecting flocculation induction, it is obvious that the inhibitor somehow affected the ability of the cells to maintain very stable flocs—perhaps not directly, but via faulty mural synthesis. In turn, the instability of the floc structure would impair the ability of a pair of cells in the floc to forge covalent linkages prior to conjugation. The formation of such linkages constitutes copulation. It is this developmental stage that is initially compromised as a consequence of the failure of 2DG-inhibited cells to form very stable flocs. In *Saccharomyces cerevisiae*, 2DG inhibits cell fusion during the mating reaction by interfering with the synthesis of mural polysaccharides (Shimoda and Yanagishima, 1974).

Like most cellular inhibitors, if not all, 2DG probably acts at many sites. Its many effects at low concentrations appear to stem from a common source: a deranged wall with deranged activities. This derangement is manifested as a failure of a floc to hold itself together, thereby causing premature deflocculation. It is indicated as a disability of a pair to effect normal sexual activity, thereby inhibiting copulation, conjugation, and sporulation. It is shown as a tendency of a cell to grow into a drumstick or a monster. It is demonstrated dramatically as a rupture at the growing ends or at the septal site of a vegetative cell, at the copulating end of a sex cell, or at the site of contact and

10. Cell Wall as Sex Organelle

fusion of a sexual pair. It arises from faulty synthesis of mural material, which may be used for the vegetative functions of growth and cell division or for the sexual function of conjugation tube formation. By interfering with mural synthesis, the inhibitor frustrates the emergence of the cell wall from a vegetative structure into an effective sex organelle.

B. The Ratio of P Cells to d Cells in a Floc

A sexual pair, the smallest cell–cell interacting unit, consists of a P cell and a d cell. Would the cell types then be equally represented in a floc? One would think that the answer to the question is readily derivable by purifying flocs, dispersing them with Pronase into free cells, plating the freed cells, and then scoring for P and d colonies. However, this approach is not fruitful, because copulation and conjugation occur before flocculation is complete, and therefore a sexual pair will not be scored as two separate colonies. Moreover, conjugation-induced lysis interferes with plating efficiency. A more useful way is to start with an uninduced culture in stationary phase, determine its P/d colony ratio by plating, then induce it to flocculate, separate the free cells from the flocs, and plate the free cells. Subtracting the number of free cells from the total number of cells yields the floc population. Table II shows the result of such an experiment.

The experimentally derived d/P ratio of 1.27 is fairly close to unity. It

TABLE II

Ratio of P Cells to d Cells in a Floc[a]

Population	Average number of colonies per plate			d/P
	Total	P	d	
Whole	262.0	94.9	167.1	1.76
(%)	(100)	(36.2)	(63.8)	
Free	115.9	30.5	85.4	2.80
(%)	(100)	(26.3)	(73.7)	
Floc	146.1	64.4	81.7	1.27
(%)	(100)	(44.1)	(55.9)	

[a] A 24-hr still culture, grown to stationary phase, was sampled for plating and then aerated. At 5 hr after the start of aeration, when flocculation had reached a maximum, the free population was separated from the flocs by gravity sedimentation and then sampled for plating. A total of 8094 colonies on 42 plates were counted. P and d colonies were scored according to Calleja et al. (1979). The floc population is derived from the difference between the whole population and the free population. Flocculation was about 56%.

confirms the one-to-one relationship of P and d. The ratio of α to a in the agglutinated mass of *Saccharomyces cerevisiae* has been reported to remain consistently close to unity, even when the input ratios are varied (Kawanabe et al., 1979; see also Chapter 11 by Yanagishima and Yoshida). On the other hand, the ratio of donor cell to recipient cell in mating aggregates during conjugation in *E. coli* can vary by a factor of four, depending upon input ratio, cell population density, and the physiological state of the cells (Achtman, 1975).

A floc may be viewed as a random collection of competent cells. The interaction between two complementary cells is decidedly specific, but the overall arrangement is stochastic. A floc of many thousands of cells is made up of about equal numbers of P cells and d cells, d cells having the slight plurality only because they constitute the majority in the interconverting population at equilibrium. The periphery of a floc is assumed to be occupied mostly by the excess d cells in the population. Because the ratio deviates significantly from unity, it may be taken as additional evidence that flocculation is not the coming together of committed pairs, confirming the more simple observation that copulation occurs only among cells in flocs.

C. Sexual Preferences, Positions, and Orientations

Electron microscopy does not reveal any mural substructure that is specifically designed as a copulatory site, making us believe that the entire wall constitutes the sex organelle. A floc is so compactly constructed that any cell, except those at the periphery, is neighbor to almost as many cells (about 12) as can be accommodated on its surface. Nonetheless, conjugation seems to be confined mainly to the poles of the cylindrical cells.

The two ends of the cylindrical cell can be differentiated from each other by their relative age. Except in a germinating spore, one end must be older than the other. The new end is generated by the last fission. The old end is that which was already there when the last fission occurred and therefore by Mitchison's rule (Mitchison, 1957), it is now the growing end, if the cell is still in the mitotic cycle. Both ends are available for conjugation. Nevertheless, one end is preferred: we observed among 277 sex cells that 171 (61.7%) conjugate by their old end, which is that end distal to the last fission, and 106 (38.3%) by their new end. Some of the data of Streiblová and Wolf (1975), for a different homothallic strain of *S. pombe*, show otherwise: the secondary pole (which is the new end in our terminology) is preferred. However, their other data show that there is a preference for the growing pole, which corresponds, by our convention, to the old end. Despite the contradiction, their data and ours tend toward one conclusion: either end of a cell is a potential site for copulation.

Fission yeast cells bear on their walls the record of past cell divisions in the form of fission scars. Serving as concomitant and associated record are the fuscannels, which are electron-dense structures in the cell wall. The age of a cell is indicated by its number of scars or fuscannels (Calleja et al., 1980a). Would sex be limited to cells of a particular number of scars? Table III shows the distribution of fuscannels in populations of vegetative cells and sex cells. (We use the fuscannel as marker because it is easier to score than the scar in ultrathin sections; there is one fuscannel for every scar.) The two populations are very similar to each other and to what is expected, except perhaps for the absence of sex cells with more than four fuscannels. But the sample is small compared to the rarity of cells with a fuscannel number greater than four. Hence, we would like to think that age is no barrier to sex. Indeed, pairs appear to be a random assortment of fuscannel classes. A cell of one fuscannel class may mate with any of the classes, including its own, with the probability being dictated mainly by the frequencies of the classes in the population.

TABLE III

Comparison of the Distributions of Fuscannels among Vegetative Cells and among Sex Cells[a]

Fuscannel class	Vegetative cells		Sex cells[b]	
	Number	%	Number	%
1	429	34.99	198	36.00
2	532	42.97	237	43.09
3	218	17.61	98	17.81
4	53	4.28	17	3.09
5	7	0.57	0	0
6	1	0.08	0	0
>6	0	0	0	0
Total counted	1238	100	550	100
Recovery of fuscannels[c]		96.84		94.00

[a] Data from transmission electron microscopy of ultrathin sections.

[b] Sex cells include copulant pairs, conjugant pairs, and asci, all counted as two cells each.

[c] In theory, the total number of fuscannels (or of scars) in a population is exactly twice the number of cells (Calleja et al., 1980a). Thus, % recovery of fuscannels = $100F/2N$, where F is the total number of fuscannels counted and N is the total number of cells counted.

Although conjugation is usually polar, the very tip of a cell is not necessarily the most likely region of contact. Indeed, even polar pairs are not usually connected coaxially. More often, they are eccentrically positioned. There are pairs that may be described as not end-to-end but end-to-side. Such a position is infrequently observed (2.21% of sex cells). Rarer still is side-to-side (0.95% of sex cells). These unconventional sexual positions may be called lateral copulation or conjugation. They seem more prone to abortion in that their resultant asci contain on the average fewer spores (ca. 43% of normal). However, they are no more lysis-prone. These observations suggest that there is more ligating activity about the poles than the rest of the wall. The infrequency of lateral conjugation may also mean that the lytic apparatus is more concentrated at the poles.

In heterothallic systems, incestuous matings (or matings between sibs) are inconceivable. Not so in homothallic systems. In agar colonies of the homothallic h^{90}, incestuous matings are common occurrences (Leupold, 1950). They are common, perhaps, because there is less freedom of movement in an agar colony and because the rate of sex interconversion in h^{90} is extremely high. In NCYC 132, or its derivative 360-2, incestuous matings are apt to be very rare events under the conditions of our experiments. There are a number of reasons why we think so. First and foremost, the P $\rightleftharpoons$ d interconversion system is rather sluggish ($<10^{-2}$ per cell division). Therefore, the probability that sibs will be of complementary mating types is very small. Besides, the interconversion must take place just prior to flocculation induction and before the sibs are completely separated from each other. Otherwise, they will have to find each other in a crowd of 5×10^7 cells ml^{-1}. Our experimental conditions do not exactly favor finding a lost sib. The cells are grown as a shaken broth culture, not as immobilized colonies on agar. Matings are likely to be random matches, because the average aggregation number of a culture in stationary phase, prior to flocculation inducation, approaches unity (Calleja and Johnson, 1977). Flocculation also discourages incestuous matings, because it is a social arrangement by which sibs are introduced to other cells. Nonetheless, let us assume that sibs which are not completely separated from each other and are of complementary mating types have been induced and are now two neighbors in a floc. But we know that the new ends, by which the sibs hang on to each other, are not the preferred sites for conjugation. Meanwhile, their old ends, the preferred sites, are hydrogen-bonded to other cells, which are of the opposite mating types. Taken together, these conditions and situations make incestuous matings highly improbable.

10. Cell Wall as Sex Organelle

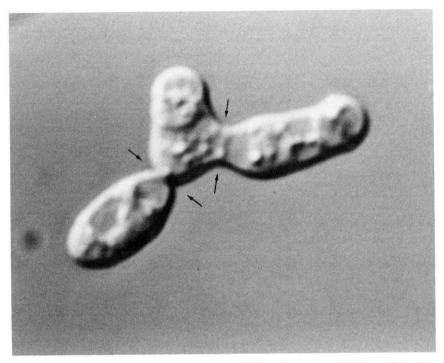

Fig. 8. A *ménage à trois* seen through the light microscope with Nomarski optics. The central cell is laterally fused with the right cell and polarly conjugating with the left cell. Arrows point to sites of fusion and conjugation.

Sexual activity is not confined to pairs (Fig. 8). Up to about 3% of copulant cells may be found in groups of greater than two cells (Table IV). We call this supernumerary sexual arrangement a *ménage* (Calleja *et al.*, 1977b). Multiple copulation, of course, is known in other yeasts (Brock, 1965; Poon *et al.*, 1974). It is not unknown elsewhere in the biological world (Freedman, 1977).

A *ménage à trois* is usually a linear chain, that is, a middle cell has either of its poles connected to a cell attached to no other. There may be two cells attached to one pole of a middle cell, making it bifunctional at one pole only. Trifunctional cells, among *ménages* of more than three cells, are very rare. Although not represented in Table IV, *ménages à cinq* have been observed by us on several different occasions. But we have observed none that is greater.

The sporulation efficiency of a *ménage* is much lower than that of a conventional pair (Table IV). The lower efficiency is not merely due to the failure of a third cell to partake of the meiotic events, thus need-

TABLE IV

Frequency and Sporulation Efficiency of *Ménages à Trois*[a]

	Number counted	Per sex cell	Per cell in *ménage*
Total sex cells[b]	11,293	1	
Total spores	11,533	1.021	
Ménage:			
3-cell	104		
4-cell	13		
5-cell	0		
>5-cell	0		
Total cells in *ménage*	364	0.032	1
Total spores in *ménage*	129		0.354

[a] Data from phase-contrast microscopy of whole cells.
[b] Sex cells include copulant pairs, conjugant pairs, and asci, all counted as two cells each.

lessly enlarging the denominator when efficiency is computed on a per-cell basis. Even when a *ménage* is considered as a pair, rather than made up of three or four cells, the computed efficiency still remains only about half of that of a pair. Not only then is a third cell usually nonproductive, it is counterproductive—it distracts the pair to which it attaches itself as the other cell. Although asci may be derived from three or more cells, we have observed, no more than a couple of times, asci with more than four spores. If a three-ring meiotic circus is possible, it must be a very rare event indeed.

The *ménage* is interesting in itself. But in a floc that consists of many thousands of cells and in which any given cell is apt to be hydrogen-bonded to a dozen or so cells, its rarity is more intriguing.

At equilibrium of the P/d sex interconversion system, there are about twice as many d cells as there are P cells in a culture. If there is no preferential induction to competence, a given P cell has a better chance of finding a mate than an average d cell. We can assume from the work on heterothallic strains that specific covalent linkages are forged only between complementary cells. This restriction surely cuts down the number of supernumerary copulants. However, there is a good chance that a given cell is neighbor to more than one complementary cell. Indeed, because a sex cell may form hydrogen bonds only with complementary cells, a P cell in the inner sanctum of a floc is presumably surrounded by d cells, and a d cell by P cells. Whereas any region of the wall is available for copulation, the ends are favored over the sides. But again, there are two ends, both capable of becoming links in a chain of cells. Furthermore, several cells can be accommodated at each end, because head-on connections are not required. Steric hindrance is not likely to prevent a *ménage*. That one end is preferred over the other, however, may discourage a third cell—especially if the preferred end of that cell is already engaged.

Still, some mechanism, other than the stickiness of ends, polar discrimination, and the required combination of P and d, must be operative to restrict a third participant. We call it the principle of the excluded third (Calleja *et al.*, 1977b). When one end of a cell is already spoken for, that is, in covalent linkage with a complementary cell, the leftover end is no longer available to a third cell. The pair are already committed to become one. It may simply be a matter of timing. The first comer takes all, as in the fertilization of an egg by a sperm. The principle of the excluded third is possibly executed by mural modulation taking place at the end not activated sexually. A communication system between the ends of one cell must then be operative. It may involve the wall, alone or with the mediation of the cytoplasm. A *ménage* may result when the bipolar communication system breaks down so that the ends are unaware of each other's activity. On the other hand, a cell may come out empty-ended, if a miscommunication somehow makes either end feel that the other is already spoken for.

Whatever the molecular mechanism, mural modulation or some other, the exclusion of a third cell does not lead to a loss of hydrogen-bonding activity of the walls, because deflocculation does not take place soon after conjugation is completed. On the contrary, the floc is made more stable by the many covalent linkages that supplant the hydrogen bonds, which until now almost solely bore the responsibility of holding the floc together. Postdevelopmental deflocculation occurs much later. It is not a factor in the sexual process, not even in sporulation.

Once sporulation is over, the social arrangement that is the floc is no longer of value to the species. By postdevelopmental deflocculation, the participants are released from the bondage and the security of the floc. Perhaps, outside the laboratory, postdevelopmental deflocculation is a means to population dispersal.

IV. PERSPECTIVES

A. Summing Up

The life cycle of the homothallic *Schizosaccharomyces pombe* is simple enough to serve as a manipulable model for morphogenesis, differentiation, and development of a eukaryote. The stable haploid cells retain the option of remaining in the vegetative state or, if the conditions become right, of moving on to sex. A cell in the presence of an energy source becomes ready for sexual interaction when it becomes mitochondrially sufficient and when released from catabolite repression, which otherwise imprisons it in the mitotic cycle. Simple aeration of a culture in stationary phase enables the cells to escape the drudgery of repetitive mitoses. Cells become induced to competence to form social structures called flocs. Induction brings about changes in the properties of the cell wall, which is the sex organelle. The changes require cytoplasmic synthesis of proteins, including the putative protein molecules that must be covalently attached to the cell wall. It is these mural proteins, possibly assembled into sex hairs, which mediate the hydrogen-bonded interaction between two cells of complementary mating types. Among the numerous cells in complementary contact in the floc, covalent linkages between two walls are forged. This activity is presumed to be catalyzed by wall-ligating enzymes. Mural activity involving both synthetic and lytic enzymes brings about the deformation of copulating ends to form the conjugation tube between two cells. Soon after mural fusion, erosion takes place until the two-layer crosswall between the two cells is gone, thereby making two cells functionally and morphologically one cell. The erstwhile two bags, through the miracle of sex, become one. After sex comes the aftermath. The resulting dikaryon becomes a zygote. Ordinarily, the diplontic state is a transient condition. The zygote is committed to undergo the pains of two meiotic divisions. During the sporulation process that follows, the four meiotic products become separately enclosed by spore walls to develop into spores. The ascus, derived from the copulating pair, is now merely an assembly line. The spores may be liberated or may remain inside the ascus. In fresh medium, they germinate and divide by fission as vegetative haploid cells, thereby completing the life cycle.

B. Directions of Things to Come

There remains much to be unraveled. We are undertaking the isolation of the putative protein molecules involved in sexual interaction. We use two approaches. Because the putative molecules are presumed to be covalently attached to the wall, we use proteolytic enzymes, trypsin and clostripain in particular, to clip them from the wall and then look for fragments by the technique of polyacrylamide gel electrophoresis. Preliminary results indicate the presence of protein bands that are derived from induced cells but not from the uninduced. It is hoped that in the bands may be found fragments with biological activity. The other approach makes use of postdevelopmental deflocculation. We test for biological activity of concentrated supernatant fluids of cultures allowed to incubate past their peak of flocculation. Biological activity is tested by a flocculation-inhibition assay, which calls for adding the test material to heat-dispersed purified flocs and measuring its reversible inhibitory activity against reflocculation. Preliminary results indicate an activity from supernatant fluids of cultures undergoing postdevelopmental defloccuation. The activity appears to derive from molecules smaller than 3000 daltons.

We are also pursuing the study of the $P \rightleftharpoons d$ sex interconversion system. We are looking into the physiology and the genetics of the interconversion. The quantitation of the rates of interconversion and the comparison of mutants should lead to insights as to the mechanism of the switch at the level of the gene. Likewise, we are attempting the isolation of the enzyme that brings about the covalent linkage of two cells, the enzymes that specifically catalyze conjugation tube formation, as well as the enzymes that cause mural erosion. It would be interesting to find out if the very same synthetic and lytic enzyme systems used for the vegetative functions of growth and cell division are components of the sexual equipment.

The phenomenon of cell clumping, initially an accidental observation and a source of annoyance, has become many things to us over the years: flocculation analogous to flocculation in the brewery, cell aggregation and a model for macromolecular denaturation, sex-directed flocculation and sexual interaction, a catabolite-repressible induction system, a model for cell differentiation, morphogenesis, and development, a genetic system, and a model for sex interconversion. It is, in the present treatment, the development of the cell wall as a sex organelle. The system, of course, is any of these and all these—indeed, more than these, as we learn more and more about it. Except that now it is studied by design and no longer, to us at least, a source of annoyance.

ACKNOWLEDGMENT

We thank Donna McLeish, Isabelle Boisclair, Teena Walker, and Harold Atkins for technical help, and Harry Turner for preparing the photographic plates.

REFERENCES

Achtman, M. (1975). Mating aggregates in *Escherichia coli* conjugation. *J. Bacteriol.* 123, 505–515.
Alfoldi, L., Jacob, F., and Wollman, E. L. (1957). Zygose létale dans les croisements entre

souches colicinogènes et non colicinogènes d'*Escherichia coli*. *C. R. Acad. Sci.* **244**, 2974–2976.
Berliner, M. D., and Reca, M. E. (1970). Release of protoplasts in the yeast phase of *Histoplasma capsulatum* without added enzyme. *Science* **167**, 1255–1259.
Biely, P., Krátký, Z., Kovařik, J., and Bauer, Š. (1971). Effect of 2-deoxyglucose on cell wall formation in *Saccharomyces cerevisiae* and its relation to cell growth inhibition. *J. Bacteriol.* **107**, 121–129.
Biely, P., Kovařik, J., and Bauer, Š. (1973). Lysis of *Saccharomyces cerevisiae* with 2-deoxy-2-fluoro-D-glucose, an inhibitor of the cell wall glucan synthesis. *J. Bacteriol.* **115**, 1108–1120.
Bowman, D. H. (1946). Sporidial fusion in *Ustilago maydis*. *J. Agric. Res.* **72**, 233–243.
Brock, T. D. (1965). Biochemical and cellular changes occurring during conjugation in *Hansenula wingei*. *J. Bacteriol.* **90**, 1019–1025.
Bücking-Throm, E., Duntze, W., Hartwell, L. H., and Manney, T. R. (1973). Reversible arrest of haploid yeast cells at the initiation of DNA synthesis by a diffusible sex factor. *Exp. Cell Res.* **76**, 99–110.
Calleja, G. B. (1970). Flocculation in *Schizosaccharomyces pombe*. *J. Gen. Microbiol.* **64**, 247–250.
Calleja, G. B. (1972a). Studies on the sexuality of *Schizosaccharomyces pombe* NCYC 132 and its derivatives. *Genetics* **71**, s9.
Calleja, G. B. (1972b). The effects of cyclic AMP and actinomycin D on an inducible system in a fission yeast. *Kalikasan Philipp. J. Biol.* **1**, 63.
Calleja, G. B. (1973). Role of mitochondria in the sex-directed flocculation of a fission yeast. *Arch. Biochem. Biophys.* **154**, 382–386.
Calleja, G. B. (1974a). On the nature of the forces involved in the sex-directed flocculation of a fission yeast. *Can. J. Microbiol.* **20**, 797–803.
Calleja, G. B. (1974b). Glucose effect and the effect of cyclic AMP on flocculation and sporulation of *Schizosaccharomyces pombe*. *Proc. Int. Symp. Yeasts, 4th, Hochsch. Bodenkult., Vienna* Part 1, pp. 73–74.
Calleja, G. B., and Johnson, B. F. (1971). Flocculation in a fission yeast: An initial step in the conjugation process. *Can. J. Microbiol.* **17**, 1175–1177.
Calleja, G. B., and Johnson, B. F. (1977). A comparison of quantitative methods for measuring yeast flocculation. *Can. J. Microbiol.* **23**, 68–74.
Calleja, G. B., and Johnson, B. F. (1979). Temperature sensitivity of flocculation induction, conjugation and sporulation in fission yeast. *Antonie van Leeuwenhoek; J. Microbiol. Serol.* **45**, 391–400.
Calleja, G. B., and Johnson, B. F. (1980). The effects of nutritional factors on microbial aggregation. *In* "CRC Handbook of Nutrition and Food" (M. Rechcigl, ed.). CRC Press, West Palm Beach, Florida (in press).
Calleja, G. B., Yoo, B. Y., and Johnson, B. F. (1977a). Conjugation-induced lysis of *Schizosaccharomyces pombe*. *J. Bacteriol.* **130**, 512–515.
Calleja, G. B., Yoo, B. Y., and Johnson, B. F. (1977b). Fusion and erosion of cell walls during conjugation in the fission yeast (*Schizosaccharomyces pombe*). *J. Cell Sci.* **25**, 139–155.
Calleja, G. B., Johnson, B. F., Zuker, M., and James, A. P. (1979). The mating system of a homothallic fission yeast. *Mol. Gen. Genet.* **172**, 1–6.
Calleja, G. B., Zuker, M., Johnson, B. F., and Yoo, B. Y. (1980a). Analyses of fission scars as permanent records of cell division in *Schizosaccharomyces pombe*. *J. Theor. Biol.* **84**, 523–544.
Calleja, G. B., Johnson, B. F., and Yoo, B. Y. (1980b). Macromolecular changes and

10. Cell Wall as Sex Organelle

commitment to sporulation in the fission yeast *Schizosaccharomyces pombe*. *Plant Cell Physiol.* **21** (in press).
Conti, S. F., and Naylor, H. B. (1960). Electron microscopy of ultrathin sections of *Schizosaccharomyces octosporus*. II. Morphological and cytological changes preceding ascospore formation. *J. Bacteriol.* **79,** 331–340.
Crandall, M. (1978). Mating-type interactions in yeasts. *In* "Cell–Cell Interaction" (A. Curtis, ed.), pp. 105–119. Cambridge Univ. Press, London and New York.
Crandall, M., Egel, R., and MacKay, V. L. (1977). Physiology of mating in three yeasts. *Adv. Microb. Physiol.* **15,** 307–398.
Day, A. W., and Poon, N. H. (1975). Fungal fimbriae. II. Their role in conjugation in *Ustilago violacea*. *Can. J. Microbiol.* **21,** 547–557.
Day, A. W., Poon, N. H., and Stewart, G. G. (1975). Fungal fimbriae. III. The effect on flocculation in *Saccharomyces*. *Can. J. Microbiol.* **21,** 558–564.
Doi, S., and Yoshimura, M. (1977). Temperature-sensitive loss of sexual agglutinability in *Saccharomyces cerevisiae*. *Arch. Microbiol.* **114,** 287–288.
Doi, S., and Yoshimura, M. (1978). Temperature-dependent conversion of sexual agglutinability of *Saccharomyces cerevisiae*. *Mol. Gen. Genet.* **162,** 251–257.
Eddy, A. A. (1955). Flocculation characteristics of yeasts. II. Sugars as dispersing agents. *J. Inst. Brew., London* **61,** 313–317.
Egel, R. (1971). Physiological aspects of conjugation in fission yeast. *Planta* **98,** 89–96.
Egel, R. (1973). Commitment to meiosis in fission yeast. *Mol. Gen. Genet.* **121,** 277–284.
Egel, R. (1976). The genetic instabilities of the mating type locus in fission yeast. *Mol. Gen. Genet.* **145,** 281–286.
Egel, R. (1977a). Frequency of mating-type switching in homothallic fission yeast. *Nature (London)* **266,** 172–174.
Egel, R. (1977b). "Flip-flop" control and transposition of mating-type genes in fission yeast. *In* "DNA Insertion Elements, Plasmids, and Episomes" (A. I. Bukhari, J. A. Shapiro, and S. L. Adhya, eds.), pp. 447–455. Cold Spring Harbor Lab., Cold Spring Harbor, New York.
Egel, R., and Egel-Mitani, M. (1974). Premeiotic DNA synthesis in fission yeast. *Exp. Cell Res.* **88,** 127–134.
Fowell, R. R. (1969). Sporulation and hybridization of yeasts. *In* "The Yeasts" (A. H. Rose and J. S. Harrison, eds.), Vol. 1, pp. 303–383. Academic Press, New York.
Freedman, H. (1977). "Sex Link: The Three-Billion-Year-Old Urge and What the Animals Do About It." M. Evans, New York.
Geilenkotten, I., and Nyns, E. J. (1971). The biochemistry of yeast flocculence. *Brew. Dig.* **46,** 64–70.
Gutz, H., and Doe, J. F. (1973). Two different h^- mating types in *Schizosaccharomyces pombe*. *Genetics* **74,** 563–569.
Gutz, H., Heslot, H., Leupold, U., and Loprieno, N. (1974). *Schizosaccharomyces pombe*. *In* "Handbook of Genetics" (R. C. King, ed.), Vol. 1, pp. 395–446. Plenum, New York.
Hartwell, L. H. (1973). Synchronization of haploid yeast cell cycles, a prelude to conjugation. *Exp. Cell Res.* **76,** 111–117.
Hawker, L. E. (1966). Environmental influences on reproduction. *In* "The Fungi: An Advanced Treatise" (G. C. Ainsworth and A. S. Sussman, eds.), Vol. 2, pp. 435–469. Academic Press, New York.
Hirsch, H. M. (1954). Environmental factors influencing the differentiation of protoperithecia and their relation to tyrosinase and melanin formation in *Neurospora crassa*. *Physiol. Plant.* **7,** 72–97.

Johnson, B. F. (1968a). Lysis of yeast cell walls induced by 2-deoxyglucose at their sites of glucan synthesis. *J. Bacteriol.* 95, 1169-1172.

Johnson, B. F. (1968b). Dissolution of yeast glucan induced by 2-deoxyglucose. *Exp. Cell Res.* 50, 692-694.

Johnson, B. F., and Rupert, C. M. (1967). Cellular growth rates of the fission yeast, *Schizosaccharomyces pombe,* and variable sensitivity to 2-deoxyglucose. *Exp. Cell Res.* 48, 618-620.

Johnson, B. F., Yoo, B. Y., and Calleja, G. B. (1973). Cell division in yeasts: Movement of organelles associated with cell plate growth of *Schizosaccharomyces pombe. J. Bacteriol.* 115, 358-366.

Johnson, B. F., Calleja, G. B., and Yoo, B. Y. (1977). A model for controlled autolysis during differential morphogenesis of fission yeast. *In* "Eucaryotic Microbes as Model Developmental Systems" (D. H. O'Day and P. A. Horgen, eds.), pp. 212-229. Dekker, New York.

Kawanabe, Y., Yoshida, K., and Yanagishima, N. (1979). Sexual cell agglutination in relation to the formation of zygotes in *Saccharomyces cerevisiae. Plant Cell Physiol.* 20, 423-433.

Kohli, J., Hottinger, H., Munz, P., Strauss, A., and Thuriaux, P. (1977). Genetic mapping in *Schizosaccharomyces pombe* by mitotic and meiotic analysis and induced haploidization. *Genetics* 87, 471-489.

Kröning, A., and Egel, R. (1974). Autolytic activities associated with conjugation and sporulation in fission yeast. *Arch. Microbiol.* 99, 241-249.

Leupold, U. (1950). Die Vererbung von Homothallie und Heterothallie bei *Schizosaccharomyces Pombe. C. R. Trav. Lab. Carlsberg, Ser. Physiol.* 24, 381-480.

Leupold, U. (1970). Genetical methods for *Schizosaccharomyces pombe. Methods Cell Physiol.* 4, 169-177.

Lindner, P. (1893). *Schizosaccharomyces Pombe* n. sp., ein neuer Gährungserreger. *Wochenschr. Brau.* 10, 1298-1300.

Lu, B. C. (1974). Genetic recombination in *Coprinus*. IV. A kinetic study of the temperature effect on recombination frequency. *Genetics* 78, 661-677.

Meade, J. H., and Gutz, H. (1976). Mating-type mutations in *Schizosaccharomyces pombe. Genetics* 83, 259-273.

Megnet, R. (1965). Effect of 2-deoxyglucose on *Schizosaccharomyces pombe. J. Bacteriol.* 90, 1032-1035.

Mill, P. J. (1964). The nature of the interaction between flocculent cells in the flocculation of *Saccharomyces cerevisiae. J. Gen. Microbiol.* 35, 61-68.

Mitchison, J. M. (1957). The growth of single cells. I. *Schizosaccharomyces pombe. Exp. Cell Res.* 13, 244-262.

Mitchison, J. M. (1970). Physiological and cytological methods for *Schizosaccharomyces pombe. Methods Cell Physiol.* 4, 131-165.

Ottow, J. C. G. (1975). Ecology, physiology, and genetics of fimbriae and pili. *Annu. Rev. Microbiol.* 29, 79-108.

Poole, R. K., and Lloyd, D. (1973). Effect of 2-deoxy-D-glucose on growth and cell walls of *Schizosaccharomyces pombe* 972h$^-$. *Arch. Mikrobiol.* 88, 257-272.

Poon, N. H., and Day, A. W. (1975). Fungal fimbriae. I. Structure, origin, and synthesis. *Can. J. Microbiol.* 21, 537-546.

Poon, N. H., Martin, J., and Day, A. W. (1974). Conjugation in *Ustilago violacea*. I. Morphology. *Can. J. Microbiol.* 20, 187-191.

Prescott, D. M. (1976). "Reproduction of Eukaryotic Cells." Academic Press, New York.

Robinow, C. F. (1977). The number of chromosomes in *Schizosaccharomyces pombe:* Light microscopy of stained preparations. *Genetics* 87, 491-497.

Shimoda, C., and Yanagishima, N. (1972). Mating reaction in *Saccharomyces cerevisiae.* III. Changes in autolytic activity. *Arch. Mikrobiol.* 85, 310-318.

Shimoda, C., and Yanagishima, N. (1974). Mating reaction in *Saccharomyces cerevisiae.* VI. Effect of 2-deoxyglucose on conjugation. *Plant Cell Physiol.* 15, 767-778.

Skurray, R. A., and Reeves, P. (1973). Physiology of *Escherichia coli* K-12 during conjugation: Altered recipient functions associated with lethal zygosis. *J. Bacteriol.* 114, 11-17.

Slooff, W. C. (1970). Genus 19. *Schizosaccharomyces* Lindner. *In* "The Yeasts: A Taxonomic Study" (J. Lodder, ed.), pp. 733-755. North-Holland Publ., Amsterdam.

Sokatch, J. R. (1978). Roles of appendages and surface layers in adaptation of bacteria to their environment. *In* "The Bacteria: A Treatise on Structure and Function" (J. R. Sokatch and L. N. Ornston, eds.), Vol. 7, pp. 229-289. Academic Press, New York.

Stewart, G. G., Garrison, I. F., Goring, T. E., Meleg, M., Pipasts, P., and Russell, I. (1976). Biochemical and genetic studies on yeast flocculation. *Kem.-Kemi* 3, 465-479.

Streiblová, E. (1976). Control of conjugation during the cell cycle of *Schizosaccharomyces pombe. Folia Microbiol. (Prague)* 21, 194.

Streiblová, E., and Wolf, W. (1975). Role of cell wall topography in conjugation of *Schizosaccharomyces pombe. Can. J. Microbiol.* 21, 1399-1405.

Svoboda, A., and Smith, D. G. (1972). Inhibitory effect of 2-deoxyglucose on cell wall synthesis in cells and protoplasts of *Schizosaccharomyces pombe. Z. Allg. Mikrobiol.* 12, 685-699.

Tsuboi, M., Ohashi, K., Takahara, M., and Hayashibe, M. (1978). Sexual process in eight-spored ascus-forming yeast, *Schizosaccharomyces japonicus.* I. Culture medium for the synchronous sexual process. *Plant Cell Physiol.* 19, 1327-1332.

Wickerham, L. J. (1969). Yeast taxonomy in relation to ecology, genetics, and phylogeny. *Antonie van Leeuwenhoek; J. Microbiol. Serol.* 35, Suppl. Yeast Symp., Part 1, 31-58.

Yanagishima, N. (1978). Sexual cell agglutination in *Saccharomyces cerevisiae:* Sexual cell recognition and its regulation. *Bot. Mag., Spec. Issue* 1, 61-81.

Yoo, B. Y., Calleja, G. B., and Johnson, B. F. (1973). Ultrastructural changes of the fission yeast (*Schizosaccharomyces pombe*) during ascospore formation. *Arch. Mikrobiol.* 91, 1-10.

11

Sexual Interactions in *Saccharomyces cerevisiae* with Special Reference to the Regulation of Sexual Agglutinability

NAOHIKO YANAGISHIMA AND KAZUO YOSHIDA

I.	Introduction...	261
II.	Survey of the Literature	263
III.	Current Research......................................	266
	A. What Is Sexual Agglutination?	266
	B. Regulation of Sexual Agglutinability	275
IV.	Perspectives..	289
	References	291

I. INTRODUCTION

The ascosporogenous yeast *Saccharomyces cerevisiae* is a unicellular eukaryotic microorganism. The life cycle of the yeast is shown in Fig. 1. The yeast has two mating types, **a** and α, which are controlled by codominant allelic genes (*MAT*a and *MAT*α) which are located on chromosome 3. In heterothallic strains, **a** and α haploid strains can be obtained by isolating single spores from asci, each of which usually contains 2 **a** and 2 α spores which have been produced by meiosis. These **a** and α haploid cells proliferate mitotically when cultured separately. However, when **a** and α cells are mixed under favorable conditions, the mating reaction occurs. The mating reaction that is thus induced is called a mass mating and consists of four successive stages:

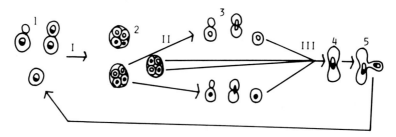

Fig. 1. The life cycle of heterothallic *S. cerevisiae*. I, sporulation; II, single spore isolation; III, mating reaction caused by mixing **a** and α spores or vegetative cells. 1, vegetative diploid cells; 2, asci (spores); 3, vegetative haploid cells; 4, zygote, 5, diploid bud.

(1) sexual cell agglutination; (2) formation of **a** and α pairs; (3) cytogamy; and (4) nuclear fusion (Fig. 2). Finally diploid cells come from the zygotes by budding. Since we can induce the mating reaction at any time under various conditions, using the mass mating system, an analytical assessment of the mating reaction of the yeast has been undertaken.

Although *S. cerevisiae* has the complex cell structure of a eukaryote and a life cycle comparable to higher organisms, molecular biological methods developed mainly with bacteria can be used with the organism. Hence, we can perform analytical studies on the regulation of the expression of sexuality and the mating process, taking advantage of these characteristics of the yeast cells.

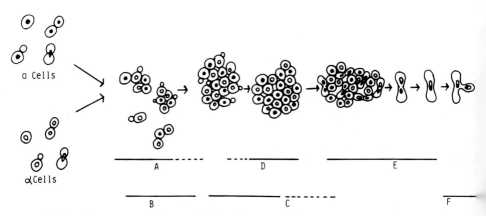

Fig. 2. The sequence of events during the process of mass mating in *S. cerevisiae*. A, induction or enhancement of agglutinability by sex pheromone of opposite mating type; B, formation of small aggregates; C, formation of large aggregates; D, cell cycle arrest in aggregates; E, formation of zygotes; F, formation of diploid bud.

The yeast expresses its sexuality and recognizes the opposite mating type through the sex-specific glycoproteins which are termed agglutination substances (Yanagishima, 1978b). In this chapter, we will deal with the mechanism, the molecular basis, the biological significance, and the regulation of sexual agglutination of *S. cerevisiae*. The strains used in our experiments were derived from the original strains given to us by Dr. D. C. Howthorne, University of Washington, Seattle, and Dr. T. Takahashi, Central Research Laboratory, Asahi Brewery Limited, Tokyo.

We generally cultured cells in YHG medium [5 gm KH_2PO_4, 2 gm $MgSO_4 \cdot 7 H_2O$, 4 gm yeast extract (Difco), 5 gm peptone and 50 gm glucose in 1000 ml distilled water] with gentle shaking (53 76-mm strokes/min) on a reciprocating shaker at 28°C. We also employed two modified YHG media, VHG, which contained traces of vitamins in place of yeast extract, and VHG-BII, which was made by enriching VHG with 40 mg of adenine sulfate and uracil per liter. Synthetic medium was employed only as a selection medium which was used for the genetic analyses, because high production of the agglutination substances, both inducible and constitutive, occurs in such rich media. The high concentration of glucose, gentle shaking, and the incubation temperature, 28°C, are all favorable to both constitutive and induced production of the agglutination substances responsible for sexual agglutination. Phosphate buffer, pH 5.5, 10^{-2} M, was used as the standard buffer (PBS).

II. SURVEY OF THE LITERATURE

Sexual agglutination in ascosporogenous yeast was first observed in *Hansenula wingei* (Wickerham, 1956). After Brock (1958) published the first analytical results on the sexual agglutination of this yeast, the molecular basis of the sexual agglutination was actively studied (Taylor, 1964, 1965; Taylor and Orton, 1971; Yen and Ballou, 1973, 1974; Crandall et al., 1974; Burke et al., 1980). The chemical characterization of the mating type-specific substances responsible for sexual agglutination (agglutination substances) in *H. wingei* has been previously reviewed (Crandall and Brock, 1968; Crandall, 1978). In spite of the progress in the chemical characterization of the agglutination substances in *H. wingei,* only a few studies have been performed on the regulation and biological significance of the sexual agglutination (Crandall and Brock, 1968; Crandall, 1978). Crandall and Caulton (1973) showed that the repression of the production of the agglutination substances in diploid cells was reversed by environmental factors.

This finding may give a clue to the direction future workers should pursue in the analysis of the genetic and physiological regulation of the synthesis of the agglutination substances.

In *Schizosaccharomyces pombe,* the chemical characterization of the agglutination substances has not been performed, although a considerable number of physiological studies on the induction of sexual agglutinability have been carried out. Egel (1971) suggested the importance of the interaction of cells of the opposite mating types, h^+ and h^-, but failed to obtain evidence for the involvement of sex pheromones in the sexual agglutination process. Later Calleja and Johnson (1971) performed physiological analyses of the induction of sexual agglutination in a homothallic strain of *S. pombe,* defining the culture conditions required for the induction process. The details of these studies are described by Calleja, Johnson, and Yoo in Chapter 10. Itoh et al. (1976) have found that a homothallic strain of *Schizosaccharomyces japonicus* requires light for the induction of sexual agglutination which can occur in the presence of yeast extract. This finding may provide a system for the study of photomorphogenesis at the cellular level.

The interaction between opposite mating types during mass mating in *S. cerevisiae* was first observed nearly 40 years ago (Lindegren and Lindegren, 1943). Although since then mass mating has been generally used to obtain hybrids for genetic studies, only a few analytical studies on the regulation of the mating process have been performed. *S. cerevisiae* has a pheromonal system for the regulation of the mating reaction (Levi, 1956; Duntze et al., 1970; Sakai and Yanagishima, 1972). In addition, the genetic and biochemical characterization of *S. cerevisiae* has been pursued more than in any other of the ascosporogenous yeasts. This allows us to analyze more readily the regulation of sexuality and the mating process using this yeast. In 1971, Sakai and Yanagishima first reported the quantitative analysis of sexual agglutination in *S. cerevisiae*. Since then we have been studying the pheromonal regulation of sexual agglutinability. Sakai and Yanagishima (1971, 1972) revealed the involvement of a sex pheromone in the regulation of the sexual agglutinability in **a** cells of *S. cerevisiae* and also found that two types of **a** cells existed: constitutive and inducible. The former strains had sexual agglutinability even when cultured separately, but the latter strains became sexually agglutinable only in response to the sex pheromone of the opposite mating type. Since the sexual agglutination in *S. cerevisiae* is rather weak compared with that of *H. wingei,* it has not been easy to quantitatively express the degree of sexual agglutination in this species. At first, we expressed the degree of agglutination by the ratio of the absorbance at 530 nm of the lower part of an agglutinating cell suspension to that of the upper part after the cell suspension had been left to settle. This method was based on the concept that cell aggregates would sediment more rapidly than single cells (Sakai and Yanagishima, 1972). Later, we used an agglutinating index which was the ratio of the A_{530} of a given agglutinating suspension after sonication to the A_{530} before sonication since sonication dispersed cells in the aggregates completely without disrupting them (Shimoda et al., 1976a).

Sakurai et al. (1976a,b) succeeded in determining the peptide structure of the agglutinability-inducing pheromone produced by α cells called α substance IA. Stötzler et al. (1976) reported the chemical structure of the α factor peptides which specifically inhibited DNA synthesis in a cells. From the results of these two research groups it was shown that these two molecules, which were characterized on the basis of different functions, are identical. Thus, the α mating type peptide pheromone probably has two kinds of actions: the specific inhibition of DNA synthesis and the induction of sexual agglutinability in a cells. A little later, Tanaka et al. (1977) isolated α factor peptides and tried to cause cell agglutination by adding α factor to a cells with negative results by their misunderstanding of our results (Sakurai et al., 1976b). Recently, Betz et al. (1978) have confirmed that α pheromone induced sexual agglutinability of a cells. Yanagishima et al. (1976) discovered the pheromone produced by a cells that induced agglutinability of α cells and called this pheromone a substance I. As mentioned earlier it seemed probable that a substance I was identical with the a factor which inhibited DNA synthesis of α cells (Shimoda and Yanagishima, 1973; Wilkinson and Pringle, 1974). Indeed, recently Betz and Duntze (1979) succeeded in isolating the a factor and clearly showed that the factor had this dual function. This work is discussed in detail in Chapter 2 by Manney et al. In parallel with the above experiments, we have performed a chemical characterization of the sex-specific substances that are responsible for sexual agglutination (Yoshida et al., 1976; Hagiya et al., 1977). At first we tried to solubilize the agglutination substances by treatment of wall fractions with Glusulase (Shimoda et al., 1975). However, Yoshida et al. (1976) and Hagiya et al. (1977) have devised a new autoclave method which solubilized a and α agglutination substances easily and allowed us to prepare the agglutination substances in large quantities in a short time. Yoshida et al. (1976) and Hagiya et al. (1980) have succeeded in the purification of the agglutination substances from autoclave extracts.

The physiological conditions for the constitutive production of the agglutination substances have also been analyzed, and both temperature and the kind of fermentable sugar employed in the culture were shown to be important factors (Yanagishima et al., 1976; Doi and Yoshimura, 1977, 1978; Tohoyama et al., 1979; Nishi and Yanagishima, 1979). Physiologically repressed sexual agglutinability was shown to be reversed by the action of opposite mating type pheromone (Yanagishima et al., 1976; Tohoyama et al., 1979). Carboxypeptidase Y was known to destroy a agglutination substance specifically, although the enzyme was contained in a, α and a/α cells, which suggests that this enzyme is involved in the regulation of sexual agglutinability in a cells (Matsushima et al., 1976; Shimoda et al., 1976b; Shimoda, personal communication). The α substance I (α pheromone)-binding substance has been extracted from a cells and the role of this substance in the induction of sexual agglutinability was proposed (Yanagishima et al., 1977; Shimizu et al., 1977).

It has been found that the mode of action in inducing agglutinability is different for the a and α pheromones (Nishi and Yanagishima, 1978; Yanagishima, 1979). The process of repression of the synthesis of the agglutination substances during the formation of diploid cells through sexual conjugation was also analyzed (Tohoyama et al., 1979). The mating type-nonspecific genes responsible for the regulation of the production of the agglutination substances were found and the genetic and physiological characterization of them was performed in an attempt to reveal the regulatory mechanism of the synthesis of the agglutination substances (Shimoda et al., 1976a; Nakagawa and Yanagishima, 1978; Yanagishima, 1978a; Yanagishima and Nakagawa, 1980). The results of these experiments will be detailed later.

III. CURRENT RESEARCH

A. What Is Sexual Agglutination?

1. Biological Significance

Sexual agglutination is a general phenomenon observed in various eukaryotic microorganisms (Wickerham, 1969). Since sexual agglutination is a sex-specific cell–cell recognition, the mechanism of sexual agglutination must be related to the fundamental mechanism of cell–cell recognition in all eukaryotic cells.

In *S. cerevisiae* we have isolated two single-gene, α mutants which have lost the abilities to undergo sexual agglutination and to form zygotes when mixed with a cells (Matsushima *et al.*, unpublished). In addition, in mating mixtures of wild type a and α cells, zygotes were found almost exclusively in cell aggregates which consisted of approximately equal numbers of a and α cells irrespective of the a to α ratio at the initial mixing (Kawanabe *et al.*, 1979). Although these results indicate the necessity of sexual agglutination for the formation of zygotes, there are some experimental results which seem to be inconsistent with this suggestion. Sena *et al.* (1973) found cell pairs at the early stage of mating reaction before the formation of detectable cell aggregates and Fehrenbacher *et al.* (1978) reported that mating type-specific adhesion of a–α cells took place shortly after the cells were mixed. These results although inconsistent do not disprove the necessity of agglutination for zygote formation since the results were observed after centrifugation of mating mixtures, which artificially forces cells of opposite mating types to adhere to each other. Furthermore, even in the case where cell aggregates are formed through random assembly of a and α cells, transient sexual pairing can occur at the initiation of the formation of cell aggregates. We have already proposed that a and α cells form cell aggregates through random assembly without forming a–α cell pairs before the formation of cell aggregates (Kawanabe *et al.*, 1979). Since concanavalin A (400–500 μg/ml) completely inhibits the formation of large aggregates with little inhibition of the formation of zygotes, large cell aggregates (> 100 cells) are not necessary for zygote formation. On the other hand, the formation of small aggregates (5–50 cells) seems to be necessary for the formation of zygotes (Kawanabe *et al.*, 1978). Hence, the intensity of sexual agglutination is not necessarily correlated with frequency of the occurrence of zygotes. From the above facts the event of sexual agglutination is thought to give heterothallic haploid cells the chances to come directly into contact with a cell of the opposite mating type, and more important to set up physiological conditions favorable to the formation of zygotes, e.g., the local retention of a high concentration of sex pheromones necessary for cell cycle arrest (Fig. 2). Wickerham (1969) reported that many heterothallic ascosporogenous yeasts showed sexual agglutination when opposite mating type cells were mixed. It is of interest then that this sexual cell–cell recognition is not only mating type-specific but also species-specific. Hence, it may be possible to gain some insight into phylogenic relationships from the degree of sexual agglutination between cells of different species. Furthermore, we may be able to determine the biological affinity on the basis of the chemical nature of the agglutination substances. The autoclave method that we have devised allows us to perform such experiments.

2. Quantitation

In order to measure the degree of the sexual agglutinability of a given cell population, we have to mix them with opposite mating type cells which reveal a high degree of sexual agglutinability. When we use living cells for the measurement of sexual agglutinability, it is impossible to know the sexual agglutinability of a given sample *in situ*, because the sexual agglutinability of a strain is affected by environmental factors and is especially sensitive to the sex pheromone of the opposite mating type. For example, in the case of inducible a cells, α sex pheromone (α substance I) shows significant induction of sexual agglutinability in 20 min at a concentration of 1 ng/ml. Indeed, if we use living cells, we can hardly detect a difference in sexual agglutination between inducible strains and constitutive ones. However, some authors (Manney and Meade, 1977; Fehrenbacher *et al.*, 1978) feel that our heat-treatment method might give anomalous results. We have already reported that the heat treatment (5 min at 100°C in PBS) has practically no effect on the sexual agglutination reaction of the treated cells (Yoshida and Yanagishima, 1978; Tohoyama *et al.*, 1979) (Fig. 3; see also Fig. 6) and the isolated agglutination substances are also resistant to the heat treatment (Hagiya *et al.*, 1977). By using the heat treatment, we could discriminate inducible strains from constitutive ones, by measuring the *in situ* sexual agglutinability of a given sample. Typical examples of a sexual agglutination of intact and heat-killed cells of *H. wingei*, *S. cerevisiae*, and *S. pombe* are shown in Fig. 3. The order of the intensity of sexual agglutination of these three yeasts is *H. wingei*, *S. cerevisiae*, *S. pombe*. Since the sexual agglutination of *S. cerevisiae* is weaker than that of *H. wingei*, we have to slowly shake the mixed culture.

In order to express the intensity of the sexual agglutination reaction, we have been using an agglutination index (AI_x) which is a ratio of attenuance at x nm after and before sonication of the agglutination assay mixtures. AI_x is a ratio of attenuance of the nonagglutinating cell suspension to that of the agglutinating cell suspension, since sonication almost completely disperses the cells in the aggregates. For convenience, we use either AI_{260} or (mostly) AI_{530}. When the absorption spectrum (210–370 nm) of reaction mixture (e.g., 5 and 21 cells of *H. wingei* or a and α of *S. cerevisiae*) was scanned at successive intervals after the initiation of the mixture, the absorbance (or attenuance) of

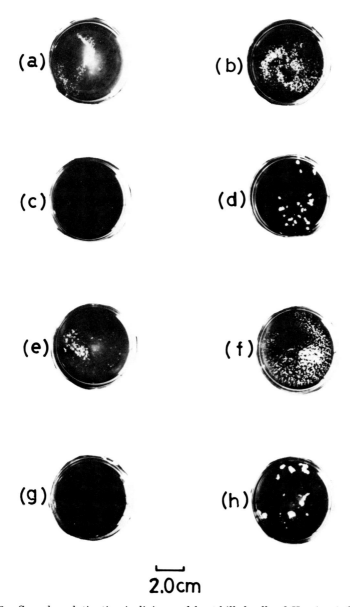

Fig. 3. Sexual agglutination in living and heat-killed cells of *H. wingei*, *S. cerevisiae*, and *S. pombe* (Yoshida and Yanagishima, 1978). The same numbers of cells of opposite mating types or diploid cells were mixed in PBS and shaken at 28°C. After shaking for 3 hr, photographs of the reaction mixtures were taken. Living cells were used in (a), (b), (c), and (d). In (e), (f), (g), and (h), boiled cells were used. (a), (e), *S. pombe* M216(h^+)+L972(h^-); (b), (f), *S. cerevisiae* T27(a^c)+T26(α^c); (d), (h), *H. wingei* 5+21; (c), *S. cerevisiae* T27(a^c)+diploid cells D8; (g), *H. wingei* 21 cells only.

the mixture decreased with time. This attenuance of the reaction mixture decreases concomitantly with the progression of aggregate formation. As a result, we tried to establish a quantitative relationship among absorbance, AI and the mean radius ($\bar{R}$) of cell aggregates in the agglutinating mixture (Yoshida and Yanagishima, 1978). $\bar{R}$ is one of the parameters which express the intensity of agglutination. AI and $\bar{R}$ were measured with a photometer and an image analyzing computer, respectively. The obtained experimental data were plotted in Fig. 4 (see ○ in the figure). The functional equation between AI and $\bar{R}$ was obtained with the aid of computer analysis and on the basis of the flattening effect theory, first introduced by Duysens (1956) in the analysis of *Chlorella* cells. He found that the absorption band of aggregates of pigment particles or light-absorbing cells was flattened compared with the absorption band of the same concentration of the pigments or cells when dispersed in solution. He named this phenomenon the flattening effect and decided that it was

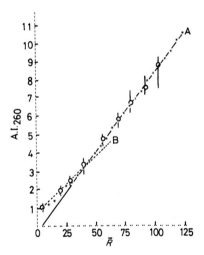

Fig. 4. Plotting of the relationship between AI_{260} and mean radius $\bar{R}$ of sexual aggregates with a computer (Yoshida and Yanagishima, 1978). The mean radius $\bar{R}$ (μm) of sexual aggregates was estimated with an image analyzing computer (Shimadzu-Bausch & Lomb, Omnicon PAS system). The $\bar{R}$ and photometrically estimated AI_{260} were plotted (○). Vertical bars indicate deviations. The equation

$$AI_{260} = \frac{k(\bar{R}-r)}{1 - e^{-k(\bar{R}-r)}}$$

was plotted with a minicomputer (Nova 2, Data General) (∗) at $k = 0.088$ and $r = 4$ μm. The best fitted equations for the curves A (---) and B (-----) were obtained by the least square method with the computer.

11. Sexual Interactions in Saccharomyces cerevisiae

caused by the heterogenous distribution or localization of pigments in particles. Through the quantitative analysis of the flattening effect, Itoh et al. (1963) successfully estimated the size of spinach chloroplasts and the action spectra of grana. The yeast cell suspension reveals an absorption optimum around 260 nm (Yoshida and Yanagishima, 1978). This indicates that the yeast cells can also be regarded as particles containing pigments just like the grana of chloroplasts. In other words, sexual agglutination may be quantified on the basis of the flattening effect theory. We have introduced the following Eqs. (1) and (2). Equation (3) was introduced modifying the equations from Itoh et al. (1963) (see Yoshida and Yanagishima, 1978).

$$AI_{260} = \frac{E_c}{E_a} \tag{1}$$

$$= \frac{1}{\text{flattening coefficient}} \tag{2}$$

$$= \frac{k(\bar{R}-r)}{1-e^{-k(\bar{R}-r)}} \tag{3}$$

where E_a, E_c, $\bar{R}$, k, and r indicate the extinction of the aggregate-forming solution, the extinction of the cell suspension, the mean radius of aggregates, a constant, and the mean radius of cells, respectively. The mean radius of aggregates and that of cells were measured with the image analyzing computer and the data was calculated [Eq. (3)] and was plotted with a computer as shown in Fig. 4 (see ∗). The calculated values (∗) match the experimental values (○) quite well. This indicates that the model of aggregate formation we have described is very useful for the quantitative analysis of agglutination according to the flattening effect theory. When $\bar{R}$ is large enough ($\bar{R} > 40$ or $AI_{260} > 3.4$), Eq. (3) becomes close to curve A, $AI_{260} = 0.0869 \bar{R} - 0.15$. When $\bar{R} < 40$ or $AI_{260} < 3.4$, Eq. (3) becomes similar to curve B, $AI_{260} = 0.0645 \bar{R} + 0.85$. Agglutination of $H.$ $wingei$ is roughly expressed by curve A. On the other hand, agglutination of $S.$ $cerevisiae$ and $S.$ $pombe$ is roughly expressed by curve B since their AI_{260} is almost always below that of $H.$ $wingei$ even under the best conditions. The other parameter to express the intensity of agglutination, mean cell number per aggregate ($\bar{N}$), was roughly estimated by the equation $\bar{N} = (4/3)\pi \bar{R}^3 (4/3)\pi r^3 = (\bar{R}/r)^3$. The analysis of this parameter supported the above relationship between AI and $\bar{R}$ (Yoshida and Yanagishima, 1978).

The relationship between AI_{260} and AI_{530} is expressed as follows:

$$AI_{260} = 1.199 \, AI_{530} + 0.479 \tag{4}$$

k in Eq. (3) and constant parameters in Eq. (4) are changeable depending on the assay conditions and photometer used (Yoshida and Yanagishima, 1978).

3. Molecular Basis

To solubilize the mating type-specific cell wall substances responsible for sexual agglutination, two kinds of methods (enzymatic and autoclaving) have been used. At first, cell wall fractions obtained by disrupting cells with a Vibrogen Cell Mill were treated with the snail gut enzyme, Glusulase, followed by heating at 100°C for 5 min (Shimoda et al., 1975). The resulting enzyme digests were known to have a masking action on cells of the opposite mating type, indicating that both a- and α-specific substances are univalent (Shimoda and Yanagishima, 1975; Shimoda et al., 1975). However, even the partially purified substances released by the enzymatic method had extremely high molecular weights near 1×10^6 daltons. The enzymatically released substances probably consisted mainly of nonfunctional wall components which caused a difficulty in further purification of the substances and in the detection of molecular relationship between a and α agglutination substances.

In order to overcome these problems, we devised a convenient autoclaving method (Yoshida et al., 1976; Hagiya et al., 1977). Cells were washed and suspended in PBS buffer, then boiled for 5 min, followed by immediate chilling in crushed ice. The cell suspensions were put in an autoclave (HL 36, gas operating type Hirayama, Tokyo) after which the pressure was quickly raised to 1 kg/cm² (about 120°C) and kept for 3 min; the pressure was quickly lowered by opening the exhaust valve. Then, the cell suspension was immediately (within 2 min) transferred to crushed ice. If rapid autoclaving and quick chilling steps are not adhered to, the released agglutination substances are inactivated. Therefore, the electric type of autoclave is unsuitable for this method. The autoclaved cell suspension was centrifuged to remove cells. The resultant supernatants containing respective agglutination substance were used as the starting materials for further purification. The length of time for autoclaving is a critical condition, especially for the release of α agglutination substance. At first, we added urea to the autoclaving buffer to prevent the aggregate formation of agglutination substances themselves (Yoshida et al., 1976). However, urea itself was converted to an agglutination-inhibitory substance by prolonged autoclaving (in longer than 10 min). Thereafter, we omitted urea from the autoclaving buffer (Hagiya et al., 1977). Our results indicated that only haploid strains released agglutination substances when the autoclave method was applied to various strains of *S. cerevisiae*. This method has been shown to be applicable to yeast species other than *S. cerevisiae*, such as *S. globosus, S. kluyveri, H. wingei, H. holstii,* and *H. fabianii* (Yoshida *et al.*, 1980). We have also found agglutination substances in cytoplasms of **a** and α cells (Hagiya *et al.*, 1977). The substances released by autoclaving primarily came from the cell surface, since cytoplasmic agglutination substances were almost completely recovered from the autoclaved cells by disrupting with the Vibrogen Cell Mill (Hagiya *et al.*, 1977). Nevertheless, to completely avoid contamination by cytoplasmic substances, extensive washing of the cell walls should be carried out.

In order to purify the agglutination substances, the supernatants of autoclaved cell suspensions were subjected successively to the following: (a) acid precipitation of contaminating materials, (b) DEAE-cellulose column chromatography, (c) ultrafiltration, (d) gel filtration, (e) affinity chromatography on Con A–Sepharose, and (f) isoelectrofocusing. The purified substances showed single band on polyacrylamide gel electrophoresis (Yoshida *et al.*, 1976; Hagiya *et al.*, 1980). Important characteristics of **a** and α-agglutination substances thus purified are summarized in Table I.

TABLE I

Characterization of Agglutination Substances Released from *S. cerevisiae* Using the Autoclave Method

	Agglutination substances	
	a	α
Molecular weight[a]	23,000	130,000
Carbohydrate content (%)[b]	61	47
pI[b]	4.5	4.3
Biological activity	\multicolumn{2}{l}{Masking action of sexual agglutinability of opposite mating type cells; formation of a molecular complex with opposite mating type agglutination substance.}
Binding activity	Univalent	
Protein character	Glycoprotein	

[a] Estimated by gel filtration.
[b] Approximate values.

Since it was possible that the a- and α-agglutination substances were in fact complementary molecules, tests were carried out to assess this. In order to detect if such molecular complex as resulted (i.e., aAS + αAS → a-αAS complex, AS = agglutination substance), partially purified (by lowering pH) a and α substances released by the autoclave method were mixed and this mixture was then successively shaken with a and α cells to remove free α and a substances but leave any a-α agglutination substance complexes. After confirmation of the lack of the biological activity in the mixture, the pH of the mixture was shifted to pH 9.5 in order to break linkages of the a-α agglutination substance complex. The mixture was then applied onto a DEAE-cellulose column. As shown in Fig. 5, the activity of a and α agglutination substances was recovered on the corresponding positions at which

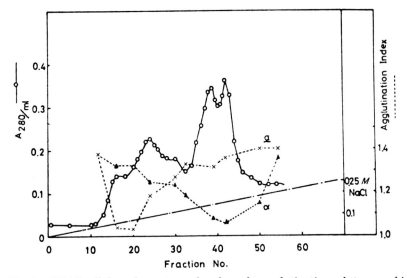

Fig. 5. DEAE-cellulose chromatography of a and α agglutination substances which were recovered from a-α agglutination substance complex (Yoshida et al., 1976). The partially purified a and α agglutination substances released by the autoclave method were mixed and shaken to form a complex. After shaking, the mixture was successively shaken with a and α cells to remove free substances. After confirmation of the lack of biological activity in the mixture, the pH of the mixture was shifted to 9.5 in order to break linkages of the a-α agglutination substance complex. Then the mixture was applied onto DEAE-cellulose chromatography. Biological activity of each fraction was assayed and expressed by masking action on opposite mating type tester cells, the lowest agglutination index, being the highest activity of each agglutination substance. ○, absorbance at 280 nm; ×, activity of a agglutination substance which is expressed in terms of agglutination index; ▲, activity of α agglutination substance in terms of agglutination index; •, NaCl gradient.

independently applied a or α agglutination substance was eluted. The fact that the biological activity of both a- and α agglutination substances disappeared in mixing but reappeared after subsequent treatment indicates that a molecular complex between a and α agglutination substances was actually formed *in vitro*. This finding strongly supports the notion that the agglutination substances released from a and α cell surfaces are directly involved in the sexual agglutination process. We have also detected the molecular complex of a and α agglutination substances on an affinity column (Hagiya *et al.*, 1980).

Recently, we have observed that agglutination substances which had been partially purified (by DEAE-cellulose chromatography followed by ultrafiltration) after extraction by the autoclave method specifically inhibited not only sexual agglutination, but also zygote formation in VHG-BII medium, when we pretreated living a and α cells with the opposite mating type agglutination substance before mixing (Hagiya *et al.*, 1978). These results indicate that the sexual agglutination substances released by the autoclave method play an important role not only in the sexual agglutination but also in the formation of zygotes.

B. Regulation of Sexual Agglutinability

In order to have insight into the regulation mechanism of sexual agglutinability, it is necessary to know the relationship between genetic control and physiological regulation. Since the physiological regulation system is complicated in eukaryotes, it has been difficult to find the relationship between genetic control and physiological regulation in eukaryotic systems. *Saccharomyces cerevisiae* is one of the most suitable eukaryotic materials for the study of the mechanism of regulation of complicated biological processes such as the expression of sexuality and the ability to recognize opposite mating type cells.

1. Physiological Regulation

First, we found, in our stock strains, two a strains which were inducible for sexual agglutinability (Sakai and Yanagishima, 1972; Shimoda *et al.*, 1976a). These inducible a strains have been used for the physiological and biochemical studies on α pheromone (Yanagishima, 1978b). We found that the a and α strains, T55(a) and T56(α), derived from a diploid strain by single spore isolation, produced mutants inducible for sexual agglutinability. We designated the indicible mutant from T55(a^c) as T55s-41(a^i) and that from T56($α^c$) as T562s-161($α^i$). These inducible mutants have been known to carry the same inducible gene, *saa1*, that will be described in detail later. The superscripts c and i given to the mating type symbols a and α indicate constitutive and inducible agglutinability, respectively. None of the inducible strains produced detectable amounts of agglutination substance when cultured at 28°–30°C, the optimum growth temperatures (Shimoda et al., 1976a; Yanagishima and Nakagawa, 1980). Boiled cells of these strains, contrary to cells of the constitutive strains, do not show sexual agglutination when mixed with boiled tester cells of respective opposite mating type which have high sexual agglutinability. However, when living cells of the inducible strains are mixed with living tester cells of the opposite mating type in the presence of nutrients (nitrogen and carbon sources), sexual agglutination takes place. These results indicate that some stimuli coming from the cells of the opposite mating types induces sexual agglutinability in these inducible strains. The stimuli have recently been identified as the same peptide pheromones as those causing cell cycle arrest of the opposite mating type.

a. Culturing Conditions. Incubation with strong aeration or in the absence of fermentable sugars results in the loss of sexual agglutinability even in constitutive strains (Yanagishima et al., 1976). On the contrary, respiration deficiency caused by a cytoplasmic mutation did not result in the loss of sexual agglutinability (Yanagishima, unpublished results). The sexual agglutinability that is lost by incubating without fermentable sugars is recovered by sex pheromone of opposite mating type (Yanagishima et al., 1976).

Incubation temperature has a distinct effect on sexual agglutinability in the constitutive strains T55(a^c) and T56(α^c) such that above 28°C the agglutinability decreases as the temperature increases (Tohoyama et al., 1979). Incubation at a temperature higher than 35°C results in the complete loss of sexual agglutinability in both a^c and α^c strains. In Figs. 6 and 7, the sexual agglutination of T55 and T56 cells cultured at 28° and 36°C is shown. In the 36°C grown cells, only living cells showed sexual agglutination, as the result of induction after mixing with opposite mating type cells. When the cell extracts from cell wall and cytoplasm were tested for their ability to mask the sexual agglutinability of tester cells it was apparent that both cell wall agglutination substances and cytoplasmic ones were lost in cells incubated at 36°C. Although T55s-41(a^i) and T562s-161(α^i) produce no detectable amount of agglutination substances when cultured at 28°C, these strains show sexual agglutinability, that is, they constitutively produce agglutination substances when cultured at 22°C. In both the inducible and constitutive strains mentioned above, the temperature-repressed sexual

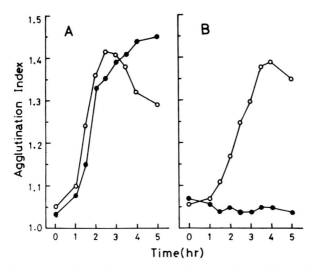

Fig. 6. Time course of sexual agglutination of living and heat-killed cells cultured at 28° and 36°C (Tohoyama *et al.*, 1979). Cells of T55 (a^c) and T56(α^c) cultured until the middle logarithmic phase at 28°C (A) or 36°C (B) were washed, mixed in YHG before or after boiling, and shaken at 28°C. Samples were taken at intervals and agglutination indexes of them were measured. Living cells (○); boiled cells (●).

agglutinability is derepressed by the sex pheromone of the opposite mating type. Hence, it is concluded that T55(a^c) and T56(α^c) are constitutive at 28°C and inducible at 36°C, while T55s-41(a^i) and T562s-161(α^i) are constitutive at 22°C and inducible at 28°C.

We have isolated some inducible mutants which are inducible at all the incubation temperatures examined (Nakagawa and Yanagishima, 1979). We found that when cultured at 28°C in the presence of Triton X-100 T55(a^c) and T56(α^c) showed only low level of sexual agglutinability. The Triton X-100-repressed sexual agglutinability was also recovered by the treatment with opposite mating type sex pheromones. The temperature dependency and the effect of Triton X-100 suggest that the cell membrane system plays an important role in the constitutive production of the agglutination substances in T55 and T56.

In spite of the diversity of physiological factors representing sexual agglutinability, the repressed sexual agglutinability is derepressed by the action of the sex pheromones of the opposite mating type (Yanagishima *et al.*, 1976; Tohoyama *et al.*, 1979), indicating that the pheromonal regulation of sexual agglutinability is most fundamental in the physiological regulation of sexual agglutinability. In Fig. 8, the

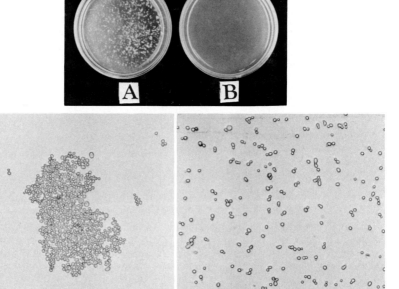

Fig. 7. Sexual agglutination of cells cultured at 28° and 36°C (by courtesy of H. Tohoyama). Cells of T55(a^c) and T56(α^c), cultured at 28°C (A) and 36°C (B) were boiled, mixed in PBS buffer, and shaken at 28°C for 2 hr to measure agglutination index. (A) gave an agglutination index of 1.42 and (B) that of 1.07. (C) and (D): microphotographs of (A) and (B), respectively.

action of the sex pheromones in relation to inducibility of cells is shown schematically.

Changes in sexual agglutinability during the cell cycle were observed with synchronous cultures of T55 and T56. As shown in Fig. 9, small immature cells including buds have lower agglutinability than mature large cells in a mating type (T55), but the difference in sexual agglutinability is not so significant in α mating type (T56) (Kawanabe et al., 1979). In general, α cells are known to be more stable than a cells concerning sexual agglutinability.

b. Pheromonal Regulation. Induction of sexual agglutinability by sex pheromones seems to play an essential role in the first step of the

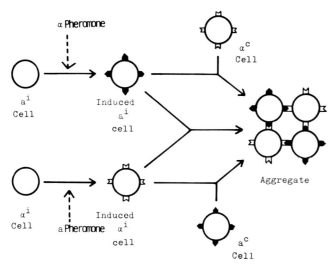

Fig. 8. Schematic expression of induction of sexual agglutinability by sex pheromones. ●, a agglutination substance; ⋈, α agglutination substance. (From Yanagishima, 1978, with modification.)

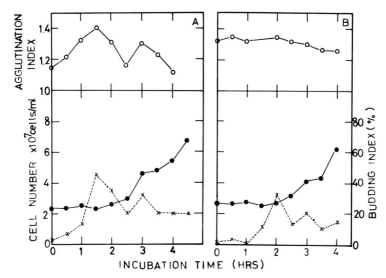

Fig. 9. Changes in sexual agglutinability during synchronous culture (Kawanabe et al., 1979). Small cells without buds separated by sucrose density gradient centrifugation were washed with PBS buffer, inoculated into YHG at a cell density of about 10^7 cells/ml, and shaken. At intervals, cell number (●), budding index (×), and the ability to agglutinate with tester cells (agglutinability) (○) were measured. Agglutinability was measured after boiled and expressed in terms of the agglutination index. A, T55(a^c) cells; B, T56($α^c$) cells.

mating reaction. From the theoretical point of view, the induced synthesis of agglutination substances provides a good model for the analysis of the action mechanism of hormone-like substances. The pheromones which induce cell cycle arrest are called α factor and a factor (Duntze et al., 1970; Wilkinson and Pringle, 1974) and those inducing sexual agglutinability are called α substance I and a substance I (Sakai and Yanagishima, 1972; Yanagishima et al., 1976). Since these two kinds of actions are known to be due to the dual action of the same substances, we will use the terms a and α pheromones to express a factor or a substance I, and α factor or α substance I, respectively.

i. Mode of Action of Sex Pheromones. It is of importance that the time necessary for the maximum induction of sexual agglutinability is about 1 hr for α pheromone and about 2 hr for a pheromone (Fig. 10), irrespective of concentrations of pheromones and strains used (Nishi and Yanagishima, 1978; Tohoyama et al., 1979). Both the pheromones needed carbon and nitrogen sources for the induction of sexual

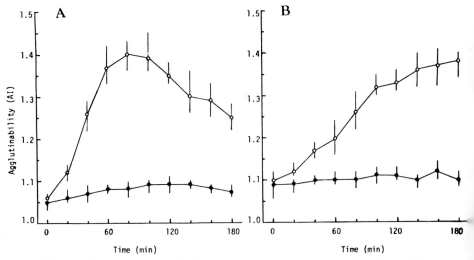

Fig. 10. Time course of induction by sex pheromones of sexual agglutinability in cells cultured at 36°C (by courtesy of H. Tohoyama). Treated with (○) or without (●) opposite mating type pheromone. Cells of (A) T55(a^c) or (B) T56($α^c$) cultured at 36°C were treated with α pheromone or a pheromone (each about 25 units/ml) in YHG at 28°C. At intervals treated cells were taken, boiled, and examined for sexual agglutinability by mixing with boiled tester cells of opposite mating type with high agglutinability after boiling. One unit of each pheromone was defined as the minimum amount of each pheromone which caused visually detectable sexual agglutinability of 10^7 cells of inducible strains of the opposite mating type. A. T55(a^c). B. T56($α^c$).

TABLE II

Comparison of a and α Pheromones[a]

Pheromone	Approximate time till maximum induction (hr)	Mating type-specific adsorption to opposite mating type cells	Mating type-specific inactivation by opposite mating type cells	Requirement for induction	Cycloheximide inhibition (1 μg/ml)
α	1	+	+	Nutrients and[b] physiological temperature	+
a	2	−	−[c]	Nutrients and physiological temperature	+

[a] +, Detected; −, not detected.
[b] Glucose and peptone, 28°C.
[c] A slight inactivation may occur.

agglutinability and the inducing action was inhibited by the addition of cycloheximide (1 μg/ml) or by cold temperature (0°C), indicating that metabolic activity, especially protein synthesis, is needed for the induction (Shimoda et al., 1976a; Yanagishima, 1978a). In Table II, the characteristics of the a and α pheromones are compared. We observed changes in autoclave-extractable or Glusulase-solubilizable agglutination substances by the action of the sex pheromones. Inducible cells or physiologically repressed cells treated with sex pheromone of the same or opposite mating type were subjected to the extraction of agglutination substances and the extracts were tested for biological activity of agglutination substances to mask agglutinability of opposite mating type tester cells. In these experiments, a and α agglutination substances were produced in response to α and a pheromone, respectively (Shimoda et al., 1976a; Tohoyama et al., 1979; Yanagishima and Nakagawa, 1980). It is of note that agglutination substances in both cell wall and cytoplasm fractions are produced in response to the opposite mating type pheromones. The preliminary results suggest that the production of soluble agglutination substances in the cytoplasm is induced first and then the appearance of the cell wall agglutination substances follows (H. Tohoyama, personal communication). Thus, it appears that the synthesis of the substances occurs in cytoplasm and then they are incorporated into the cell wall. The degree of the induced sexual agglutinability expressed by the agglutination index in the inducible a strain, T55s-41(a^i), was found to be proportional to the logarithm of the concentrations of α pheromone used for the induction (Yanagishima et al., 1977). Hence, we can estimate the relative amount of α pheromone from the agglutination index of the induced a^i cells.

ii. Mechanisms of Agglutinability-Inducing Action. In a cells carrying the inducible gene, *saa1* (see Section III,B,2,c), sexual agglutinability is not induced by α pheromone at 38°C, but it is induced at 28°C (Nishi and Yanagishima, 1979; Yanagishima and Nakagawa, 1980). To determine the temperature-sensitive period during the induction, temperature shift experiments were performed (Nishi and Yanagishima, 1979). In the following experiment, the induction was initiated by transferring α pheromone-absorbed cells to the nutrient medium, VHG, to separate the absorption process from the induction process. At various intervals after the initiation of induction of sexual agglutinability by VHG incubation, the temperature was shifted from 28°C to 38°C or from 38°C to 28°C. The cells kept at 38°C for 5 min did not show any induced sexual agglutinability after incubation at 28°C for 1 hr, but the cells kept for 10 min at 28°C and transferred to 38°C showed sexual agglutinability as intensively as cells incubated continuously at 28°C. These results indicate that the temperature-sensitive period occurs within 10 min after transference to VHG, suggesting that the induction of sexual agglutinability is triggered during the first 5 min and the completion of the temperature-sensitive event occurs over the next 5 min, although the maximum induction was observed after 1 hr. Cycloheximide inhibits the induction of sexual agglutinability at 28°C only when added to VHG containing α pheromone within 20 min after the addition of a^i cells, indicating that the protein(s) necessary for the induction was synthesized during the first 20 min (K. Nishi and N. Yanagishima, unpublished). It is highly probable that during the temperature-sensitive period the trigger is pulled for the synthesis of the protein(s) necessary for the induction process. It is interesting that when the T55s-41 (a^i) strain is treated with 1 M sorbitol or 0.2% Triton X-100 followed by washing with PBS buffer it becomes insensitive to the agglutinability-inducing action of α pheromone (Yanagishima and Nakagawa, 1980). These results suggest that the membrane system plays an important role in the induction of sexual agglutinability by α pheromone in a cells carrying *saa1* gene.

Yanagishima et al. (1977) have found that not only a cells but also cell-free culture medium of a cells inactivate α-pheromone activity and the inactivating action was observed in both inducible and constitutive a cells. Even when a cells were suspended in α pheromone at 0°C, significant inactivation of the pheromone was observed in 30 to 60 min. Hence, it is possible that **a** cells take up α pheromone by binding the pheromone.

To confirm this possibility, we have tested the pheromone-inactivating action of boiled **a** cells (constitutive and inducible) with positive results (Yanagishima *et al.*, 1977). Boiled or living T55(**a**^{c}) or T55s-41 (**a**^{i}) cells were incubated with α pheromone for 1 hr, then the **a** cells were removed and the remaining solutions were tested for α pheromone activity on the basis of their sexual agglutinability-inducing action on the **a**^{i} strain. In all cases the agglutination index was reduced by this procedure, indicating that the **a** cells take up α pheromone (Yanagishima *et al.*, 1977). We further found that the biological activity of α pheromone inactivated by treatment with cell-free culture medium of **a** cells was recovered by 5 min heat treatment at 100°C, significantly though not completely (Yanagishima *et al.*, 1977). The heat treatment did not inactivate α pheromone but inactivated the activity of the culture medium of **a** cells to inactivate the pheromone. Thus, the inactivating action of the culture medium of **a** cells was, at least partially, attributed to the binding action of the culture medium of **a** cells. We have further succeeded in extracting the inactivating substance from both inducible and constitutive cells of **a** mating type by hot water (Yanagishima *et al.*, 1977). The extracted inactivating substance was partially purified by DEAE-cellulose column chromatography (Shimizu *et al.*, 1977). We call the inactivating factor, which was present in the cell extract and the culture medium of **a** cells, "binding substance." The biological activity of the α pheromone inactivated by the partially purified binding substance from the **a** cell extract was recovered by the heat treatment at 100°C for 5 min, significantly though incompletely (Shimizu *et al.*, 1977). The partially purified binding substance is protenacious.

When **a** cells were suspended in PBS buffer containing α pheromone at 0°C for 5 min followed by washing, the induction of sexual agglutinability occurred after 1 hr incubation of the treated cells in VHG at 28°C, suggesting the physiological role of the binding substance for α pheromone (Yanagishima and Inaba, 1980). Next, the characteristics of the binding substance for **a** pheromone were studied (Yanagishima, 1979). Although α cells lowered the biological activity of **a** pheromone specifically under a condition where cell growth was allowed, the decrease in activity was slower and less when compared with the inactivation of α pheromone by **a** cells. In addition, the decrease of **a** pheromone activity caused by the α cells was not observed at 0°C or in the absence of nutrients. The cell extract obtained from α cells by the

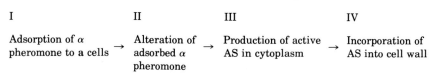

Fig. 11. Proposed sequential events in the induction of sexual agglutinability of the inducible a strain, T55s-41, by α pheromone. AS, agglutination substance.

same method used for the extraction of the binding substance for α pheromone from a cells had no inactivating action on a pheromone (Yanagishima, 1979). These results do not necessarily mean that there is no binding substance for a pheromone, but at least it appears that α cells do not produce a binding substance for a pheromone which is comparable to the binding substance for α pheromone from a cells. Although inactivation of α pheromone by a cells was reported, no binding substances have been shown (Hicks and Herskowitz, 1976; Chan, 1977). Recently, it has been reported that a cells inactivated the α pheromone by enzymatic degradation (Maness and Edelman, 1978; Finkelstein and Strausberg, 1979; Ciejek and Thorner, 1979). The relationship between the binding substance that we proposed and these findings has not yet been made clear. We are now trying to purify the binding substance from a cells in an attempt to characterize their chemical nature, in the hope of clarifying this relationship.

From the above results, it is possible to propose an action process of α pheromone on the inducible mutant carrying *saa1*, as is summarized in Fig. 11.

TABLE III

Comparison between Agglutinability-Inducing Action and Cell Cycle-Arresting Action of α Pheromone

Action	Induction after washing of treated cells	Cell cycle dependency	Time for[a] action	Concentration[a] needed
Agglutinability inducing	Yes[b]	No[b]	Short	Low
Cell cycle-arresting	No[c]	Yes[c]	Long	High

[a] A random population of H1-0(a^i) was used. Induced agglutinability was detected within 20 min and reached maximum after 1 hr at 1 ng/ml α pheromone (the optimum concentration for the induction). Only 9% inhibition of DNA synthesis was detected after 1 hr at 1 μg/ml α pheromone and 40% inhibition after 2 hr. Purified α substance IA (α2) peptide was used as α pheromone.

[b] T55s-41 (a^i) was used.

[c] Generally accepted conclusion (Bücking-Throm *et al.*, 1973; Hereford and Hartwell, 1974).

iii. Dual Action of Sex Pheromones. The optimum concentration of purified α pheromone for the induction of sexual agglutinability of an inducible a strain was 1 ng/ml and the inducing action was detected in 20 min at this concentration. On the other hand, DNA synthesis of the same strain was inhibited by 9% in 1 hr and 40% in 2 hr at 1 μg/ml (Shimoda *et al.*, 1978). In addition, the induction of sexual agglutinability occurred independent of stage of the cell cycle (Yanagishima and Nakagawa, 1980), but the inhibition of DNA synthesis was observed only before the initiation of DNA synthesis (Hereford and Hartwell, 1974). As already mentioned, 5 min of contact with α pheromone at 0°C is enough to cause the induction of sexual agglutinability after 1 hr of subculturing in growth medium at 28°C (Yanagishima and Inaba, 1980). On the contrary, in the case of inhibition of DNA synthesis washing of pheromone-treated cells results in the recovery of the inhibition (Bücking-Throm *et al.*, 1973; Hereford and Hartwell, 1974). These results indicate a difference in the action mechanism for the two functions of α pheromone. In Table III these two functions are compared.

2. Genetic Control

a. Mating Type Locus. Since sexual agglutination is a sex-specific phenomenon, the mating type alleles must have the fundamental regu-

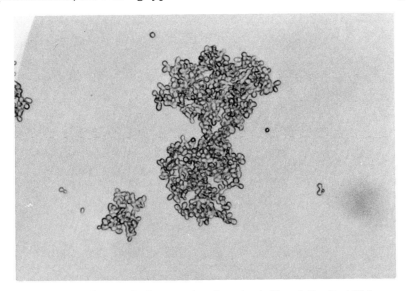

Fig. 12. Microphotograph of aggregates of zygotes (self-agglutination) (Tohoyama *et al.*, 1979). Zygote fraction obtained from the 3.5-hr-old mating mixture of T55(a^c) and T56(α^c) cells by sucrose density gradient centrifugation after boiling was shaken for 2 hr in PBS.

latory role in controlling the sexual agglutinability. We have confirmed that, although diploid cells carry the information for both a and α agglutination substances, no sexually specific agglutination substances are detected in diploid cells (Hagiya et al., 1977). We have observed changes in sexual agglutinability during sexual conjugation in connection with changes in agglutination substances, to find a clue toward knowing the role of the mating type alleles in the regulation of the production of the agglutination substances (Tohoyama et al., 1979). As shown in Fig. 12, significant self-agglutination was observed even

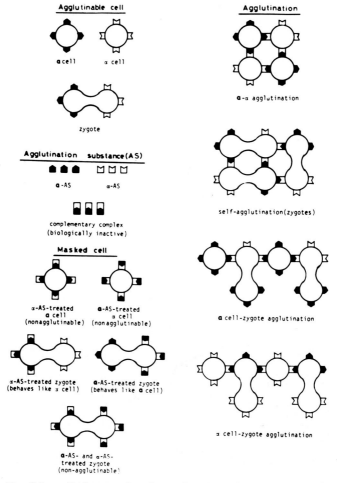

Fig. 13. Schematical expression of sexual agglutination of a and α cells and zygotes (Tohoyama et al., 1979). Zygotes treated with a and α agglutination substances behave like a and α cells, respectively. Symbols as in Fig. 8.

in zygotes. To confirm that the sexual agglutination in zygotes is brought about through the complementary binding of the agglutination substances, the effect of haploid cells of opposite mating type, a and α agglutination substances, and some substances which destroy a or a and α agglutination substances on the self-agglutination of zygotes was studied (Tohoyama et al., 1979). The results can be explained as shown in Fig. 13. Finally, the presence of the agglutination substances in zygotes was confirmed by the extraction of agglutination substances (Fig. 14). When the a–α agglutination substance linkage was broken by shifting the pH (see legend to Fig. 5), the recovery of both a and α agglutination substances was possible, indicating the presence of the agglutination substances in zygotes. The above facts suggest that the agglutination substances in the cell walls of a and α cells were retained even after the formation of zygotes. Tohoyama et al. (1979) have further shown that the first diploid daughter cells coming from zygotes show self-agglutination caused by a and α agglutination substances on the cell surface. However, diploid daughter cells coming from old zygotes which had already born diploid buds showed no sexual agglutinability. These phenomena can be explained in the following way. The activity of a and α genes is suppressed when a and α cells are fused, resulting in the switch-off of the production of agglutination substances in the cytoplasm. However, remaining soluble agglutination substances which exist in the cytoplasm may move into the cell surface of the daughter diploid cells. The agglutination substances which are incorporated into the wall of diploid cells cause sexual agglutinability. Since new agglutination substances probably are not synthesized in zygotes and diploid cells, preexisting agglutination substances may be diluted out through successive cell division, eventually producing nonagglutinable diploids.

b. Nonagglutinable Mutants. Matsushima et al. (unpublished) have isolated four nonmating α mutants, all of which were found to be nonagglutinable. They used these mutants in an attempt to reveal the relationship between sexual agglutination and cell wall agglutination substances.

Among the four mutants, two mutants were known to be controlled by single genes; one not linked to the mating-type locus, and the other closely linked to the mating type locus. The mutation not linked to the mating type locus is not specific to α mating type, since it causes sexual nonagglutinability in a cells when it is introduced through mating. No agglutination substances were detected in Glusulase digests of the cell walls of the above four α mutants.

From these results, it may be concluded that the agglutination sub-

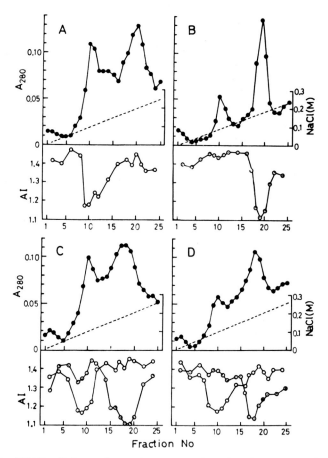

Fig. 14. DEAE-cellulose column chromatography of agglutination substances extracted from zygotes by the autoclave method (Tohoyama et al., 1979). (A) Agglutination substance extracted from T55(a^c); (B) agglutination substance extracted from T56(α^c); (C) mixture of agglutination substances from T55 and T56; (D) agglutination substances from zygote fraction. ●, absorbance of 280 nm; - - -, NaCl concentration; ○, activity of a agglutination substance (AI of boiled tester α cells treated with each fraction and mixed with boiled a tester cells); ⊙, activity of α agglutination substance (AI of boiled a tester cells treated with each fraction and mixed with boiled α tester cells). Since the activity of a and α agglutination substances was expressed in terms of the masking action on the opposite mating type, the lowest agglutination index indicates the highest activity (the peak of the activity) of each agglutination substance (see Fig. 5). Agglutination substances from zygotes were extracted from the zygote fraction isolated from 3-hr-old mating mixture of T55 and T56 in YHG after boiling, the ratio of zygotes to total cells in the zygote fraction being 91.8%.

stances that we have isolated are responsible for the sexual agglutination, one of the essential stages in the mating reaction. The functional relationship between nonagglutinable genes and mating type alleles remains to be studied.

c. **Inducible Mutants.** As described already, we have physiologically characterized some inducible mutants. In an attempt to reveal the genes responsible for their physiological characteristics, we performed genetic analysis of the inducibility (Nakagawa and Yanagishima, 1978; Yanagishima and Nakagawa, 1980). In three of the inducible mutants [T55s-41 (a^i), T562s-161 (α^i), and H22 (a^i)] single genes control the inducibility. The inducible genes carried by T55s-41 and T562s-161 are not linked to the mating type locus while the inducible gene carried by H22 is. The inducible strain H1-0(a^i) carries two inducible genes, at least one of which is linked to the mating-type locus. At least one of these inducible genes was shown to be dominant by constructing a^c/a^i diploid by the protoplast fusion method. The inducible gene carried by T55s-41 and that by T562s-161 are thought to be identical, because recombination and complementation did not occur between the two genes. The inducible gene is denoted as *saa1* and known to be recessive.

Besides the genetic differences the inducible a strains, T55s-41 and H1-0, are different physiologically in that the former is temperature-sensitive for inducibility but the latter not while the induction of sexual agglutinability by α pheromone is sensitive to Triton X-100 in the former but not in the latter (Yanagishima, 1978a; Yanagishima and Nakagawa, 1980). Recently, Nakagawa and Yanagishima (1979) have isolated various inducible a mutants which differ in their sensitivity to α pheromone and temperature. These mutants should aid our analysis of the regulation mechanism of sexual agglutinability in *S. cerevisiae*.

IV. PERSPECTIVES

The results described in this chapter may be summarized as follows:
1. Sexual cell aggregates, consisting of essentially equal numbers of randomly assembled a and α cells are formed at various stages in the cell cycle. These aggregations give cells the chance to contact an opposite mating type cell and maintain physiological conditions favorable for zygote formation.
2. The theoretical basis for the quantitative measurement of sexual agglutination by a spectrophotometric method has been given. The agglutination substances responsible for sexual agglutination have been chemically characterized and the formation of a complementary

complex between the opposite mating type agglutination substances was shown. The agglutination substances, extracted by the newly devised autoclave method, are active in living cells, not only in the masking of sex-specific agglutinability of opposite mating types, but also in the inhibition of the formation of zygotes. High incubation temperature and the lack of fermentable sugar cause loss of sexual agglutinability even in constitutive cells.

3. Both physiologically and genetically repressed sexual agglutinability is reversed by the addition of pheromones from the opposite mating type. The mode and mechanism of actions of these pheromones have been described.

4. The a cells produce the binding substance for α pheromone which seems to play an important role in the first step of α pheromone induction. In α cells we could not detect the binding substance for a pheromone. The involvement of the cell membrane in the mechanism of agglutinability induction by α pheromone was proposed.

5. Zygotes and, at least, the first diploid cells from zygotes still retain a and α sexual agglutination substances which cause self-agglutination. Inducible mutants for sexual agglutinability have been isolated and genetically and physiologically characterized.

The following problems remain.

1. The molecular structure of a and α agglutination substances in relation to the molecular mechanism of the formation of a complementary complex must be clarified. The intracellular distribution and the quantitative changes of these substances during sexual conjugation and spore germination should be pursued. We are going to pursue the mechanism of mating type conversion in homothallic strains through the use of cytological detection methods for these substances.

2. It is important to reveal the biochemical role of cell membrane and enzymatic alteration of α pheromone in the process of the induction of sexual agglutinability to completely understand the mechanism of action of α pheromone. In this regard, the chemical characterization of the binding substances for α pheromone is also essential.

3. The autoclave method we have devised allows us to perform comparative studies on sexual agglutination in yeasts at the biochemical level. First, the validity of the application of the autoclave method to various species of yeasts must be established. In parallel with this, comparative tests of sexual agglutinability among different but related yeasts through the use of intact or boiled cells should be carried out. These experiments have already been started (Yoshida and Yanagishima, 1978).

4. Inducible mutants which differ in their physiological characters

have been isolated. Comparative studies of these mutants should reveal the general features of the mechanism of induction of sexual agglutinability by sex pheromones. The mechanism of repression of sexual agglutinability in diploid cells and that of the derepression in haploid cells are two areas yet to be pursued.

5. The role of cell-cell recognition through agglutination substances and cell-to-cell interaction through sex pheromones during the germination of spores in both heterothallic and homothallic strains should be clarified to understand the biological significance of sexual agglutination.

6. Finally, we would like to point to the following problems. Since sexual agglutination has close relations with various fields of biological science such as molecular biology, biochemistry, physiology, taxonomy, and even ecology, it is important to deal with sexual agglutination, coordinating the viewpoints of all these fields.

ACKNOWLEDGMENT

A part of the work described in this chapter was supported by a Grant-in-Aid from the Ministry of Education, Science and Culture in Japan.

REFERENCES

Betz, R., and Duntze, W. (1979). Purification and partial characterization of a factor, a mating hormone produced by mating-type-a cells from *Saccharomyces cerevisiae*. *Eur. J. Biochem.* 95, 469–475.

Betz, R., Duntze, W., and Manney, T. R. (1978). Mating-factor-mediated sexual agglutination in *Saccharomyces cerevisiae*. *FEMS Microbiol. Lett.* 4, 107–110.

Brock, T. D. (1958). Mating reaction in *Hansenula wingei*. Preliminary observation and quantitation. *J. Bacteriol.* 75, 697–701.

Bücking-Throm, E., Duntze, W., Hartwell, L. H., and Manney, T. R. (1973). Reversible arrest of haploid yeast cells at the initiation of DNA synthesis by a diffusible sex factor. *Exp. Cell Res.* 76, 99–110.

Burke, D., Mendonça-Previato, L., and Ballou, C. E. (1980). Cell–cell recognition in yeast: Purification of *Hansenula wingei* 21 cell sexual agglutination factor and comparison of the factors from three genera. *Proc. Nat. Acad. Sci. U.S.A.* 77, 318–322.

Calleja, G. B., and Johnson, B. F. (1971). Flocculation in a fission yeast: An initial step in the conjugation process. *Can. J. Microbiol.* 17, 1175–1177.

Chan, R. K. (1977). Recovery of *Saccharomyces cerevisiae* mating type a cells from G_1 arrest by α factor. *J. Bacteriol.* 130, 766–774.

Ciejek, E., and Thorner, J. (1979). Recovery of *S. cerevisiae* a cells from G_1 arrest by α factor pheromone requires endopeptidase action. *Cell* 18, 623–635.

Crandall, M. (1978). Mating type interactions in yeasts. *In* "Cell-Cell Recognition" (A. S. G. Curtis, ed.), pp. 105-123. Cambridge Univ. Press, London.

Crandall, M. A., and Brock, T. D. (1968). Molecular basis of mating in the yeast *Hansenula wingei*. *Bacteriol. Rev.* **32**, 139-163.

Crandall, M., and Caulton, J. H. (1973). Induction of glycoprotein mating factors in diploid yeast of *Hansenula wingei* by vanadium salts or chelating agents. *Exp. Cell Res.* **82**, 159-167.

Crandall, M., Lawrence, L. M., and Saunders, R. M. (1974). Molecular complementarity of yeast glycoprotein mating factors. *Proc. Nat. Acad. Sci. U.S.A.* **71**, 26-29.

Doi, S., and Yoshimura, M. (1977). Temperature-sensitive loss of sexual agglutinability in *Saccharomyces cerevisiae*. *Arch. Microbiol.* **114**, 287-288.

Doi, S., and Yoshimura, M. (1978). Temperature-dependent conversion of sexual agglutinability in *Saccharomyces cerevisiae*. *Mol. Gen. Genet.* **162**, 251-257.

Duntze, W., MacKay, V., and Manney, T. R. (1970). *Saccharomyces cerevisiae*: A diffusible sex factor. *Science* **168**, 1472-1473.

Duysens, L. N. M. (1956). The flattering of the absorption spectrum of suspension, as compared to that of solutions. *Biochim. Biophys. Acta* **19**, 1-12.

Egel, R. (1971). Physiological aspects of conjugation in fission yeast. *Planta* **98**, 89-96.

Fehrenbacher, G., Perry, K., and Thorner, J. (1978). Cell-cell recognition in *Saccharomyces cerevisiae*: Regulation of mating-specific adhesion. *J. Bacteriol.* **134**, 893-901.

Finkelstein, D. B., and Strausberg, S. (1979). Metabolism of α-factor by a mating type cells of *Saccharomyces cerevisiae*. *J. Biol. Chem.* **254**, 796-803.

Hagiya, M., Yoshida, K., and Yanagishima, N. (1977). The release of sex-specific substances responsible for sexual agglutination from haploid cells of *Saccharomyces cerevisiae*. *Exp. Cell Res.* **104**, 263-272.

Hagiya, M., Yoshida, K., and Yanagishima, N. (1978). Chemical characterization and physiological action of agglutination substances. *Proc. Annu. Meet. Bot. Soc. Jpn., 43rd* p. 194. (In Jpn.)

Hagiya, M., Yoshida, K., and Yanagishima, N. (1980). Purification and physicochemical characterization of α agglutination substance in comparison with a agglutination substance in *Saccharomyces cerevisiae*. In preparation.

Hereford, L. M., and Hartwell, L. H. (1974). Sequential gene function in the initiation of *Saccharomyces cerevisiae* DNA synthesis. *J. Mol. Biol.* **84**, 111-117.

Hicks, J. B., and Herskowitz, I. (1976). Evidence for a new diffusible element of mating pheromones in yeast. *Nature (London)* **260**, 246-248.

Itoh, M., Izawa, S., and Shibata, K. (1963). Disintegration of the chloroplasts with dodecylbenzene sulfonate as measured by flattering effect and size distribution. *Biochim. Biophys. Acta* **69**, 130-142.

Itoh, S., Takahashi, S., Tsuboi, M., Shimoda, C., and Hayashibe, M. (1976). Effect of light on sexual flocculation in *Schizosaccharomyces japonicus*. *Plant Cell Physiol.* **17**, 1355-1358.

Kawanabe, Y., Hagiya, M., Yoshida, K., and Yanagishima, N. (1978). Effect of concanavalin A on the mating reaction in *Saccharomyces cerevisiae*. *Plant Cell Physiol.* **19**, 1207-1216.

Kawanabe, Y., Yoshida, K., and Yanagishima, N. (1979). Sexual cell agglutination in relation to the formation of zygotes in *Saccharomyces cerevisiae*. *Plant Cell Physiol.* **20**, 423-433.

Levi, J. D. (1956). Mating reaction in yeast. *Nature (London)* **177**, 753-754.

Lindegren, C. C., and Lindegren, G. (1943). A new method for hybridizing yeast. *Proc. Natl. Acad. Sci. U.S.A.* 29, 306–308.

Manes, P. F., and Edelman, G. M. (1978). Inactivation and chemical alteration of mating factor α cells and spheroplasts of yeast. *Proc. Natl. Acad. Sci. U.S.A.* 75, 1304–1308.

Manney, T. R., and Meade, J. H. (1977). Cell–cell interactions during mating in *Saccharomyces cerevisiae*. *In* "Receptors and Recognition" (J. L. Reissig, ed.), Ser. B, Vol. 3, pp. 281–321. Chapman & Hall, London.

Matsushima, Y., Shimoda, C., and Yanagishima, N. (1976). A factor in cell-free extracts inactivating sexual agglutinability of a mating type in *Saccharomyces cerevisiae*. *Plant Cell Physiol.* 17, 621–625.

Nakagawa, Y., and Yanagishima, N. (1978). Mutants inducible for sexual agglutinability II. Genetic analysis. *Proc. Annu. Meet. Bot. Soc. Jpn., 43rd* p. 193. (In Jpn.)

Nakagawa, Y., and Yanagishima, N. (1979). Gene control of sexual agglutinability in yeast. *Proc. Annu. Meet. Bot. Soc. Jpn., 44th* p. 75. (In Jpn.)

Nishi, K., and Yanagishima, N. (1978). Physiological characterization of the agglutinability-inducing pheromones, a and α substances. *Proc. Annu. Meet. Jpn. Soc. Plant Physiol.* p. 35. (In Jpn.)

Nishi, K., and Yanagishima, N. (1979). Temperature-sensitivity of the agglutinability-inducing action of α pheromone in yeast. *Proc. Annu. Meet. Jpn. Soc. Plant Physiol.* p. 138. (In Jpn.)

Sakai, K., and Yanagishima, N. (1971). Mating reaction in *Saccharomyces cerevisiae* I. Cell agglutination related to mating. *Arch. Mikrobiol.* 75, 260–265.

Sakai, K., and Yanagishima, N. (1972). Mating reaction in *Saccharomyces cerevisiae* II. Hormonal regulation of agglutinability of a type cells. *Arch. Mikrobiol.* 84, 191–198.

Sakurai, A., Tamura, S., Yanagishima, N., and Shimoda, C. (1976a). Isolation and chemical characterization of the peptidyl factor inducing sexual agglutination in *Saccharomyces cerevisiae*. *Agric. Biol. Chem.* 40, 255–256.

Sakurai, A., Tamura, S., Yanagishima, N., and Shimoda, C. (1976b). Structure of the peptidyl factor inducing sexual agglutination in *Saccharomyces cerevisiae*. *Agric. Biol. Chem.* 40, 1057–1058.

Sena, E. P., Radin, D. N., and Fogel, S. (1973). Synchronous mating in yeast. *Proc. Natl. Acad. Sci. U.S.A.* 70, 1373–1377.

Shimizu, T., Yoshida, K., and Yanagishima, N. (1977). Specific binding substance for the agglutinability-inducing pheromone, α substance-I in *Saccharomyces cerevisiae*. *Proc. Annu. Meet. Jpn. Soc. Plant Physiol.* p. 77. (In Jpn.)

Shimoda, C., and Yanagishima, N. (1973). Mating reaction in *Saccharomyces cerevisiae* IV. Retardation of deoxyribonucleic acid synthesis. *Physiol. Plant.* 29, 54–59.

Shimoda, C., and Yanagishima, N. (1975). Mating reaction in *Saccharomyces cerevisiae* VIII. Mating type specific substances responsible for sexual cell agglutination. *Antonie van Leeuwenhoek; J. Microbiol. Serol.* 41, 521–532.

Shimoda, C., Kitano, S., and Yanagishima, N. (1975). Mating reaction in *Saccharomyces cerevisiae* VII. Effect of proteolytic enzymes on sexual agglutinability and isolation of crude sex-specific substances responsible for sexual cell agglutination. *Antonie van Leeuwenhoek; J. Microbiol. Serol.* 41, 513–519.

Shimoda, C., Yanagishima, N., Sakurai, A., and Tamura, A. (1976a). Mating reaction in *Saccharomyces cerevisiae* IX. Regulation of sexual cell agglutinability of a type cells by a sex factor produced by α type cells. *Arch. Microbiol.* 108, 27–33.

Shimoda, C., Matsushima, Y., and Yanagishima, N. (1976b). Mating reaction in *Saccharomyces cerevisiae* X. Agglutinability-inactivating factor: A factor which destroys sexual agglutinability of a mating-type cells. *Antonie van Leeuwenhoek; J. Microbiol. Serol.* 42, 511–521.

Shimoda, C., Yanagishima, N., Sakurai, A., and Tamura, S. (1978). Induction of sexual agglutinability of a mating-type cells as the primary action of the peptidyl sex factor from α mating-type cells in *Saccharomyces cerevisiae*. *Plant Cell Physiol.* 19, 513–517.

Stötzler, D., Kiltz, H.-H., and Duntze, W. (1976). Primary structure of α factor peptides from *Saccharomyces cerevisiae*. *Eur. J. Biochem.* 69, 397–400.

Tanaka, T., Kita, H., Murakami, T., and Narita, K. (1977). Purification and amino acid sequence of mating factor from *Saccharomyces cerevisiae*. *J. Biochem (Tokyo)* 82, 1681–1687.

Taylor, N. W. (1964). Specific, soluble factor involved in sexual agglutination of the yeast *Hansenula wingei*. *J. Bact.* 87, 863–866.

Taylor, N. W. (1965). Purification of sexual agglutination factor from the yeast *Hansenula wingei* by chromatography and gradient sedimentation. *Arch. Biochem. Biophys.* 111, 181–186.

Taylor, N. W., and Orton, W. L. (1971). Cooperation among the active binding sites in the sex-specific agglutinin from the yeast, *Hansenula wingei*. *Biochemistry* 10, 2043–2049.

Tohoyama, H., Hagiya, M., Yoshida, K., and Yanagishima, N. (1979). Regulation of the production of the agglutination substances responsible for sexual agglutination in *Saccharomyces cerevisiae:* Changes associated with conjugation and temperature shift. *Mol. Gen. Genet.* 174, 269–280.

Wickerham, L. J. (1956). Influence of agglutination on zygote formation in *Hansenula wingei*, a new species of yeast. *C. R. Trav. Lab. Carlsberg, Ser. Physiol.* 26, 423–443.

Wickerham, L. J. (1969). Yeast taxonomy in relation to ecology, genetics and phylogeny. *Antonie van Leeuwenhoek; J. Microbiol. Serol.* 35, Suppl. Yeast Symp., Part 1, 31–58.

Wilkinson, L. E., and Pringle, J. R. (1974). Transient $\bar{G}_1$ arrest of *S. cerevisiae* cells of mating type a. *Exp. Cell Res.* 89, 175–187.

Yanagishima, N. (1978a). Mutants inducible for sexual agglutinability I. Physiological characterization. *Proc. Annu. Meet. Bot. Soc. Jpn., 43rd* p. 193. (In Jpn.)

Yanagishima, N. (1978b). Sexual cell agglutination in *Saccharomyces cerevisiae:* Sexual cell recognition and its regulation. *Bot. Mag., Spec. Issue* 1, 61–81.

Yanagishima, N. (1979). Mode of action of a pheromone in *Saccharomyces cerevisiae*. *Yeast Genet. News—Jpn.* No. 12, p. 23 (In Jpn.).

Yanagishima, N., and Inaba, R. (1980). Induction of sexual agglutinability by absorbed α peptidyl factor in a mating type cells of *Saccharomyces cerevisiae*. In preparation.

Yanagishima, N., and Nakagawa, Y. (1980). Mutants inducible for sexual agglutinability in *Saccharomyces cerevisiae*. *Mol. Gen. Genet.* 178, 241–251.

Yanagishima, N., Yoshida, K., Hamada, K., Hagiya, M., Kawanabe, Y., Sakurai, A., and Tamura, S. (1976). Regulation of sexual agglutinability in *Saccharomyces cerevisiae* of a and α types by sex-specific factors produced by their respective opposite mating types. *Plant Cell Physiol.* 17, 439–450.

Yanagishima, N., Shimizu, T., Yoshida, K., Sakurai, A., and Tamura, S. (1977). Physiological detection of a binding substance for the agglutinability-inducing pheromone, α substance-I in *Saccharomyces cerevisiae*. *Plant Cell Physiol.* 18, 1182–1192.

Yen, P. Y., and Ballou, C. B. (1973). Composition of a specific intercellular agglutination factor. *J. Biol. Chem.* **248,** 8316–8318.

Yen, P. Y., and Ballou, C. B. (1974). Partial characterization of the sexual agglutination factor from *Hansenula wingei* Y-2340 type 5 cells. *Biochemistry* **13,** 2428–2437.

Yoshida, K., and Yanagishima, N. (1978). Intra- and intergeneric mating behaviour of ascosporogenous yeasts I. Quantitative analysis of sexual agglutination. *Plant Cell Physiol.* **19,** 1519–1533.

Yoshida, K., Hagiya, M., and Yanagishima, N. (1976). Isolation and purification of the sexual agglutination substance of mating type a cells in *Saccharomyces cerevisiae*. *Biochem. Biophys. Res. Commun.* **71,** 1085–1094.

Yoshida, K., Kabuto, H., and Yanagishima, N. (1980). Intra- and intergeneric mating behaviour of ascosporogenous yeasts. II. In preparation.

12

Sexual Interactions in the Green Alga *Chlamydomonas eugametos*

H. VAN DEN ENDE

I.	Introduction	297
II.	Description of the Sexual Process	299
III.	Cultivation and Gametogenesis	301
IV.	The Flagellar Surface	302
	A. Morphology	302
	B. Isoagglutinins	303
	C. Composition of the Flagellar Membrane	304
V.	Nature of the Agglutination Process	311
VI.	Prospects	315
	References	316

I. INTRODUCTION

Sexual reproduction in *Chlamydomonas* deserves attention as being one of the best experimental systems in the plant world to study intercellular recognition. Sexually active gametes can be produced easily and in large numbers; recognition is visible as clumps of agglutinating cells and can be quantified by determining the number of fusion products; the parts of the cell involved are also obvious: gametic contact is established by naked flagella, without the complication of a cell wall barrier. Also, in the reactions that follow primary cell contact, which ultimately lead to cell fusion, the flagella play a central role, and can therefore be considered as genuine sensory organelles.

The universal recognition of this system is largely due to the work of one man, Dr. L. Wiese, who worked *inter alia* with three species,

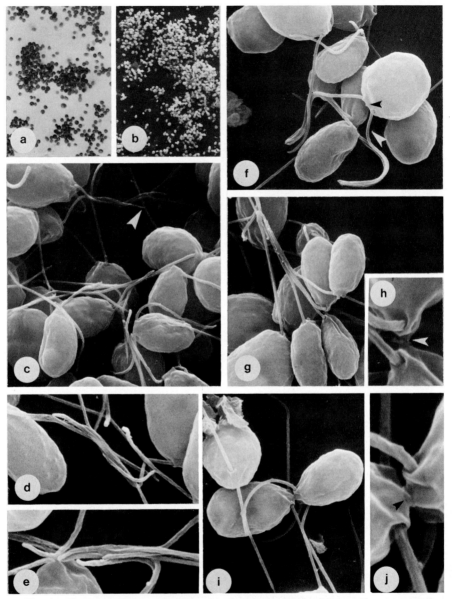

Fig. 1. Scanning electron micrographs of various stages of the mating process of *Chlamydomonas eugametos* (from Mesland, 1976, with permission). (a) Typical light microscopic picture of aggregates of mating gametes (200×); (b) low magnification of scanning electron micrograph of aggregates of mating gametes (180×); (c) flagellar contacts 15 sec after mixing of the gametes; note typical "T" contact (arrow) and lack of

Chlamydomonas eugametos, C. moewusii, and *C. reinhardti* (for reviews, see Wiese, 1969, 1974; Wiese and Wiese, 1978). The first two species are closely related and as far as syngen I of *C. moewusii* is concerned sexually compatible (Gowans, 1976; Wiese and Wiese, 1978), but incompatible with *C. reinhardti.* These species are heterothallic and isogamous, morphologically indistinguishable gametes of two mating types (mt^+ and mt^-) interacting to produce progeny. Recently, work has concentrated on *C. reinhardti,* mainly due to the efforts of the group of Dr. U. W. Goodenough (for review, see Goodenough, 1977).

This chapter, however, will deal with some recent results obtained with *C. eugametos.* The main emphasis will be on the flagella and particularly on the characterization of the flagellar surface. Additional information about *C. reinhardti* will be presented as far as it is applicable to *C. eugametos* or could be useful for comparative reasons.

II. DESCRIPTION OF THE SEXUAL PROCESS

The following picture of sexually interacting *C. eugametos* cells was obtained by Mesland (1976) using the scanning electron microscope, in accordance with earlier light microscopical studies (e.g., Lewin, 1952). The mixing of suspensions of mt^+ and mt^- gametes leads to the formation of many clumps of cells in which practically all cells participate, caused by an apparently sex-specific adhesiveness of the flagella (Fig. 1a and b). A highly characteristic feature that distinguishes clumps of sexually interacting cells with homotypic cell aggregates caused by antibody or lectin cross-linking (see below) is a violent twitching of the flagella, giving the clumps a vibrating appearance. While originally the contacts seem to be of a random nature (Fig. 1c and d), the flagella

---

tip-to-tip contacts (4500×); (d) flagellar bundle with flagella oriented both tip-to-tip and tip-to-base, 1.5 min after mixing (7800×); (e) flagellar bundle, almost completely lined up tip-to-tip (9000×); (f) aggregate of three cells; flagellar tips lie adjacent to each other although the upper gamete has longer flagella (white arrow); The upper gamete shows an extending papilla (black arrow) (4500×); (g) a cell pair just after papillar fusion; note the visible plasma bridge, the right-over-left position of the pair's flagella and their equal lengths; two other cells are visible which have an outgrown papilla and one of their flagella joining the same flagellar bundle; the other flagella have contacts with other cells in the aggregate (not shown) (1800×); (h) higher magnification of the plasma bridge shown in (g) (18,000×); (i) vis-à-vis pair; the mt^- gamete has shortened its flagella which exhibit a typical posture around the mt^+ cell body; the flagella still adhere to flagella not belonging to the pair (4500×); (j) higher magnification of (i); the cell walls are closely appressed and the plasma bridge is no longer visible (arrow) (18,000×).

gradually associate over their whole length into bundles. A continuous reassociation and sorting process leads within a few minutes to a pattern in which the flagella are aligned over their whole length, with the tips adjacent to each other and the cell bodies positioned pairwise with the anterior ends facing each other. (Fig. 1e). Notable features at this stage are (1) that of each pair of cells the right-hand flagellum as seen from the posterior end of each cell always overlies the base of the left-hand flagellum of its partner, and (2) that the flagellar tips are always associated (Fig. 1e and f). At this stage both cells are seen to extend a short pointed plasma papilla, also called fertilization tubule (Friedmann et al., 1968) or gamosomal tubule (Cavalier Smith, 1975) that protrudes through the anterior ridge of the cell wall between the flagellar collars. By fusion of these papillae a plasma connection is established between each pair of cells (Triemer and Brown, 1975) (Fig. 1f and g). Directly after fusion, the flagella lose adhesiveness, so that the resulting vis-à-vis pair can escape from the clump (Fig. 1h, i, and j). The mt^+ pair of flagella resumes the swimming action, while the mt^- flagella now appear to have contracted by 30% of their original length, and are held bowed around the partner cell. Notwithstanding the extreme rapidity of the agglutination reaction, the rate of vis-à-vis pair formation is relatively slow. Usually the maximal number of pairs is reached after about 45 min. The paired cells fuse completely after approximately 12 hr in the light after which a zygote wall is produced (Brown et al., 1968). Only then do the original cell walls disappear. In the dark, the vis-à-vis pair stage is maintained indefinitely (Lewin, 1952; Gowans, 1960).

In C. reinhardti, this process differs in some major respects. In the first place, agglutinating cells lose their walls, due to the action of a lytic factor (Claes, 1971; Schlösser et al., 1976). Flagellar interaction between the resulting naked gametes occurs much faster than in C. eugametos, and shows mainly adhesion at the flagellar tips. Subsequently, a fertilization tubule is produced in the mt^+ partner by which the cytoplasmic connection between pairs is established (Weiss et al., 1977). This connection rapidly widens, leading to complete fusion of the cells. As in the former species, the four flagella of the resulting zygote immediately lose their adhesive properties. No vis-à-vis pair stage is apparent in C. reinhardti (Cavalier Smith, 1975; Martin and Goodenough, 1975).

During sexual agglutination, activation of papillar outgrowth in C. eugametos is not observed until both flagella are agglutinated along most of their length. This suggests a causal relationship which is confirmed by the fact that papillar extension can be evoked by treating gametes with high concentrations of isolated flagella or membranes derived from flagella of the opposite mating type (Mesland and van den Ende, 1978a). Low concentrations, although sufficient to induce strong isoagglutination of cells by cross-linking cells of the same mating type, do not have this effect. This implies that agglutination per se is not sufficient to activate the cells. Perhaps a large area of the flagella has to be associated with the surface of opposite flagella. In *C. reinhardti*, association of the flagellar tips ("tipping," see below) is required for activation (Goodenough *et al.*, 1979). In deflagellated cells, normal mating occurs only when the regenerating flagella have attained their normal length (Mesland, 1977). Similar observations have been made by Solter and Gibor (1977) for *C. reinhardti*. So the possibility that gamete activation is effected by some soluble factor, secreted by the interacting gametes, can be ruled out.

Also the fusion of the papillae in a cell pair and the subsequent disengagement of the flagella are causally linked processes. When fusion is inhibited by maintaining the mixed gametes at 4°C or by treating them with a fusion inhibitor, or treating the mt^+ gametes with thermolysin or chymotrypsin, the flagella remain associated indefinitely (Mesland and van den Ende, 1978b; Wiese and Wiese, 1978). This is also seen in the *imp*-1 mutant of *C. reinhardti*, which agglutinates normally, but is unable to fuse. The flagella therefore do not lose their adhesiveness with the consequence that agglutinating clumps grow in size into huge clusters (Goodenough and Weiss, 1975; Goodenough *et al.*, 1976).

The research described in this chapter is mainly directed at elucidating the mechanism of flagellar agglutination, which indeed shows a number of remarkable features, common to both *C. eugametos* and *C. reinhardti*: its extreme mating type and species specificity, its instantaneous disappearance after plasma bridge formation and the way it leads to tip-to-tip arrangements of *both* flagella in *each* gamete pair; and finally its function in the activation of the mating cells.

III. CULTIVATION AND GAMETOGENESIS

Gametogenesis in a liquid suspension of actively growing vegetative cells is induced by imposing a nitrogen deficiency. As described by Kates and Jones (1964), synchronous cultures exposed to a 12 hr light/12 hr dark regimen are most sensitive to this treatment after about 6 hr in the light period. Such cells, when placed in nitrogen-free medium, develop quantitatively into gametes after 24 hr (Wiese *et al.*,

1979). Alternatively, a convenient method to obtain a high yield of gametes is to cultivate the alga on agar plates (Sager and Granick, 1954; Schmeisser et al., 1973; Wiese, 1965). When such plates are flooded overnight with distilled water, a dense suspension of flagellated gametes is obtained. It is not clear, however, whether the cells in the agar cultures are present as gametes (Martin and Goodenough, 1975) or whether they differentiate from (stationary) vegetative cells into gametes on flooding with water (Wiese et al., 1979).

Using these methods, no significant physiological differences between the two types of gametes have been observed, except for a considerable difference in cell size (Martin and Goodenough, 1975; Wiese et al., 1979). This point has not been studied systematically for *C. eugametos*, however. Also, caution is needed with respect to vegetative cells, which in a study of the chemical background of mating should serve as control cells. They can be obtained from nitrogen-containing liquid cultures, or by flooding agar plates with low concentrations of an ammonium salt (Wiese, 1965). In the latter case, the sexually inactive cells could be dedifferentiated gametes.

The present study was carried out with gametes and vegetative cells obtained from 10- to 14-day-old agar plates, on a medium described by Wiese (1965) and maintained at 15°–20°C, 4000 lx on a 12 hr light/12 hr dark cycle. When flooded with 10 ml water per plate the yield was $1-5 \times 10^7$ cells/ml, 70–100% of which would participate in vis-à-vis pair formation. When flooded with 0.4% NH_4Cl, equally viable cells were obtained which did not exhibit any sexual reaction.

IV. THE FLAGELLAR SURFACE

A. Morphology

The flagellar axoneme is enclosed by a membrane, the outer surface of which is covered by a fuzzy coat, 17–34 nm thick (McLean et al., 1974), also called the flagellar sheath. It is assumed that it consists of carbohydrate material associated with or being part of the integral membrane constituents. On the flagella hairlike projections called mastigonemes are present, about 1 μm long (Brown et al., 1968; Bergman et al., 1975; Snell, 1976; Mesland, 1977). Freeze fracture studies showed particle-studded P and E fracture faces on the membrane of *C. reinhardti* flagella. Gametes tended to carry more particles than vegetative cells in the E fracture face, but there was no change in particle distribution in agglutinating flagella (Bray et al., 1974;

Bergman et al., 1975; Snell, 1976). Otherwise, no differences between gametic and vegetative flagella were distinguished or between mt^+ and mt^- flagella. Mesland (1977) observed in *C. eugametos* that glutaraldehyde-fixed flagella carried rather elaborate disk-bearing appendages which were continuous with the membrane. The presence of these structures on regenerating flagella was correlated with mating competence of the gametes involved. These appendages could be artifacts of glutaraldehyde fixation since they were not observed in OsO_4-fixed flagella, but nevertheless could reflect structural changes related to the adhesiveness of the membrane. That they might be bona fide structures at the flagellar surface, however, is suggested by the fact that by a shock treatment of gametes with acetic acid, particles were released very similar in shape and size to the appendages (Mesland and van den Ende, 1978a). Because of their agglutinative and papilla-inducing properties, it is reasonable to assume that they were derived from the flagellar membrane and are in fact identical to the appendages present *in situ*.

B. Isoagglutinins

Conditioned culture medium of gametes contains material that causes homotypic agglutination of gametes of the opposite mating type (isoagglutination). Thus, material derived from mt^+ gamete suspensions isoagglutinates mt^- gametes, and vice versa. Förster and Wiese (1954), who investigated the so-called mt^- isoagglutinin of *C. eugametos* and *C. moewusii*, reported that it could be sedimented by centrifugation and precipitated by ammonium sulfate, acetone, or alcohol. It contained carbohydrate and protein. Wiese (1965) concluded that it consisted of glycoprotein components of the flagellar membrane which for some reason were easily released into the culture field. Particularly, they would be involved in the sexual agglutination process, a view based on the equal specificities of the agglutination and isoagglutination reaction, and on the equal susceptibilities toward a range of hydrolytic enzymes (cf. also Wiese, 1974; Wiese and Wiese, 1975). Vegetative cells did not produce active isoagglutinins and were not isoagglutinated by them, nor did mt^+ gametes of *C. eugametos*, which were sexually incompetent by dark exposure (Wiese, 1965).

McLean et al. (1974) found by electron microscopy that the isoagglutinin particles are vesicles of varying size and shape. They observed a difference in buoyant density between isoagglutinins derived from mt^+ and mt^- gametes of *C. moewusii*, reflecting a difference in molecular weight and/or composition of the membrane constituents.

No such difference was found for *C. eugametos* or for *C. reinhardti* isoagglutinins which also were shown to be membrane vesicles (Bergman et al., 1975; Snell, 1976; Musgrave et al., 1979). Snell (1976) compared by SDS polyacrylamide gel electrophoresis membrane constituents of flagellar membranes and isoagglutinins of *C. reinhardti*, and found no significant differences. Similar results were obtained by other authors (Bergman et al., 1975; Musgrave et al., 1979). So the conclusion is that isoagglutinins consist of vesicles derived from the flagellar membrane, with identical properties. Isoagglutination could then be envisaged as being due to the multivalency of these vesicles. They would adhere to and act as crosslinkers of flagellar surfaces in a sex-specific way.

One would consequently expect that isoagglutinin, present in the medium of mt^+ as well as mt^- gametes, would inhibit the normal sexual agglutination between the mates, due to their isoagglutinative action, and competition for adhesive sites. Intriguingly, this appears not to be the case (Mesland and van den Ende, 1978a). Another remarkable fact is that mt^+ and mt^- isoagglutinins when mixed, do not form precipitating aggregates of vesicles, although they are allegedly derived from flagella which readily interact with one another (Wiese and Wiese, 1978). They do not, or very slightly, neutralize each others activities.

The mechanism as to how and why isoagglutinin vesicles are produced by the flagellar membrane is not well understood, either. Bergman et al. (1975), in negatively stained preparations of flagella, observed a row of vesicles blebbing off the flagellar tip. Mesland (1977), on the contrary, suggested that the appendages, seen on flagella in glutaraldehyde-fixed preparations, could be a source of isoagglutinin vesicles. In either case, it seems probable that this shedding of membranous material is the consequence of a normal turnover process of the mature flagellar membrane, as is also observed in other systems, like erythrocytes (Lutz et al., 1977). An active turnover of flagellar proteins involved in the mating process has been clearly demonstrated by Ishiura and Iwasa (1973) with cycloheximide as inhibitor of protein synthesis (see also Wiese and Wiese, 1978).

C. Composition of the Flagellar Membrane

1. Membrane Glycoconjugates

Agglutination clearly involves components exposed at the outside of the flagellar or isoagglutinin membranes. We may therefore assume that either lipids or membrane proteins are involved. Membrane proteins are invariably glycosylated when they are exposed outside the cell and research is rapidly establishing that glycoproteins or glycolipids are regularly involved in cell–cell recognition. Wiese and co-workers have shown that the agglutinability of mt^+ and mt^- gametes and isoagglutinins of *C. eugametos* is susceptible to proteolytic enzyme activity and that the agglutinability of mt^+ gametes and isoagglutinin is also susceptible to α-mannosidase and concanavalin A treatment (Wiese and Shoemaker, 1970; Wiese and Wiese, 1975). This, together with the fact that isoagglutinins only contain glycoproteins, strongly suggests that glycoproteins are involved in cell recognition by *C. eugametos* and consequently the attention has been focused on this fraction.

Flagella can be separated from the cell bodies by the pH shock procedure, described by Witman *et al.* (1972), by which cells are exposed for 30 sec at 4°C to pH=4. They are purified by centrifugation on a discontinuous sucrose gradient. They effectively adhere to flagella of intact gametes of the other mating type, so that at least part of the membrane components functional in adhesion have remained in the membrane after the shock procedure (Musgrave *et al.*, 1979).

Figure 2a shows a diagram of 5% polyacrylamide gels of flagellar extracts of *C. eugametos,* stained with the periodic acid–Schiff reagent (PAS). No difference is detectable between mt^+ and mt^- flagella, either derived from gametes or vegetative cells. Generally, PAS-stainable bands are visible, reflecting glycoconjugate fractions present in the flagellar membrane. The major fractions are labeled with the I_2 lactoperoxidase method of intact flagella (Musgrave *et al.*, 1979) indicating that they are exposed at the exterior of the flagellar membrane. From several experiments it appears that two bands are consistently predominant, a heavy band near the origin (PAS 1) and a band with

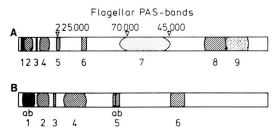

Fig. 2. Diagram showing patterns of glycoprotein bands in SDS polyacrylamide gels of flagellar material, stained with the periodic acid–Schiff procedure (PAS). The gels were run overnight at 1 mA/gel. (A) 5% acrylamide according to Weber *et al.* (1972); the positions of molecular weight markers is indicated. (B) 4% acrylamide according to Laemmli (1970).

apparent molecular weight of approximately 225,000 (PAS-5). However, PAS-5 is the major cell wall glycoprotein and appears to be a contaminant in flagella and isoagglutinin preparations. It can be removed by centrifuging the flagella or isoagglutinins on a cesium chloride cushion at 20,000 g (Musgrave *et al.*, 1979).

When *C. eugametos* is compared with *C. reinhardti* (Witman *et al.*, 1972; Snell, 1976; Bergman *et al.*, 1975) some differences are observed. Gels of the latter contain only one major PAS-stainable band with an estimated molecular weight of 2.5×10^5. In addition there is no mention of cell wall contaminants. Instead, in both *C. reinhardti* and *C. moewusii* free mastigonemes are the only conspicuous contaminants in crude membrane preparations (McLean *et al.*, 1974; Bergman *et al.*, 1975).

2. Sex-Specific Flagellar Surface Components

From the fact that sexual adhesion is a rapid and highly specific phenomenon, one would predict that components of the flagellar membrane carrying determinants for this specificity would be predominant at the flagellar surface. Several lines of evidence suggest that such components are present.

An approach that led to the detection of mating type-specific components was to raise specific antibodies by injecting rabbits with either mt^+ or mt^- flagella (Lens, 1980). When crude sera or the immunoglobulin fractions were tested with live gametes, they showed a remarkable specificity in that anti-mt^+ antibody isoagglutinated predominantly mt^+ gametes and not mt^- gametes, while anti-mt^- antibody was active mainly on mt^- gametes. This specificity was confirmed by indirect immunofluorescence. A direct comparison could be made using vis-à-vis pairs in which the mt^+ flagella can be distinguished from mt^- flagella. It is evident from Fig. 3 that the mt^+ pair of flagella fluoresce more strongly than the mt^- pair. Immunoglobulin fractions could be adsorbed to absolute specificity with flagella of the opposite mating type, which procedure eliminates all antigens common to both. The resulting preparations then acted on one single mating type in the isoagglutination and the immunofluorescence test.

It was investigated which components of the flagellar membrane were antigenic with these mating type-specific antibodies by using Triton extracts of flagella as antigenic material in cross-immunoelectrophoresis. When adsorbed immunoglobulin fractions were used, only one precipitation line was present (compared with seven lines using nonadsorbed fractions) and only when antibody and antigen were of the same mating type. For further identification this

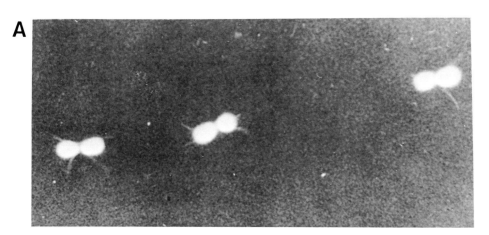

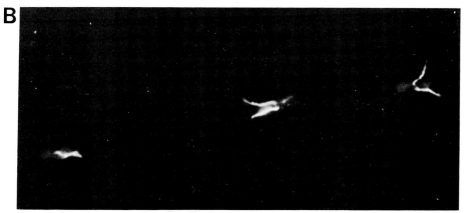

Fig. 3. Photograph showing that anti-mt^+ serum binds preferentially to the flagella of one of the gametes in a vis-à-vis pair. (A) The natural fluorescence of the cells under blue light. (B) The presence of bound rabbit anti-mt^+ antibodies on the same cells made visible via fluorescein labeled horse anti-rabbit Ig antibodies. Vis-à-vis pairs, fixed with 1.25% glutaraldehyde were washed with water and spread on glass slides covered with a layer of gelatin. They were then washed with acetone for 30 min, and in phosphate buffered saline (PBS), pH 7.6 for 15 min. The cells were then incubated with antiserum for 30 min in a humid chamber, after which the slides were washed three times with PBS to remove unbound immunoglobulins. Then they were incubated for 30 min with horse anti-rabbit immunoglobulin conjugated with fluorescein isothiocyanate. The slides were again washed with PBS and the cells embedded in a few drops of PBS–glycerol (1:1). A cover glass was sealed over the cells with nail varnish and the cells examined using a Zeiss fluorescence microscope.

precipitation line was cut out from the agarose gels and after extraction subjected to gel electrophoresis. After PAS or CBB staining the PAS-1 and the PAS-4 bands appeared. The same result was obtained with mt^+ and mt^- gametic material.

Another approach to identifying the sex-specific components was to incubate SDS gels of flagellar proteins with anti-mt^+ and anti-mt^- immunoglobulin fractions (Rostas et al., 1977) and visualizing the presence of adsorbed antibody with the peroxidase/3,3'-diaminobenzidine/H_2O_2 technique. When nonadsorbed immunoglobulin fractions were used, most of the PAS bands bound antibody. Although anti-mt^+ immunoglobulin tended to react better with gels of mt^+ flagella and mt^- immunoglobulin with gels of mt^- flagella, the staining patterns were more or less similar. When the gels were incubated with adsorbed fractions, only the bands PAS-1 and PAS-4 were stained, and only when the flagella and antibodies were of the same mating type. It can be concluded then that these two high molecular weight glycoconjugates present at the flagellar membrane possess antigenic characteristics which are specific for both mating types.

Preliminary results with antisera raised against the PAS-1 fraction, obtained by injecting homogenized gel sections, gave comparable results. The first two collections showed extreme specificity even without adsorption in the immunofluorescence and agglutination test. As shown in Table I only mt^+ gametes were isoagglutinated by anti-mt^+ PAS-1, and mt^- gametes by anti-mt^- PAS-1. Also cross-immunoelectrophoresis with the immunoglobulin fractions and Triton extracts of flagella were highly specific in producing only one single precipitation line in the appropriate combinations. However, when SDS gels of flagella were incubated with the anti-PAS-1 sera, no specificity was evoked. This can be explained by the fact that these nonabsorbed sera still contained nonspecific antibodies to antigens exposed only in denatured material, and not in intact flagella or Triton extracts, and thus obscured the specific reactions.

Thus, if the PAS-1 and PAS-4 fraction of the flagellar surface show differences along with the mating type in gametes, at least in the clones used in this research, the question arises whether they are really sex-linked. A study to answer this question has not yet been performed. A second question is whether components of the PAS-1 and PAS-4 fractions are functional in sexual agglutination. Some results pertaining to this question can be reported. First, in all the immune reactions described so far, vegetative cells also showed mating type specificity, although in lesser degree (see Table I). So sex specificity is not necessarily restricted to the binding sites postulated to be present

TABLE I

Isoagglutination of *C. eugametos* Cells Mediated by Antisera Directed against the PAS-1 Fraction of Gamete Flagella[a]

Cell type	Anti-PAS-1 sera		
	Anti-mt^+	Anti-mt^-	Preimmune serum
mt^+ gametes	128	0	0
mt^+ vegetative cells	16	0	0
mt^- gametes	0	32	0
mt^- vegetative cells	0	8	0

[a] Antisera were obtained by injecting rabbits with the portions of SDS polyacrylamide gels containing the PAS-1 fraction of gamete flagella which previously had been homogenized and washed with 25% isopropanol, 10% methanol, and phosphate-buffered saline (PBS), in succession, after which the gel pellets were resuspended in PBS. Each rabbit was injected with 2 ml antigen suspension derived from approximately 10^{10} gametes; 1 ml was injected intramuscularly in the hind leg and 1 ml was injected subcutaneously in the back. Collections were made every other week. The data in the table refer to the second collection. Agglutinative titers were determined by preparing binary dilution series and incubating an aliquot of each dilution with the same volume of cell suspension, containing about 10^7 cells/ml. Represented are the highest dilutions in which isoagglutinative action was observed by light microscopy.

at the flagellar surface of gametes only. Also the flagella of vis-à-vis pairs reacted efficiently with high specificity, indicating that loss of agglutinability in these flagella is not correlated with a loss of specific antigenic sites. Second, anti-PAS-1 antisera or Fab fragments derived therefrom inhibited the formation of vis-à-vis pairs only moderately, as is demonstrated in Fig. 4. Thus, presently there is not much support to designate a role for the specific antigenic sites in the sexual interaction process. A second line of evidence, however, points strongly to an involvement of a component of the PAS-1 fraction in sexual agglutination.

The high molecular weight part of the gel electrophoresis was expanded by prolonged electrophoresis in 4% Laemmli gels in SDS. It then appeared that the PAS-1 fraction could be split into two fractions, PAS-1a and PAS-1b (Fig. 2B). Particularly, the band with the highest molecular weight (PAS-1a) is interesting because it was not found in preparations of vegetative flagella, and thus might be the first indication of a gamete-specific flagellar component. In the next section some results are described which strongly suggest that this component is involved in agglutination.

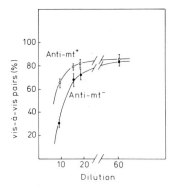

Fig. 4. Antisera directed against the PAS-1 fraction of gamete flagella were produced as described in Table I. 100 μl of diluted anti-mt^+ serum were added to 100 μl of mt^+ gamete suspensions and anti-mt^- serum to mt^- gametes. After 15 min, 100 μl of a suspension of gametes of the opposite mating type were added. The mixtures were incubated for 1 hr in the light, after which 300 μl 2.5% glutaraldehyde was added. Vis-à-vis pair percentages were determined by hemocytometer countings.

3. The mt^- Agglutination Factor

As stated in Section IV,B, the action of isoagglutinin vesicles can be envisaged to be due to cross-linking of flagellar surfaces of the opposite mating type and therefore should also carry the components determining specific agglutinability. That this is the case is shown by the effects of various treatments which release these components into solution (Homan et al., 1980).

An isoagglutinin suspension derived from mt^- gametes contains vesicles which can be collected quantitatively by centrifugation at 100,000 g, and can be resuspended into fresh buffer without loss of agglutinative action. However, a sonication treatment or extraction with guanidine thiocyanate (GTC) resulted in a reduction of isoagglutinative activity of the vesicles with a concomitant increase of proteinaceous and carbohydrate but lipid-free material in the high-speed supernatant. This supernatant showed a characteristic effect on the behavior of mt^+ gametes. It can best be described as a heavy twitching movement of the flagella, reminiscent of that observed in sexual agglutination, but now the cells remained separate and tended to accumulate at the surface of the fluid. This effect was sex-specific: mt^- cells were not influenced in their behavior. A possible explanation for this "twitch" effect is that some of the solubilized membrane components adsorbed to the flagellar surface in a specific way and exerted an influence on flagellar behavior.

On SDS gel electrophoresis of the soluble material in an mt^- sonicate, the only components present were the high molecular weight PAS-1, -2, -3, and -4 bands. Since this fraction only caused twitching and not isoagglutination, it is possible that it contained monovalent factors, adsorbing to the flagellar membrane. Isoagglutinin could then be considered as particles, deriving their multivalency from a multiplicity of such factors in or on the lipid bilayer of each vesicle. This is supported by the fact that reconstitution could take place between a fraction exhibiting only twitch effect and vesicles, rendered nonagglutinative by extensive sonification. Vesicles of mt^+ and mt^- origin both resulted on incubation with mt^- sonicate in a preparation that only isoagglutinated mt^+ gametes. Thus, the specificity was only determined by the soluble high molecular weight components. This is also indicated by the fact that on cross-linking this material with glutaraldehyde or with a combination of anti-mt^- serum and glutaraldehyde, a precipitate was formed which had isoagglutinative action on mt^+ gametes.

On fractionation of a GTC extract of mt^- isoagglutinin by gel filtration on a Bio-Gel A-150M column in 1 M GTC, activity was confined to a single fraction which on SDS gel electrophoresis showed only the PAS-1a band. From the retention in this column, a molecular weight of $30-50 \times 10^6$ was estimated.

In conclusion, it seems that at least for mt^- flagella the active component in agglutination is PAS-1a. Its absence in vegetative cells immediately explains their inability to agglutinate and presents us for the first time with a parameter at the molecular level to study gametogenesis.

So far, these results are valid only for mt^- material. When mt^+ isoagglutinin was extracted with GTC or sonicated, the same high molecular weight glycoproteins, including PAS-1a, were solubilized as were found in mt^- material, but the extracts had no effect on mt^- gametes. Although several explanations are possible, the one most at hand is that mt^+ isoagglutinin is very susceptible to inactivation, for example by freezing/thawing procedures or prolonged storage, as was already mentioned by Förster and Wiese (1954).

V. NATURE OF THE AGGLUTINATION PROCESS

Wiese has postulated that flagellar adhesion involves the interaction of complementary molecules in which noncovalent binding would occur between protein and carbohydrate, or carbohydrate and carbohydrate

TABLE II

Inactivation of ^{32}P-Labeled mt^- Isoagglutinin after Addition to Suspensions of mt^+ Gametes[a]

Volume of mt^+ gamete suspension (ml)	Mt^- iso-agglutinin added (cpm)	Radioactivity recovered in cell-free supernatant (cpm)	Biological activity recovered in cell-free supernatant (titer)	
			Found	Calculated
0.5	43,300	42,770	32	93
1.0	43,300	43,100	2	48
1.5	43,300	42,340	1	33
2.0	43,300	41,330	0	25
2.5	43,300	40,260	0	20
3.5	43,300	39,480	0	14
4.5	43,300	40,190	0	11

[a] Cells were cultivated on agar plates with M1 medium (Wiese, 1965) containing 0.2 mCi [^{32}P]phosphate per plate. Ten 14-day-old plates were flooded with water which after standing overnight was decanted. The cells were removed by centrifugation at 8000 g and the supernatant centrifuged at 50,000 g. The pellet was suspended in 2 ml water which after homogenization was pipetted on top of a 2.93 M CsCl cushion. After centrifugation at 20,000 g for 1 hr (Sorval RC2B, HB4 rotor) the interphase was collected and desalted by gel filtration over Sephadex G-25 followed by exhaustive dialysis. The isoagglutinative power of the resulting preparation was determined by binary dilution series and by establishing the dilution which just showed isoagglutination (titer). 50 μl containing 43,300 cpm with a titer of 1024 were added to cell suspensions of various size (1.02 × 10^7 cells/ml), and incubated for 30 min at room temp. Due to isoagglutination practically all the cells precipitated. A sample was taken from the clear supernatant which was centrifuged for 5 min at 1000 g after which the radioactivity and titer were determined.

moieties. The binding could be of a lectin–sugar, or antigen–antibody type. However, while the binding of concanavalin A and antibodies to surfaces can be easily demonstrated, the actual binding of (radioactively labeled) isoagglutinin to the flagellar surface of *Chlamydomonas* has so far not been detected, which suggests that the binding forces involved must be quite weak (Wiese and Wiese, 1978). This is in accordance with the easy and repeated partner exchange thought to occur within clusters of more than two agglutinating gametes.

The agglutination process exhibits a number of additional features worth mentioning, although explanations at the molecular level are not readily given. When a preparation of isoagglutinin is incubated with gametes of the opposite mating type, it appears to be gradually inactivated to large extent, a phenomenon first described by Wiese, and illustrated in Table II. Since adsorption of isoagglutinin to pelleted cells is not detectable, while the isoagglutinative titer dramatically deviates from the titer calculated from the added isoagglutinin and applied dilution, the conclusion seems warranted that isoagglutinin is somehow inactivated. This would explain the transitory character of cell aggregates induced by low concentrations of isoagglutinin (Wiese and Wiese, 1978). The dispersed cells can be reagglutinated by repeated additions of isoagglutinin; thus, the responsiveness of live gametes is maintained, but the added isoagglutinin is continuously inactivated. No inactivation of isoagglutinin occurs in suspensions of incompatible gametes or vegetative cells (Table III; see also Wiese and Wiese, 1978). The nature of this evidently quite specific progress is unknown. Similar results obtained with *C. reinhardti* have recently been described extensively by Snell and Moore (1980). The nature of this evidently quite specific process is unknown. A second point relevant to our understanding of agglutination was put forward by Ray *et al.* (1978), by showing that regenerating flagella in deflagellated cells of *C. reinhardti* must have a length greater than 5 μm before activation (in terms of cell wall lysis) of the gametes can be attained. How-

TABLE III

Inactivation of mt^- Isoagglutinin after Addition to Suspensions of Various Cell Types of *C. eugametos*[a]

Cell type	Agglutinative titer of cell-free supernatant determined with	
	mt^+ gametes	mt^- gametes
mt^+ gametes	0	0
mt^- gametes	32	0
mt^+ vegetative cells	16	0
mt^- vegetative cells	32	0
Water control	64	0

[a] A crude isoagglutinin preparation was obtained by centrifuging an mt^- gamete cell-free conditioned medium at 50,000 g for 1 hr and homogenizing the pellet in water. 0.5 ml of the resulting suspension was added to 0.5 ml cell suspension (5×10^8 cells/ml). After 30 min at room temperature, the cells were spun down and the titer of isoagglutinative power of the supernatant determined. Vegetative cells were obtained by flooding agar cultures with 0.2 gm/liter NH$_4$Cl (giving rise to 8% vis-à-vis pairs); gametes were obtained by flooding the plates with water (60% vis-à-vis pairs).

ever, sexual agglutination takes place already when the flagella have a length of less than 3 μm. The authors suggest that this points to the existence of two different types of receptors involved in the sexual process, one type for agglutination and another type for cell activation, i.e., cell wall lysis. Agglutination sites are on the flagellar surface as soon as the flagellum begins to regenerate, while the receptors responsible for cell wall lysis are not produced or incorporated into the membrane until the flagellum has grown to 50% of its maximum length. Additional support was obtained by showing that gametes that have lost mating competency by treatment with trypsin regain agglutinability in about 45 min after treatment, but carbohydrate release as a result of cell wall lysis is only observed after about 2 hr. The recovery of agglutinability is inhibited by cycloheximide but not by actinomycin D, while the reappearance of carbohydrate release is inhibited by both actinomycin D and cycloheximide (Solter and Gibor, 1978).

Particularly with respect to *C. eugametos* a point about gamete activation referred to in Section II is relevant, namely, that it seems that a relatively large number of binding sites must be occupied before sexual interaction can proceed beyond mere agglutination. In this organism papillar activation is only seen after association of the flagella over their whole length (Mesland, 1976) or after treatment with large doses of isoagglutinin (Mesland and van den Ende, 1978a). This quantitative aspect should be borne in mind when considering the interpretation of Gibor and co-workers.

A third important aspect of agglutination, referred to above, is the phenomenon that flagella do not adhere randomly but gradually align, tip-to-tip. There is general agreement that this association of flagellar tips is an essential feature of gamete activation. Lewin (1952) attributed this ability of the cells to arrange their flagella in such a well-defined way to the fact that flagella can creep or slide along those of the partner until the tips and the papillae are in apposition. This process was described in detail by Goodenough *et al.* (1979) using a paralyzed mutant of *C. reinhardti* which was more easily studied than vehemently twitching wild-type cells. Initial contact sites appeared to migrate along one or both participating flagella in both directions, and incidentally also moved into and out of the flagellar tip region. The impression gained by the authors, however, was that when *both* flagella of one cell became engaged in adhesive contacts, a contact site entering the tip region would remain stationary there ("tip locking"). Thus, one can imagine that eventually all the cells would be brought in a position with all flagella associated by their tips only.

In *C. eugametos,* the tip oriented movement of adhesion sites was advocated by Mesland and van den Ende (1978a), based on the fact that

mt^- isoagglutinin vesicles added to mt^+ gametes are mainly seen to concentrate at the flagellar tips. They postulated that the receptors responsible for adhesion have a certain mobility in the plane of the membrane. The occurrence of adhesion between flagella would lead to clusters of occupied receptors which as such would preferably move in the direction of the flagellar tip, analogous to the capping phenomenon described in lymphocytes, slime molds, etc. (Edelman et al., 1973; Beug et al., 1973). This implies that tip-oriented receptor transport and the resulting gamete activation could only be established if the appropriate ligands were in a multivalent state. In fact, Goodenough and Jurivich (1978) demonstrated that C. reinhardti gametes are activated by antiserum raised against the flagellar surface, but not by the corresponding monovalent Fab fragments. On the other hand, it should be noted that in C. eugametos there is presently suggested evidence for tip-oriented receptor transport only in the mt^+; mt^- gametes have been observed to be activated by isoaglutinin without tip-associated flagella (Mesland and van den Ende, 1978a).

VI. PROSPECTS

From this survey of sexual interaction in C. eugametos it is clear that a number of tasks have to be accomplished in the near future: (1) the evidence that in mt^- flagella the PAS-1a component is directly involved in sexual agglutination will have to be completed, and a similar line of arguments will have to be obtained for the corresponding agent(s) in mt^+ flagella; (2) work needs to continue to identify the primary structure of the PAS-1a component, which is the first stage in elucidating the nature of interflagellar binding; (3) the question whether sex-specific antigenic sites are involved in sexual agglutination must be answered, for the corresponding antibodies could be useful in identifying the sites responsible for the observed high specificity in the recognition process; and finally, (4) the mechanism by which flagellar adhesion proceeds from random contacts to the fully "lined-up" state remains to be elucidated. In addition, the relation between flagellar association and activation of the gamete cells must be explained.

ACKNOWLEDGMENT

The author expresses his gratitude to Wieger L. Homan, Peter F. Lens, and Alan Musgrave for permitting him to use unpublished material and for their assistance during the preparation of this manuscript; he is indebted to Dr. U. W. Goodenough and Dr. Dick Mesland for providing preprints of their papers. This work was partly supported by the Netherlands Organization for the Advancement of Pure Research.

REFERENCES

Bergman, K., Goodenough, U. W., Goodenough, D. A., Jawitz, J., and Martin, H. (1975). Gametic differentiation in *Chlamydomonas reinhardtii*. II. Flagellar membranes and the agglutination reaction. *J. Cell Biol.* 67, 606-622.

Beug, H., Katz, F. E., and Gerisch, G. (1973). Dynamics of antigenic membrane sites relating to cell aggregation in *Dictyostelium discoideum*. *J. Cell Biol.* 56, 647-658.

Bray, D. F., Nakamura, K., Costerton, J. W., and Wagenaar, E. B. (1974). Ultrastructure of *Chlamydomonas eugametos* as revealed by freeze etching: Cell wall, plasmalemna and chloroplast membrane. *J. Ultrastruct. Res.* 47, 125-141.

Brown, R. M., Jr., Johnson, C., and Bold, H. C. (1968). Electron and phase-contrast microscopy of sexual reproduction in *Chlamydomonas moewusii*. *J. Phycol.* 4, 100-120.

Cavalier Smith, T. (1975). Electron and light microscopy of gametogenesis and gamete fusion in *Chlamydomonas reinhardii*. *Protoplasma* 86, 1-18.

Claes, H. (1971). Autolyse der Zellwand bei den Gameten von *Chlamydomonas reinhardii*. *Arch. Microbiol.* 78, 180-188.

Edelman, G. M., Yahara, I., and Wang, J. L. (1973). Receptor mobility and receptor-cytoplasmic interactions in lymphocytes. *Proc. Natl. Acad. Sci. U.S.A.* 70, 1442-1446.

Förster, H., and Wiese, L. (1954). Untersuchungen zur Kopulationsfähigkeit von *Chlamydomonas eugametos*. *Z. Naturforsch., Teil B* 9, 470-471.

Friedmann, I., Colwin, A. L., and Colwin, L. H. (1968). Fine structural aspects of fertilization in *Chlamydomonas reinhardi*. *J. Cell Sci.* 3, 115-128.

Goodenough, U. W. (1977). Mating interactions in *Chlamydomonas*. In "Receptors and Recognition" (J. L. Reissig, ed.), Ser. B, Vol. 3, pp. 323-351. Chapman & Hall, London.

Goodenough, U. W., and Jurivich, D. (1978). Tipping and mating structure activation induced in *Chlamydomonas* gametes by flagellar membrane antisera. *J. Cell Biol.* 79, 680-693.

Goodenough, U. W., and Weiss, R. L. (1975). Gametic differentiation in *Chlamydomonas reinhardti*. *J. Cell Biol.* 67, 623-637.

Goodenough, U. W., Hwang, C., and Martin, H. (1976). Isolation and genetic analysis of mutant strains of *Chlamydomonas reinhardti*, defective in gametic differentiation. *Genetics* 82, 169-86.

Goodenough, U. W., Adair, W. S., Caligor, E., Forest, C. L., Hoffman, J. L., Mesland, D. A. M., and Spath, S. (1979). Membrane-membrane and membrane-ligand interactions in *Chlamydomonas* mating. *J. Gen. Physiol.* (in press).

Gowans, C. S. (1960). Some genetic investigations on *Chlamydomonas eugametos*. *Z. Vererbungsl.* 91, 63-73.

Gowans, C. S. (1976). Genetics of *Chlamydomonas moewusii* and *Chlamydomonas eugametos*. In "The Genetics of Algae" (R. A. Lewin, ed.), pp. 145-173. Blackwell, Oxford.

Homan, W. L., Musgrave, A., Molenaar, E. M., and Ende, H. van den. (1980). Isolation of monovalent sexual binding components from *Chlamydomonas eugametos* flagellar membranes. *Arch. Microbiol.* (in press).

Ishiura, M.,m and Iwasa, K. (1973). Gametogenesis in *Chlamydomonas*. II. Effect of cycloheximide on the induction of sexuality. *Plant Cell Physiol.* 14, 923-933.

Kates, J. R., and Jones, R. F. (1964). The control of gametic differentiation in liquid cultures of *Chlamydomonas*. *J. Cell. Comp. Physiol.* 63, 157-164.

Laemmli, U. K. (1970). Cleavage of structural proteins during the assembly of the head of bacteriophage T4. *Nature (London)* **227**, 680-685.
Lens, P. F., Briel, W. van den, Musgrave, A., and Ende, H. van den. (1980). Sex-specific glycoproteins in *Chlamydomonas* flagella. An immunological study. *Arch. Microbiol.* **126**, 77-81.
Lewin, R. A. (1952). Studies on the flagella of algae. I. General observations on *Chlamydomonas moewusii* Gerloff. *Biol. Bull. (Woods Hole, Mass.)* **102**, 74-79.
Lutz, H. U., Lomant, A. J., McMillan, P., and Wehrli, E. (1977). Rearrangements of integral membrane components during *in vitro* aging of sheep erythrocyte membranes. *J. Cell Biol.* **74**, 389-398.
McLean, R. J., Laurendi, C. J., and Brown, R. M., Jr. (1974). The relationship of gamone to the mating reaction in *Chlamydomonas moewusii. Proc. Natl. Acad. Sci. U.S.A.* **71**, 2610-2613.
Martin, N. C., and Goodenough, U. W. (1975). Gametic differentiation in *Chlamydomonas reinhardtii*. I. Production of gametes and their fine structure. *J. Cell Biol.* **67**, 587-605.
Mesland, D. A. M. (1976). Mating in *Chlamydomonas eugametos*. A scanning electron microscopical study. *Arch. Microbiol.* **109**, 31-35.
Mesland, D. A. M. (1977). Flagellar surface morphology of *Chlamydomonas eugametos*. *Protoplasma* **93**, 311-323.
Mesland, D. A. M., and van den Ende, H. (1978a). The role of flagellar adhesion in sexual activation of *Chlamydomonas eugametos. Protoplasma* **98**, 115-129.
Mesland, D. A. M., and van den Ende, H. (1978b). An inhibitor of cell fusion in *Chlamydomonas eugametos. Arch. Microbiol.* **117**, 131-134.
Musgrave, A., Homan, W. L., van den Briel, M. L., Lelie, N., Schol, D., Ero, L., and van den Ende, H. (1979). Membrane glycoproteins of *Chlamydomonas eugametos*. *Planta* **145**, 417-425.
Ray, D. A., Solter, K. M., and Gibor, A. (1978). Flagellar surface differentiation. Evidence for multiple sites involved in mating of *Chlamydomonas reinhardti. Exp. Cell Res.* **114**, 185-189.
Rostas, J. A. P., Kelly, P. T., and Cotman, C. W. (1977). The identification of membrane glycocomponents in polyacrylamide gels: A rapid method using ^{125}I-labeled lectins. *Anal. Biochem.* **80**, 366-372.
Sager, R., and Granick, S. (1954). Nutritional control of sexuality in *Chlamydomonas reinhardti. J. Gen. Physiol.* **37**, 729-742.
Schlösser, U. G., Sachs, H., and Robinson, D. G. (1976). Isolation of protoplasts by means of a "species-specific" autolysin in *Chlamydomonas. Protoplasma* **88**, 51-64.
Schmeisser, E. T., Baumgartel, D. M., and Howell, S. H. (1973). Gametic differentiation in *Chlamydomonas reinhardti:* Cell cycle dependency and rates in attainment of mating competency. *Dev. Biol.* **31**, 31-37.
Snell, W. J. (1976). Mating in *Chlamydomonas:* A system for the study of specific cell adhesion. *J. Cell Biol.* **68**, 48-69.
Snell, W. J., and Moore, W. S. (1980). Aggregation-dependent turnover of flagellar adhesion molecules in *Chlamydomonas* gametes. *J. Cell Biol.* **84**, 203-210.
Solter, K. M., and Gibor, A. (1977). Evidence for the role of flagella as sensory transducers in mating of *Chlamydomonas. Nature (London)* **265**, 444-445.
Solter, K. M., and Gibor, A. (1978). Removal and recovery of mating receptors on flagella of *Chlamydomonas reinhardti. Exp. Cell Res.* **115**, 175-181.
Triemer, R. E., and Brown, R. M., Jr. (1975). Fertilization in *Chlamydomonas reinhardti,* with special reference to the structure, development and fate of the choanoid body. *Protoplasma* **85**, 99-107.

Weber, K., Pringle, J. R., and Osborne, M. (1972). Molecular weight determinations and related procedures. In "Enzyme Structure," Part C (C. H. W. Hirs and S. N. Timasheff, eds.), Methods in Enzymology, Vol. 26, pp. 3-27. Academic Press, New York.

Weiss, R. L., Goodenough, D. A., and Goodenough, U. W. (1977). Membrane differentiations at sites specialized for cell fusion. J. Cell Biol. 72, 144-160.

Wiese, L. (1965). On sexual agglutination and mating type substances (gamones) in isogamous heterothallic Chlamydomonads. I. Evidence of the identity of the gamones with the surface components responsible for sexual flagellar contact. J. Phycol. 1, 46-54.

Wiese, L. (1969). Algae. In "Fertilization. Comparative Morphology, Biochemistry and Immunology" (C. B. Metz and A. Monroy, eds.). Vol. 2, pp. 135-188. Academic Press, New York.

Wiese, L. (1974). Nature of the sexspecific glycoprotein agglutinins in Chlamydomonas. Ann. N.Y. Acad. Sci. 234, 383-395.

Wiese, L., and Shoemaker, D. W. (1970). On sexual agglutination and mating type substances (gamones) in isogamous heterothallic Chlamydomonads. II. The effect of concanavalin A upon the mating type reactions. Biol. Bull. (Woods Hole, Mass.) 138, 88-95.

Wiese, L., and Wiese, W. (1975). On sexual agglutination and mating type substances in isogamous dioecious Chlamydomonads. IV. Unilateral inactivation of the sex contact capacity in compatible and incompatible taxa by α-mannosidase and snake venom protease. Dev. Biol. 43, 264-276.

Wiese, L., and Wiese, W. (1978). Sex cell contact in Chlamydomonas, a model for cell recognition. In "Cell Interactions" (A. S. G. Curtis, ed.), Symposia of the Society for Experimental Biology, Vol. 32, pp. 83-104. Cambridge Univ. Press, London and New York.

Wiese, L., Wiese, W., and Edwards, D. A. (1979). Inducible anisogamy and the evolution of oogamy from isogamy. Ann. Bot. (London) 44, 131-139.

Witman, G. B., Carlson, K., Berliner, J., and Rosenbaum, J. L. (1972). Chlamydomonas flagella. I. Isolation and electrophoretic analysis of microtubules, matrix, membranes and mastigonemes. J. Cell Biol. 54, 507-539.

13

Preconjugant Cell Interactions in *Oxytricha bifaria* (Ciliata, Hypotrichida): A Two-Step Recognition Process Leading to Cell Fusion and the Induction of Meiosis

NICOLA RICCI

I.	Introduction	319
	A. Systematics, Morphology, and Ecology	321
	B. Life Cycle	322
	C. Culturing Methods	324
II.	Literature Review and Current Research	325
	A. Preconjugant Cell Interactions	325
	B. Early Conjugant Cell Interactions	340
III.	Perspectives	343
	A. Cell–Cell Interactions in *Oxytricha bifaria*	343
	B. Future Research	346
	References	347

I. INTRODUCTION

The idea that the progress of biology largely depends on the suitability of the experimental tool(s) chosen to study a certain problem was realized as soon as the age of experimental investigation encroached upon the age of pure description of nature. It seems to me that one of the most important contributions of this volume, and of this chapter as well, could be to demonstrate that, even though a rather wide range of living tools has been already employed to investigate the problems of cell biology, an even larger number of potential tools is still to be discovered and properly used.

The ciliated Protozoa have been widely studied, not only to increase the knowledge of their own biology but also to investigate many general problems in cell biology. Though very well known and described since the last century (Bütschli, 1876; Maupas, 1889), the sexual act of ciliates, called "conjugation," did not become useful for biological research until 1937, when Sonneborn demonstrated how the sexual difference (called "mating type") was inherited and determined in these organisms. Metz (1954) was the first to suggest that conjugation could be studied as a model of fertilization. Yet, only recently (Miyake, 1974), the sexual reproduction of ciliates has been proposed as a useful tool for studying the problems of development at the cellular level.

The most distinctive characteristics of ciliates are (a) the complex and patterned cortex which is covered with cilia and (b) the two different kinds of nuclei which exist in each cell. A small diploid nucleus, called the "micronucleus," may be distinguished from a large, hyperpolyploid nucleus, called the "macronucleus." They are responsible, respectively, for sexual reproduction and control of the vegetative cell. Such nuclear dimorphism obviously recalls the difference between the germ and the somatic lines of cells in Metazoa, thus suggesting a possible analogy between ciliated and pluricellular organisms. On the other hand, one must also keep in mind that the ciliates are unicellular individuals. Once in a while, however, conjugation has to occur and it constitutes a sort of bottleneck in their life cycle (Nanney, 1977; Miyake, 1978). To accomplish conjugation, the ciliates must gain the capacity to communicate with each other. The interaction systems, although leading all the ciliates to the same final goal, show in the different groups a wide range of variation, according to their evolutionary and ecological strategies. Ciliate conjugation provides an ideal system with which the general mechanisms of cell–cell interactions may be investigated. More specifically, the conjugation of ciliates allows us to study the cell interactions occurring between only two types of cells at a time, and these cells may differ from each other only in their mating type.

Once some parameters are defined, the preconjugant cell interactions constitute an experimentally inducible biological system, with almost absolute reproducibility. Finally, since the preconjugant cell interactions of ciliates deal with sexual functions, it offers a good chance to gain a new insight into fundamental problems of cell biology such as cell union, induction of meiosis, and nuclear differentiation.

13. Preconjugant Cell Interactions in *Oxytricha bifaria*

Certain ciliates, such as *Paramecium, Tetrahymena, Stentor,* and *Blepharisma,* are very familiar to the biologists and already have been studied intensively. *Oxytricha bifaria,* however, represents an almost brand new research tool.

A. Systematics, Morphology, and Ecology

According to the most recent systematic compilation, elaborated by Corliss (1979), *O. bifaria* (Stokes, 1887) belongs to the family Oxytrichidae, of the order Hypotrichida, phylum Ciliophora.

The hypotrichs are easily recognizable. They possess a flattened body, which is usually elongated or oval in form and which permits one to distinguish the ventral from a dorsal surface. The absence of simple somatic cilia, together with the presence of complex ciliary organelles, characterizes this order of Ciliophora. The most striking of these organelles are the adoral zone membranelles (AZM) and the cirri. The AZM are very complex, highly polymeric feeding ciliary structures, located in the anterior, left third of the body. The cytostome is located on the ventral surface, right at the posterior end of the AZM (Fig. 1A,B). The cirri, though not absolutely specific of hypotrichs, are considered essential to this order because of their abundance, importance in feeding and moving, and specific positional patterns on the ventral surface. The cirral pattern is of considerable diagnostic value. The

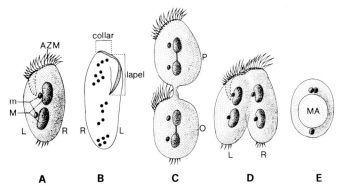

Fig. 1. The morphology of *O. bifaria*. (A) Dorsal view: M, macronucleus; m, micronucleus; AZM, adoral zone membranelles; L, left side of the cell; R, right side. (B) Ventral view: the schematic drawing shows the location of the cirri (black dots). Collar, anterior part of the AZM; lapel, central and posterior part of the AZM. (C) General morphology of a vegetative, dividing cell. P, proter; O, opisthe. (D) A pair, as old as about 1.5 hr, is shown. L, left, still partner; R, right, rotating partner. (E) An exconjugant cell is represented; MA, the developing macronuclear anlage.

flattened body, together with the cirri on the ventral surface, make the hypotrichs the only ciliates which truly walk on the substrate, and in this movement they display great mobility and skill (cf. Avoiding Reaction; Section II,A,3). These organisms, moreover, show a definite tendency to adhere to the substrate. Thus, they behave like thigmotactic organisms.

The hypotrichs are believed to represent the highest point in the phylogeny of ciliates. Well known for their ubiquitousness, the hypotrichs are usually free-living organisms, which may be easily found in fresh, brackish, and marine waters, in and/or on different substrates.

Among the hypotrichs, the Oxytrichidae are recognizable for their rows of both right and left marginal cirri and for the heavy and distinctive frontoventral and transverse cirri. A general scheme of *O. bifaria*, a hypotrich about 125 μm in length and about 50 μm in width, is shown in Fig. 1A, B, and C. This hypotrich has two diploid micronuclei and two macronuclei. Although it is composed of two different pieces, the macronucleus should be considered a unique structure, since it is a product of only one of the four nuclei that are produced by the first two mitoses of the synkaryon (cf. Section I,B; Fig. 2).

B. Life Cycle

The life cycle of *O. bifaria* is quite similar to the general one reported for other ciliates. Vegetative cells periodically undergo transverse binary fissions. The two resulting sister cells are, respectively, called "proter" and "opisthe" according to their anterior or posterior position during the cell division (Fig. 1C). At 22°C, the length of the cell cycle, i.e., the generation time, ranges between 8 to 9 hr, depending on the culture conditions (cf. Section I,C). The length of both micronuclear and macronuclear G_1, S, and G_2 phases have been studied by Dini *et al.* (1975). The micronuclear G_1 lasts about 7 hr while the S phase occurs in 1 hr just before mitosis. No G_2 is observed in the micronuclei of *O. bifaria*. The macronuclear G_1 phase lasts approximately 3.5 hr as does the S phase, which is followed by a short G_2.

The mating type system of *O. bifaria* is composed of at least nine complementary mating types, according to the data of Siegel (1956). The same results have also been obtained with the clones collected in Italy (N. Ricci, unpublished). The nine complementary types all intermate and no preferential mating was ever observed. When a clone becomes old (cf. Section II,A,2), it stops conjugating, but never undergoes selfing. Moreover, in the clones we have studied, autogamy never occurred.

13. Preconjugant Cell Interactions in *Oxytricha bifaria*

When vegetative cells of *O. bifaria,* belonging to different mating types are mixed together under the appropriate conditions (cf. Section II,A,2) the complementary cells unite in pairs. A time lag of about 1.5 to 2 hr, called the "waiting period," always elapses between mixing and pairing. It is interesting that *Paramecium* has no "waiting period" (see Chapter 14 by Hiwatashi). A stereotyped "visible reaction" (cf. Section II,A,3) occurs at the end of the waiting period leading to cell union and to the ensuing events of sexual exchange. The preconjugant cell interactions of *O. bifaria* occur during both the waiting period and the visible reaction. According to our data, we may further distinguish a "tête-à-tête" and a "rotation" phase as components of the visible reaction. A more detailed description of these phenomena is given in Sections II,A,2 and 3.

As shown in Fig. 2, the micronuclei of the two partners first undergo the two meiotic divisions followed by one mitotic division. The two haploid products of each partner are called "stationary" and "migrant" pronuclei. The migrant pronucleus of each partner then moves into the other mate and fuses with its stationary pronucleus by a sort of cross-fertilization, to produce a synkaryon. At this moment, the diploid phase of the life cycle is restored. Two successive postgamic mitoses produce four micronuclei. The third micronucleus from the anterior end always becomes the macronuclear anlage. During the meiotic

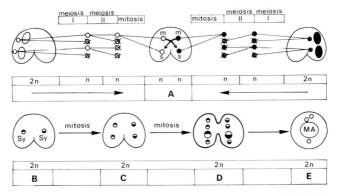

Fig. 2. Nuclear events occurring during sexual reproduction of *O. bifaria*. Only the micronuclear events are detailed. Upper part (A): the two divisions of meiosis plus one mitosis are indicated. Some products of these divisions degenerate; the other produces a migrant (m) and a stationary (s) pronucleus in each partner. Lower part: in (B) the nuclear fusion events give rise to the synkarya (Sy), which, by two successive mitoses, produce four micronuclei in each partner (C, D). The third of them, from the anterior end, will become the new macronuclear anlage (MA). At this point (E) the partners separate and two exconjugants are formed. (According to N. Ricci and R. Banchetti, submitted.)

events of conjugation, the old macronuclear pair breaks down and finally disappears. After about 19 to 20 hr, the mates separate.

A period of sexual immaturity follows conjugation. A sexually immature cell, though apparently quite normal from both the morphological and physiological points of view, divides vegetatively with a regular cell cycle, but it cannot conjugate. The cells become sexually mature and can enter the sexual phase of their life cycle only about after 170 binary fissions (R. Cetera, unpublished) after conjugation. Our results are in good agreement with what reported by Siegel (1956), who found the immaturity period to be from 1 month to 2 years. The fact that *O. bifaria's* breeding system is multipolar, together with such a long sexual immaturity and the absence of autogamy, seems to suggest that this hypotrich has possibly evolved according to outbreeding strategies (Sonneborn, 1957). The control mechanism of the immaturity period has not been defined as it has for *Paramecium* (see Chapter 14 by Hiwatashi).

Once they become sexually mature, Oxytrichae maintain their maturity for at least 2 years (Siegel, 1956). Finally, the clones become "old," and show a decreasing ability to conjugate. Beyond a certain point, a clone can only reproduce vegetatively and will eventually die (Ricci *et al.*, unpublished).

C. Culturing Methods

All the strains of *O. bifaria* used in our experiments were collected in freshwater ponds and canals near Pisa, Italy, by means of plankton nets. Because of the strong thigmotactism typical of this ciliate, it is essential to rub the net over the aquatic plants and over branches or roots dipped in the water. Although two peaks of the population density occur during late spring and early fall, specimens of *O. bifaria* may be found all year long. A minimum population density occurs during the winter (Ricci, unpublished). The cells are isolated using a dissecting microscope and then tested for their mating type. The strains are either kept in test tubes as stock cultures or grown in 1.0 liter Erlenmeyer flasks, when massive cultures are to be used for experimental purposes. The temperature is maintained at about 22°C. The general techniques are those reported by Sonneborn (1950).

The culture method for *O. bifaria* is the classic lettuce medium inoculated with *Aerobacter* (Sonneborn, 1950). One of the major problems in culturing *O. bifaria,* is that the cells are strongly thigmotactic. As a result they preferentially distribute on the walls of the flasks, leaving about 90% of the food source unexploited. As a consequence, the large quantity of living and dead bacteria in the medium seriously affects the distribution of the oxygen in the fluid, so that once the culture reaches a certain volume, the cells distribute only on the surface of the fluid and on the walls, 2–3 cm from the air–water contact zone. Preliminary experiments, in which the flasks are rotated and gently shaken, appears to help overcome both the thigmotactism and the oxygen distribution. Our initial results seem to suggest that this improves the culturing conditions (Ricci *et al.*, 1980c). The optimum ratio between the volume of the culture and of food added every time into the flasks is about 2:1. The cells must be properly starved in order for conjugation to occur. The cells are collected and then washed free from bacteria by successive mild centrifugations, and then resuspended into uninoculated lettuce medium or SMB (salt medium for *Blepharisma;* Miyake, 1968). The cells are suspended in either medium at an approximate cell density of 25,000 cells/ml and then starved for about 24 hr.

II. LITERATURE REVIEW AND CURRENT RESEARCH

A. Preconjugant Cell Interactions

The interactions occurring between mixed mating types of *O. bifaria* before they conjugate are referred to as "preconjugant cell interactions."

1. Preliminary Constraints

Among the endogenous constraints, the cells must be sexually mature and be of opposite mating type. With regard to the former, we can say that the cells must acquire some physiological capacity in order to become sensitive to the possible stimuli which lead to conjugation. Whether, as suggested by Nanney (1977), such a capacity depends on new synthesis or on the increased amount of the synthesized substances or, even, on the exposition of the same substances at the proper cortical sites is still to be ascertained. Another endogenous constraint is imposed by the cell cycle: there is a restricted period during which the cells can conjugate. As reported by Wolfe (1973) for *Tetrahymena* and by Luporini and Dini (1975) for *Euplotes crassus,* the cells of *O. bifaria* must also be in the G_1 macronuclear phase (Luporini and Dini, 1975) for the conjugation to occur. Beyond all the physical and chemical exogenous constraints that can affect the occurrence of maturation in ciliates, we must recall that the pairing was found to be related to daily rhythms in *Paramecium bursaria* (Cohen, 1964, 1965) and in

Euplotes crassus (Miyake and Nobili, 1974). In *O. bifaria* no apparent relationship between the occurrence of conjugation and any daily periodicity was found (Esposito *et al.*, 1976). Complementary cells of *O. bifaria* start conjugating any time during the day, showing no preferentiality. These results confirmed the data reported by Siegel (1956).

Finally, the influence of starvation on conjugation of *O. bifaria* was carefully investigated (Ricci *et al.*, 1975b) using as the parameter the length of the time lag between the mixture of complementary cells and the onset of pairing. Optimal reactions occur with starvation times ranging from 24 to 36 hr. Under these conditions, the cells conjugate within 45 min, while randomly starved cells need between 100 and 120 min on average to give the same reaction. Starvation times shorter than 24 hr did not significantly affect the length of the reaction time, while longer periods of starvation dramatically influenced the cells. After about 48 hr without food, the cells were irreversibly damaged and eventually died. Starving cells progressively reduce their motility, stop swimming, and walk on the bottom of the depression. It is clear that they do not depend upon any mating type-specific interaction, as they can also be observed in depressions when only one cell type is present. Thus, we can conclude that a certain degree of starvation is necessary for populations of *O. bifaria* to conjugate under optimal conditions. Whether starvation alone can induce the shift onto the sexual pathway, or whether it just represents an extremely unfavorable condition for the cells under which sexual activity can occur by overcoming its intrinsic disadvantages (Nanney, 1977) is still to be demonstrated. In favor of the first hypothesis, one can mention the data of Wolfe (1973) for *Tetrahymena,* of Mishima (1978) for *Paramecium,* and of Dini and Luporini (1979) for *Euplotes,* all of which seem to point to a major role of starvation in inducing differentiation of cells toward the sexual phase of the life cycle.

2. Long-Distance Interactions

Even if optimal starvation conditions are obtained, the cells require a time lag of at least 45 min before they can start pairing. The occurrence of a waiting period suggested that the complementary cells have to interact with each other for a while before conjugating. In general, the waiting period is defined as the time lag between the mixing of two types of cells and the onset of any distinguishable behavioral response (see Miyake, 1979; see also Chapter 5 by Miyake). Such a definition perfectly suits the *O. bifaria* situation in as much as the waiting period of this hypotrich precedes the fairly sophisticated behavioral pattern that leads to the actual cell fusion.

It is well known that during the waiting period of *Blepharisma japonicum* cellular interactions occur, which are mediated by soluble gamones (see Miyake, 1968; Miyake and Beyer, 1973; see also Chapter 5 by Miyake). On the other hand, Heckmann and Siegel (1964) reported that during the waiting period of *Euplotes crassus,* the complementary cells interact by specific contacts which generally prepare the cells for fusion. Recently, Dini and Luporini (1979) showed that the cell contacts actively induce macronuclear synchronization in G_1 phase, which is one of the preliminary conditions necessary for the conjugation. The possible role of the cortical ampules, described by Verni *et al.* (1978), in these contacts is not yet known. Also the waiting period of *Tetrahymena thermophila* has been extensively studied and it was shown that a specific cell–cell interaction, called costimulation, occurs during this period (Bruns and Brussard, 1974).

In order to ascertain whether some interaction occurs during the waiting period of *O. bifaria* or not, a careful study was conducted. It was shown that with time, the mixed cells, at first randomly distributed on the bottom, progressively aggregate in a small area; usually, there is one aggregation per depression. In this area they decrease their mobility until they lie almost still on the substrate. The first tête-à-têtes occur almost contemporaneously. Similar results have been reported by Kimball (1939) for *Euplotes patella* and, very recently, for *E. minuta,* by Heckmann (see Miyake, 1978). Moreover, Kimball demonstrated that cells of *E. patella* belonging to a certain mating type can condition their fluid so that complementary cells which are subsequently incubated in that fluid may undergo selfing. With *O. bifaria* it was shown that by treating complementary cells with cell-free fluid (CFF) obtained from different mating types, one can induce them to conjugate after a significantly shorter waiting period (30 min) than is usual (45 min) (Ricci *et al.,* 1975b). In spite of this significant activation, however, no selfing was ever observed. Though statistically different, the two results were not clearly comparable, because of the large standard deviation observable in the length of the waiting period of cells treated by the heterologous CFF.

To reduce such a wide range of variation of the reaction times, a simple diffusion apparatus was designed. Each apparatus is formed by two perspex cylinders (5 cm × 0.5 cm), closed at the outside ends by two perspex bases and held together by threaded brass rods with wing nuts. The apparatus was sealed with silicone grease. A Sartorius membrane filter (5 cm diameter) with 8 μm pores is placed between the two cylinders, in order to obtain contiguous chambers separated by the filter. Complementary mating types are placed separately into the two contiguous chambers, but may interact with each other through the filter only by means of soluble factors. As reported by Esposito *et al.* (1976), after a period of about 4–6 hr, the cells become accustomed to the apparatus and, after mixing, start conjugating without delay. Three groups of experiments were conducted to investigate several different problems.

Cells belonging to the complementary mating types X and Y were placed separately into the two contiguous chambers of an apparatus (heterologous treatment) (X/Y); two control apparatus were also prepared with the same mating type in their chambers (homologous treatment or control) (X/X; Y/Y), as shown in Fig. 3. Heterologous and control mixtures were made at various times after the beginning of the treatment. The graph in Fig. 3 shows the reaction of activated heterologous cells compared with that of homologous control cells. Since "activated" cells give many pairs within 17–20 min from the moment of the mixture, we can say that the waiting period was suppressed or, that, under these conditions, the cell–cell interactions, which must occur during the waiting period, took place across the Sartorius membrane, before the physical mixing of the complementary cells. Thus,

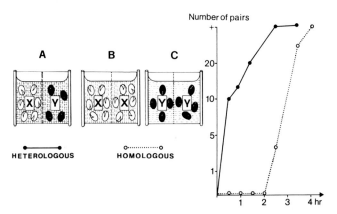

Fig. 3. Cell–cell interactions across Sartorius membranes as defined by subsequent pairing reactions. Left: schematic representation of three Sartorius apparatus. (A) "heterologous treatment": complementary mating types interact across the membrane. (B, C) "homologous treatment" only cells of the same mating type interact. Right: the typical pairing reaction of the heterologous (●) and of the homologous (○) mixtures is given. On the abscissa the times after the mixture of complementary cells and on the ordinate the number of pairs have been indicated. (Redrawn from Esposito *et al.*, 1976.)

the excretion of mating type-specific soluble factors (gamones) occurs during the mating process.

Three apparatus were prepared, as represented in Fig. 4. Cells belonging to either the X or Y mating type were put into only one chamber of a two-chamber apparatus. X and Y cells were also put into a five-chambered apparatus according to the following combination: X/Y/–/X/Y. X or Y cells were then incubated in the fluids collected from the cell-free chambers. Three different treatments were used as shown in Fig. 4. The results are shown in Fig. 5. It was demonstrated that small amounts of the gamones are excreted spontaneously by the cells in both heterologous and homologous mixtures (Fig. 5). The best reaction was, however, shown by the "plurinduced" cells, suggesting that a feedback mechanism enhances the excretion of the gamones by the cells.

In another experiment, X cells interacted with Y cells across a sartorius membrane in two-chamber apparatus and were, respectively, called X_y and Y_x. W cells similarly interacted with Z cells in another apparatus and were called, respectively, W_z and Z_w. Subsequently the following mixtures were made: (I) $X_y \times Y_x$, (II) $W_z \times Z_w$, (III) $X_y \times W_z$, and (IV) $Y_x \times Z_w$. It should be noted that, in mixtures I and II, cells that had previously interacted with each other were put

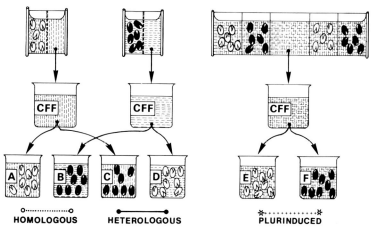

Fig. 4. Isolation of cell free fluid from various Sartorius cultures. Two Sartorius apparatus are shown on the left, upper part: from their cell-free chambers, cell-free fluids (CFF) are obtained and then used for "homologous" (A, B) and "heterologous" (C, D) treatments. On the right, a five-chamber Sartorius apparatus is shown, from which a CFF is obtained and used for "pluriinduced" (E, F) treatments. Mixtures of complementary cells, similarly treated, are made: A×B (○), C×D (●), and E×F (*). (Redrawn from Esposito et al., 1976.)

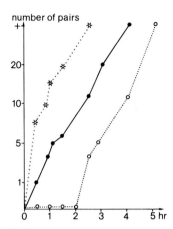

Fig. 5. The effect of cell free fluid from various Sartorius culture on the pairing reaction. The pairing reaction of plurinduced (*), heterologous (●), and homologus (○) cells. More details are given in the legend of Fig. 4. (Redrawn from Esposito et al., 1976.)

physically together, whereas in combinations III and IV the mixed cells had not experienced the reciprocal interaction, although they had been heterologously activated. The pairing reactions in these experiments were identical to those of the "plurinduced" curve shown in Fig. 5. It was therefore concluded that the activation induced by the soluble gamones is not mating type-specific; the mating readiness induced by the gamones is independent of the source of the complementary gamones themselves.

To summarize the information obtained on the role of the waiting period in the events leading to pair formation in *O. bifaria,* we can say that, during the waiting period, each mating type spontaneously excretes its own soluble gamone(s). In addition, there appears to be a positive feedback mechanism for the enhancement of gamone secretion, once two mating types start interacting. Moreover, the soluble gamones activate the complementary cells so that they can contact other similarly activated potential mates, and undergo the visible reaction regardless of the mating type which activated them. Under our standard conditions, the heterologous treatment can induce cells of the same mating type to interact with each other in a tête-à-tête position (homotypic tête-à-tête), but it cannot induce the same cells to form homotypic pairs. This result indicates that the gamone system of *O. bifaria* somehow differs from that of *Blepharisma japonicum* (see Chapter 5 by Miyake). In fact, in the same apparatus and under similar conditions, homotypic pairs of *B. japonicum* are formed, according to the classic results of Miyake (1968). Thus the action of the *O. bifaria* gamones seems to be to make the cells more ready to pair without any synchronizing effect on the populations they activate. If we compare the curves of heterologous and homologous cells in Fig. 3, we can see that the pairing of the heterologous cells simply occurs earlier than the pairing of the controls.

Is there any metabolic event, occurring during the waiting period of *O. bifaria*, which is specifically related to conjugation? Earlier studies had shown that a critical role is played by RNA and protein syntheses in the processes leading to conjugation, of *Paramecium* (Nobili, 1963; Nobili and Kotopulos De Angelis, 1963; Bleyman, 1964), of *Tetrahymena* (Tyler and Wolfe, 1972; Cleffmann et al., 1974), of *B. japonicum* (Miyake and Honda, 1976), and of *Stylonychia mytilus* (Sapra and Ammermann, 1973). Experiments were therefore carried out to study the effect (if any) of actinomycin D and cycloheximide on the conjugation of *O. bifaria*. Complementary populations were treated by these inhibitors for different periods before mixing (Esposito and Ricci, 1975). The results from these experiments showed that the RNA(s) and protein(s) which function in the cell pairing are synthesized in different periods during the preconjugant cell interactions of *O. bifaria*. Although preliminary, these results assume more relevance, when compared with similar, recent results obtained for *E. crassus* (Dini and Miyake, 1977, in prep.) and *T. pyriformis* (Bruns and Palestine, 1975; Allewell et al., 1976; Ofer et al., 1976).

In summary, the waiting period of *O. bifaria* constitutes a period of interaction between complementary cells, which is necessary but not sufficient by itself to induce pair formation. The waiting period seems to prepare the cells for the second, more specific phase of cell interactions and it induces the cells to aggregate on the bottom and to differentiate some specific parts of their cortical areas. We do not know yet whether the cell clustering is due to some specific short-range gamone, which might also direct the mating dances occurring during the visible reaction (Nanney, 1977), or to some substance released and attached directly on the substrate. The last possibility could be expected on the basis of the general thigmotactic strategy underlying all the preconjugant cell interactions of *O. bifaria*.

3. Direct Interactions

Cells of *O. bifaria*, slowly creeping in the area of a cluster, apparently contact each other by chance, usually making contact at their anterior tips (Fig. 6A). When two vegetative cells come in contact, they immediately escape from each other by means of one, or a series of avoiding reactions. On the other hand, activated cells of *O. bifaria*, slowly moving on the substrate, are capable of undergoing fruitful

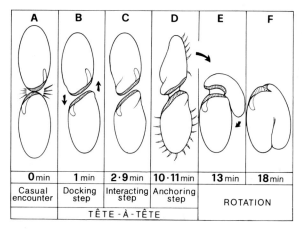

Fig. 6. A diagrammatic representation of the steps in the visible reaction. The term for each phase is indicated below the time intervals. It must be stressed that the homotypic tête-à-tête never goes beyond the interacting step, although it can last as long as 20 min before the cells separate. (Partially from Ricci et al., 1980a.)

contacts and do not avoid the potential partner. Within 30–90 sec the cells creep toward each other, shifting slightly to the right, so that their peristomial areas come in close contact and so that direct interactions may begin (Fig. 6B). Such a process has been called the "positioning step" by Ricci et al. (1980a). This process is similar to the docking step, a common feature of the short-range microbial interactions (see review in Reissig, 1977). The fact that the two cells come close to each other by shifting on their right sides and that they also stop when their peristomial areas are in a right position strongly suggests that some general topographical recognition occurs between the two partners. This kind of recognition is not mating type-specific, since it may occur between complementary or identical cells. Thus, the positioning step seems to be some kind of mechanical prerequisite for the tête-à-tête.

The next phase of the visible reaction is the "interacting step" (Fig. 6C), during which the two cells interact with each other by means of their peristomial ciliary organelles. This phase lasts approximately 7–8 min. A careful study of living specimens revealed that the cells, which are now absolutely motionless on the substrate, lift up their anterior left ends soon after the positioning step occurs. They interact in this position by means of their ciliary organelles. Based on such observations, a mating differentiation of the peristomial membranelles and cirri from the somatic cilia may be suggested, although it can not yet be said whether or not gamones are involved. It appears that the "interacting step" is the critical step for mating-type recognition.

Homotypic (X/X; Y/Y) and heterotypic (X/Y) tête-à-têtes occur, but homotypic pairing never does. This was first reported by Siegel (1956). We can therefore conclude that the recognition role suggested by Esposito et al. (1976) for the visible reaction is likely the "interacting step" resulting in the direct interactions which lead to pair formation. If this is true, then the mechanism preventing homotypic pairing (Esposito et al., 1976) can be ascribed to this same interacting step.

After two cells of the same mating type interact with each other for a while (10–20 min) they separate ("undocking step"), although still retaining the ability to contact other cells. This mechanism constitutes a good example of modulation of microbial interactions. Whether the "undocking step" is due to a loss of ciliary agglutinability of the interacting regions or to some change at the cirri substrate level is not yet known. The fact that, after the undocking step, the cells maintain their capability of contacting other cells seems to support the second hypothesis.

The "anchoring step" occurs about 10–11 min after the first casual encounter (Fig. 6D and Fig. 7A). One cell bends down its marginal cirri

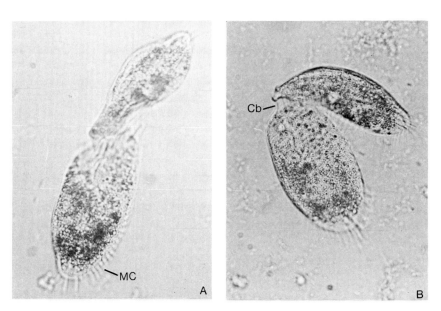

Fig. 7. Photographs showing the characteristics of the visible reaction. At this time the cells are completely motionless. (A) The "anchoring step" can be recognized because the marginal cirri (MC) bend down to anchor the still partner to the substrate, while the partner cell rotated sideways. (B) During late rotation the pairing is almost complete and the cytoplasmic bridge (Cb) is already a clear-cut morphological trait. (Partially from Ricci et al., 1980a.)

and firmly anchors itself to the substrate, while the other cell rotates sideways. The outer pellicles of the partners become fused (Ricci et al., 1975a). This stage of the visible reaction constitutes the first event which makes the two partners distinguishable from each other. The anchored cell became the so-called "still" or "left" partner, while its mate is the "rotating" or "right" partner (Fig. 1D). The two partners differ from each other in many other traits as well. By split-pair experiments (see Sonneborn, 1950), it was shown that pairing is a random event, i.e., that a cell will be the right or the left partner independently of its mating type. Thus, we may infer that, during the interacting step, the two potential mates not only recognize each other as belonging to a complementary mating type, but also "decide" their "role" in the pairing. Whether they actually differentiate into either the left or the right partners or they just recognize their relative readiness to undergo either fate is still to be ascertained.

During the "rotation step," following the interacting step, the future "right" partner rotates clockwise over the other cell, which is now firmly anchored to the substrate. The parameter by which one can distinguish the successive phases of the rotation is the angle formed by the two major axes of cells: at first it is of about 180°, but within 6–7 min it progressively becomes as narrow as 30° (Figs. 6E,F and 7B). Once the visible reaction is over, a pair is formed and the newly formed bicellular complex start moving again at an increasing speed. Moreover, the pairs start swimming again, showing that a perfect morphological and physiological coordination of the two sets of ciliary organelles has been obtained.

While it is possible experimentally to reduce the waiting period to 0 min using a Sartorius apparatus, the temporal and morphological patterns of the visible reaction cannot be affected in any way without completely blocking them. This aspect of the visible reaction suggests that during this period the reactions are so perfectly scheduled and so finely coordinated that any perturbation affecting the process will stop the total sequence of interactions.

Another major problem still unsolved is the length of time that split partners must interact a second time before they become capable of pairing. Do they have to undergo a second complete visible reaction? The answer to this question will allow us to determine to what extent the visible reaction can be considered an all-or-none process. This visible reaction, and more specifically the interacting step, mediates the recognition between the complementary partners and, at the same time, works as a kind of safety lock against homotypic pairing. These two characters make the role of the visible reaction particularly important on the pathway toward conjugation.

Finally, we have to remember that ciliary agglutination never occurs in *O. bifaria*. Miyake (1978) suggests that there are at least two roles of such an important mating reaction: (a) to supply a sufficiently strong adhesion between two cells to let them unite later, and (b) to mediate some specific recognition. If this is true for *B. japonicum*, one has to consider that *O. bifaria* (with its clear-cut thigmotactism) largely relies on the substrate to carry on the preconjugative processes. This is shown by the progressive aggregation of the cells on the bottom of the depression during the waiting period and by the critical role played by the visible reaction. On the basis of these observations, we may expect that the specific mating substances involved in the visible reaction work as recognition mediators, rather than as adhesive-type materials.

A series of experiments was carried out to evaluate the importance of the mechanical strength required to keep the cells united to the substrate and to each other (Ricci *et al.*, 1980a). A glass needle was used to separate the cells during the successive steps of the visible reaction. It was applied at the level of their interacting regions either with an upward or with a downward pressure. The results indicated that before the anchoring step the cells can be split rather easily and that, although affected seriously, they keep lying on the substrate. After the onset of the rotation step, when membrane fusion begins, the glass needle can either lift the two cells from or move them on the substrate without splitting them. Thus, they become more attached to each other than to the substrate. These results confirm that the first part of the visible reaction strongly depends upon anchoring to a substrate (thigomotactic strategy). After membrane fusion, the rotation is not affected and pairing regularly occurs even if the cells are pushed away from the substratum.

In order to study the nature of the substances responsible for the direct interactions, "clouds" of trypsin, diastase, and lipase were applied with a micropipette at the level of the interacting reactions of the partners (Ricci *et al.*, 1980a) (Fig. 8). The lowest specific activity capable of splitting the interacting partners was termed the "threshold activity." The threshold activities were plotted versus times (Fig. 8). Trypsin was found to be incapable of splitting the cells after the interacting step. Moreover, after the beginning of the rotation, neither trypsin nor diastase had any effect, while lipase was still effective in splitting the cells. These results suggested that cortical protein(s)

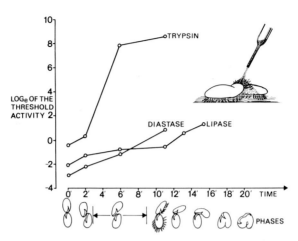

Fig. 8. Enzymatic splitting of cell couple. The upper, right scheme depicts the technique used for these experiments. On the abscissa, the times and the correspondent phases are indicated for both heterotypic and homotypic tête-à-têtes. On the ordinate, the natural logarithm of the lowest specific activity capable of splitting the successive steps is given. More details are given in the text. (Redrawn from Ricci et al., 1980a.)

plays a critical role during the first steps of the visible reaction. After the interacting step, lipase interfered with the visible reaction. These results confirm the preliminary data already obtained by Ricci et al. (1975b).

We next studied the possible correlation between cortical protein(s) and gamone-induced activation. Apparatus were set up as shown in Fig. 9. Four different combinations of cells were thus obtained: (I) "heterologous control cells," belonging to either X or Y mating type, which had been activated by complementary gamones, but which never contacted complementary cells; (II) "heterologous mixed cells," X and Y cells, which had been activated both by complementary gamones and by direct contacts with each other; (III) "homologous control cells," which never interacted with complementary cells in any way; (IV) "homologous mixed cells," not preactivated by complementary gamones, but activated by direct interactions with complementary cells. The effect of the trypsin on tête-à-tête formation of the four different groups was studied. The results, shown in Fig. 10 reveal the following: (a) the gamone activates the cells, making them more resistant to the trypsin treatment, during the first step of the visible reaction, while no additional effect is subsequently observed (Fig. 10, curve II versus IV); (b) interacting cells become increasingly resistant to the splitting action of the trypsin, only if they belong to complementary

13. Preconjugant Cell Interactions in *Oxytricha bifaria* 337

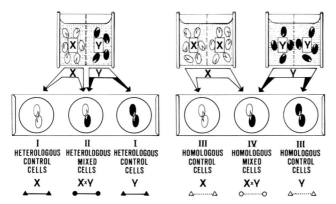

Fig. 9. The Sartorius apparatus were used to obtain visible reactions between "heterologous control cells" (I), "heterologous mixed cells" (II), "homologous control cells" (III), and "homologous mixed cells" (IV). (Redrawn from Ricci *et al.*, 1980a.)

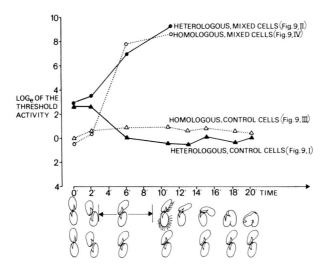

Fig. 10. The ability of trypsin to split cell pairs during the different stages of the visible reaction. On the abscissa the times and the morphological appearance of the stages are indicated: upper line refers to heterotypic visible reaction; lower line refers to homotypic tête-à-tête. On the ordinate the natural logarithm of the lowest specific activity of trypsin which is capable of splitting cell pairs is shown. The four curves are referred to according to the terminology of Fig. 9.

mating types (Fig. 10, curve IV versus III); (c) if properly preactivated cells subsequently interact with cells belonging to the same mating type, they lose their activation (i.e., their resistance to trypsin) within 4-6 min after beginning the interaction (Fig. 10, curve I). These results demonstrated a link between gamones which function during the waiting period and the cell-bound recognition substances, which work during the visible reaction. Whether the soluble gamones of the waiting period and the cell-bound materials are the same substances is not known.

The last aspect of the visible reaction which has been recently investigated (Ricci, in prep.) is the behavior that cells of *O. bifaria* show when they are separated during this period, since it is comparable to the avoiding reaction of vegetative cells. Vegetative cells of *O. bifaria* can escape unfavorable conditions and physical obstacles as well, by a single or a series of "avoiding reactions." The avoiding reaction, first described for *Paramecium* by Jennings (1906), has been recently studied in *O. bifaria* by means of a modified Dryl (1958) technique. An avoiding reaction contains a series of stereotyped elements: (a) sudden cessation of walking, (b) backward motion of about 100 μm, (c) clockwise rotation of about 63°, and (d) forward motion on the new pathway. If two cells are disturbed during the visible reaction they undergo what has been called the "exploring reaction." The split cells tended to move and to occupy a series of successive positions within the same restricted area. When studied with microphotographs, their tracks overlapped in an intriguing pattern which was difficult to study. To overcome this problem, a videotape recorder was used. This technique enabled us to measure both the backward motions and the angles of each "jerk." One jerk is formed by (a) a backward motion of about 100-120 μm, (b) a clockwise rotation of about 65 degrees, (c) a forward motion of about 100-120 μm, and (d) the sudden cessation of this motion. Soon after the stimulus separating the partners, a series of jerks begins, which forms the exploring reaction and results in up to ten or more complete round angles. The probability of the split partners of contacting each other again is apparently rather high (about 50% of the split partners met again and regularly pair).

The relationship between the avoiding reaction and a single jerk of the exploring reaction is self-evident since both the length of the motion and the width of the angle are identical. What is unknown is the cause (exogenous and/or endogenous) of the sudden stop of the forward motion, which is responsible for such a regular repetitivity of the jerks during the exploring reaction. A study of the variation of the angle of the jerks, covering the first round angle after the splitting stimulus, was made. It was shown that the width of that angle is, in

13. Preconjugant Cell Interactions in *Oxytricha bifaria*

general, related to the period of time that the cells had interacted. The longer this period, the more narrow was the angle (Fig. 11A). The initial value of about 65° is progressively reduced down to about 35°. This change in behavior occurs only when the two interacting cells belong to complementary mating types. If a homologous tête-à-tête is split, the angle of the jerks of the exploring reaction remains approximately 65° regardless of the time the cells had spent together (Fig. 11A). The parameter "reduction of the angle," may in some way measure the differentiation of the cells toward their sexual fate. How long does such a differential behavior last? The behavior of 14 couples of cells split during the late interacting step was recorded and the angles of all the jerks were carefully measured for the first seven round angles. The results (Fig. 11B) showed that the angle of the jerks covering the first round angle was around 35° and that it progressively changed to 38°, 52°, 62°, 67°, 72°, and 67° during the six successive round angles. These data suggest that the split cells can "remember" their sexual differentiation for as long as 30–60 sec.

A normal avoiding reaction, which is essential for the survival of cells of *O. bifaria,* may sometimes be used in meeting the split partner again. The advantage of such a tactic is immediately clear if one con-

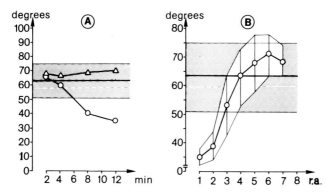

Fig. 11. The variation of the "jerk" angle as compared to the length of time cells have interacted. (A) The times of the exploring reaction (abscissa, min) have been plotted versus the width (degrees) of the angles of the jerks, by which the split cells explore the first round angle of the entire reaction (ordinate); homologous visible reaction (△), heterologous visible reaction (○). The black horizontal line has been drawn at the level of the mean value of the avoiding reaction's angle. The shadowed area covers the standard deviation distribution. (B) On the abscissa are indicated the successive round angles explored by the split cells. On the ordinate are indicated the degrees of the angles of the jerks covering the successive round angles. Each value shows the mean of the jerks of each round angle and the vertical lines the standard deviations. The black horizontal line and the shadowed area have the same meaning as in (A).

siders that, with this behavior, the species does not need to encode two different behaviors in its genetic memory.

The visible reaction, as well as the waiting period, constitutes a well-defined part of the preconjugant cell interactions of *O. bifaria*, which are necessary (although not sufficient) to allow conjugation to occur. While it is not yet clear whether the interactions mediated by soluble gamones are the result of a true recognition process or not (the activation induced by gamones is independent of the source of the gamones themselves), we can say that the two activated cells undergoing the visible reaction truly recognize each other as proper, complementary cells. This recognition is not the only major event occurring during the visible reaction. In addition to recognition, the cells also develop into either a right or a left partner.

B. Early Conjugant Cell Interactions

When two complementary partners make final contact, a cytoplasmic bridge unites them and their cell membranes fuse. It should be noted that a long series of developmental changes still must occur for a successful sexual event. However, the preconjugant cell interactions end when the cell membranes fuse. From this point on, other cell interactions guide the complex cortical and cytoplasmic integration events which occur between the two partners. This involves nuclear differentiation, the migration of the pronuclei and their union with the complementary pronuclei, and, finally, the separation of the mates. These interactions have been called "conjugant cell interactions" (Miyake, 1974), to indicate that they occur between fused cells and that they are to be distinguished from the preconjugant cell interactions.

1. Cell Union

In ciliates, the first ciliary contacts between complementary cells progressively bring them to a more intimate union eventually leading to the newly formed bicellular entity. Though the preconjugant strategies guiding the cells to the cell union may differ from one group of ciliates to another, the cell union in itself seems to be rather uniform (Miyake, 1978).

The cell union in *O. bifaria* was studied at the ultrastructural level by Ricci *et al.* (1975a). The cortical apparatus of vegetative cells of *O. bifaria* is composed of a singular external pellicle and the cell membrane. Cross section of conjugating pairs, 10–13 min after the onset of the visible reaction, revealed that protuberances of only slightly dense cytoplasm are formed at the peristomial level. They are surrounded by the pellicle and extend toward the partner's cytoplasm. The fact that they are found only within the peristomial area suggests that these protuberances could be the expression of the gamone-induced cortical changes. A few minutes later, the situation dramatically changes: couples in the early-rotation stage reveal that the external pellicles of the two partners are already fused in the anterior part of the ventral area. The two cell bodies are still well separated and a highly vacuolized sack from the left partner extends to fill the gap between the two mates, within the joined pellicles. Before rotation is over, lysis of the cell membranes occurs and a clear continuity between the two cytoplasms is observed. At the level of the fusion areas, some mitochondrial clustering occurs. Further development of the cell union involves the differential reabsorption of parts of their AZM and several other ciliary structures in order to form the new peristomial area of the pair. The right rotating partner undergoes a dramatic cortical rearrangement reabsorbing the lapel of its AZM (Fig. 1B), the frontal cirri, and the anterior area of the ventral surface. The left, still partner reabsorbs a smaller area, which includes the collar of its AZM and the most distal part of the ventral surface. This differential, well-coordinated reabsorption results in a new AZM of the pair, which is practically indistinguishable from the normal AZM complex of vegetative cells. Serial cross sections of pairs, fixed about 15 hr after their formation, revealed that the cytoplasmic bridge extends approximately over the anterior third of the length and over almost the entire dorsoventral height. No particular limiting structure can be observed, such as those described by Nobili (1967) for pairs of *Euplotes*. Finally, about 18-20 hr after pair formation, the cytoplasmic bridge progressively narrows so that the cells are united by a thin, densely vesiculated cytoplasmic cord. This connection is eventually broken by the active uncoordinated swimming of the partners, which are now referred to as exconjugants.

It is possible to compare the results of the enzymatic treatments of the visible reaction with that observed at the ultrastructural level during cell fusion. The increase in resistance to trypsin (Fig. 10, curve II) suggests an increasingly important role of protein components up to the interacting step, after which the cells can no longer be split. On the other hand, since lipase can split the cells up until the first stages of rotation, this suggests that lipids are critically involved in the later events of the mating process. These results are in good agreement with what is already known about the role played by proteins as mediators of specific cellular contacts leading to membrane fusion (Poste and Allison, 1973; Plattner *et al.*, 1973; Satir *et al.*, 1973; Heuser *et al.*,

1974; Ahkong et al., 1975; Beisson et al., 1976; Pollock, 1978), as well as with what was reported by Lawson et al. (1977) about the critical role of protein–lipid bilayers for membrane fusion.

Under natural conditions, membrane fusion constitutes the "no return" point in the processes leading the vegetative cells to conjugation. The general traits of the first stages of the *O. bifaria*'s conjugation agrees with what has been reported for *O. fallax* (Hammersmith, 1976). As the visible reaction occurs, the cells not only progressively differentiate as sexual cells, but they also become more and more different from each other from a physiological point of view (left versus right partner). This observation is in agreement with what was observed during the anchoring step of the visible reaction. The right cell rotates over its still partner (differential behavior), and also that it actively reabsorbs a larger part of its cortex during the fusion with the left partner (differential cortical rearrangement). When and how these differences are actually triggered remains to be ascertained.

2. Induction of Meiosis

The early stages of the conjugant interactions were investigated (Ricci et al., 1980) to ascertain the moment at which the cells are induced to undergo the micronuclear meiotic processes. This induction is commonly considered as the true point of no return.

Since the visible reaction of *O. bifaria* ends with the rotation of one partner over the other, such a rotation can be studied in its successive stages, if one distinguishes the angles between the partners rather then the times of the steps themselves. The constant correlation between the time of the rotation and the angle formed by the partners allows such a shift from one parameter to another. The rotation may be operationally distinguished into five successive stages, with respective angles of 180°, 150°, 90°, 60°, and 30°. Some fifty couples were split for each stage and the two partners were singly isolated in depressions in order to study their fate. The rotating and the still cell were carefully observed and distinguished from each other by their cytoplasmic differences (inclusions, food vacuoles, nuclear peculiarities). One such distinction was made—the couple was picked up, transferred into a very small droplet of SMB, and then was split by the technique of Ricci et al. (1980a). The two partners were finally isolated into depressions distinguished as left or right. Observations were made 24 hr later. Two different fates were observed: (a) "vegetative," the partner started dividing vegetatively, and (b) "autogamic," the partner underwent meiosis and a huge macronuclear anlage was observed (Fig. 1E). It was found that the partners were always asynchronous as to their induction of meiosis. At every stage, the percentage of right partners already triggered to their sexual fate was constantly and significantly higher than that of the left partners. Such results suggested that the rotating partner is the first to be induced to meiotic processes. This idea was confirmed by comparing the fates of both partners of a pair: whenever their fates differed, it was always the rotating right cell which was already induced to the autogamic process. According to these results, the physiological differences between the left and the right partners are very dramatic. In addition to the differential behavior and the differential cortical rearrangement, there is also a "differential meiotic induction." Whether these three differences are independent traits or are merely different aspects of the same physiological differentiation toward the right or the left fate is not yet known. While the "border" between the vegetative and the sexual phases of the life cycle of *O. bifaria* is the initiating step of the visible reaction, we have also to keep in mind that even beyond such a border (for 2–5 min more) the cells can shift back into a vegetative state. The results suggest that the relative time for induction in either partner depends on the extent of the cortical rearrangements that have occurred in the cells rather than upon the size of the bridge uniting the conjugants. Whereas the size of the bridge is important for *Tetrahymena* (Preparata and Nanney, 1977), in *O. bifaria* the temporal difference of the induction of meiosis is more likely dependent on the quality rather than on the quantity of the cortical rearrangements. Although it is impossible to distinguish whether the induction of meiosis depends on heterologous contacts or on membrane fusion, we may overlook this disadvantage. Considering that (a) the induction of meiosis occurs according to a defined time table and (b) the first two divisions are meiotic, the *O. bifaria* model can be used to further investigate the mechanisms associated with meiotic induction.

III. PERSPECTIVES

A. Cell–Cell Interactions in *Oxytrichia bifaria*

The life cycle of ciliates shows a periodical alternation between vegetative and sexual reproduction. In general, ciliates do not undergo conjugation. It occurs only under extreme environmental conditons and represents not only a unique social event for ciliates, but also an exceptional process which can only occur when many different restraining conditions are satisfied. For conjugation to occur an extremely accurate synchronization of cellular metabolism and physiology must be obtained.

Cell–cell interactions trigger and regulate all of the necessary differentiating events. Once the sexual pathway is triggered by the environment, a population of ciliates starts behaving in much the same way as a metazoan organism. In Metazoa, however, cell recognitions and interactions usually mediate many other processes, in addition to sexual reproduction. Conjugation in ciliates is an easily inducible developmental process, which experimentally offers an unique tool for various biological investigations (see reviews in Miyake, 1974, 1978, 1981; Reissig, 1974, 1977; Crandall, 1977; Nanney, 1977; Sonneborn, 1977).

On the basis of what is known for *O. bifaria,* a scheme of its preconjugant cell interactions can be drawn (Fig. 12). Vegetatively growing cells are displaced from asexual reproduction by the environmental message "food is no longer available." Once several preliminary constraints are satisfied, the cells shift to the sexual mode. The cells become synchronized in the G_1 macronuclear phase and the cells start excreting soluble gamones into the medium. These mediate the long-distance preconjugant cell interactions. During the waiting period, these mating type-specific soluble factors enhance their own production by means of a positive feedback mechanism and also induce those cells to come into contact with similarly prepared cells. The next interactions are directly mediated by cellular contacts during the visible reaction. In this phase, the cells not only interact with each other, but also recognize their partner as belonging either to a complementary mating type ("go ahead") or to their own mating type ("stop, start again with new contacts"). Under standard conditions a sort of cascade of reactions occurs following the "go-ahead" message. This cascade includes cell membrane fusion and the induction of meiosis. Experi-

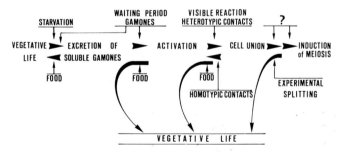

Fig. 12. A diagrammatic scheme showing the successive phases leading to pair formation, with the factors that affect the different stages.

mentally split pairs can still go back onto the vegetative reproduction, even after the cell membrane fusion (Fig. 12). Only the induction to meiosis can irreversibly commit the cells to their sexual fate.

In this scheme, the cell interactions play two different roles: (1) they facilitate cell differentiation leading to pair formation and (2) they coordinate and synchronize such a differentiation in the two partners in order to optimize the condition required for the occurrence of cell union and sexual processes. Two major interacting steps must occur before conjugation: (a) interaction from a distance, mediated by soluble gamones, and (b) the direct interaction, mediated by cell-bound substances. These two successive interactions can be referred to, respectively, as "long-range interactions" and "short-range interactions" (Reissig, 1977). "Come close," "get ready," and "tell others" are the three messages the gamones mediate. Differently from *Blepharisma* (Honda and Miyake, 1975), no clear chemotactic effect of *O. bifaria*'s gamones was shown by preliminary experiments so we cannot say whether the cells do shout "come close" with their gamones. However, the "get ready" effect of the soluble gamones of *O. bifaria* has been clearly demonstrated and the activation of the cells is independent of the source of the gamones themselves. Do the gamones mediate also some kind of "tell others" message? We can say "yes" on the basis of the positive feedback mechanism which enhances the excretion of gamone. Such a relay function can be interpreted as a sort of mechanism which progressively involves larger and larger amounts of cells in the activation.

With regard to preconjugant cell interactions of the other ciliates we can say that the kind of two-stage preconjugant cell interactions described for *O. bifaria* is similar to the "initiation-costimulation" system of *Tetrahymena pyriformis* (Bruns and Brussard, 1974). As described elsewhere in this book, complementary paramecia adhere to each other as soon as they are mixed, although the pairs are formed 2 hr later. Similarly, complementary *Blepharisma* cells have to wait about 120 min before uniting with each other.

Is the *O. bifaria*'s interacting system an exception or does it represent the more general pattern for ciliates? If the latter is true, then the interacting systems of *Paramecium* and *Blepharisma* would be the extreme examples of the general case, the former having lost the long-distance interactions and the latter the short-distance interactions. Only future research into the preconjugant cell interactions of as many ciliates as possible will give an answer to this problem, and the answer will probably reflect the extremely different ecological and evolutionary strategies followed by these ancient organisms, which were well-established a billion years ago (Nanney, 1977).

B. Future Research

The *O. bifaria* model has been extensively used to gain new insights into the problem of the nature of cell–cell interaction. While several areas have been well defined, many others are waiting to be examined. More careful and more accurate studies of the molecular mechanisms of the gamone-receptor system are certainly needed.

One of the problems of preconjugant cell interaction being investigated in our lab is the progressive aggregation of complementary cells when mixed together. Such an aggregation occurs in a single area for the two types of cells. Perhaps a particular substance that is not mating type-specific plays some important role in aggregation? Regardless of the possible specific–aspecific nature of this proposed substance, two major hypotheses have been made: (1) the reaction could be induced by a short-range soluble substance which is excreted into the medium, or (2) the reaction depends on a substance which is bound to the substrate. Gentle removal of the medium around the cellular aggregates on the bottom of the depressions would negatively affect such a clustering if any short-range soluble factor were around. Moreover, the removed fluid should have some inductory effect on other cells. On the other hand, if the substance(s) responsible for the aggregation of the cells is (are) bound to the substrate, picking up the aggregated cells and substituting that population with different cells already mixed elsewhere should induce the aggregation of this new population of cells in the same area. The testing of these two hypotheses will tell us whether the first aggregation of cells is dependent on the already known soluble gamones or, the aggregation reaction constitutes an autonomous mechanism in the cell–cell interactions.

Associated with this problem, is the study of the progressive slowing down of motility shown by the cells during the waiting period. Is such a function continuous or does it show any clear discontinuity? The first case would suggest that an aspecific, progressive differentiation of the cells occurs. A discontinuous function would indicate that, at a certain point of the waiting period, some specific reaction is triggered that in some way affects cell motility. Another aspect of the waiting period currently being investigated is whether the gamones have a positive chemotactic action on complementary cells.

One of the most interesting aspects of the cell interactions in *O. bifaria* remaining to be examined is the relatively long period of cellular contacts that is necessary during the early steps of the visible reaction. The cells lie quite still on the substrate, interacting with each other for about 10 min before the anchoring step finally occurs. During this time lag the cells are kept together by increasingly strong forces. As suggested by Ricci et al. (1980a), this aspect of the preconjugant cell interactions of *O. bifaria* should be further investigated as a model for the study of the specific adhesion of cells to other cells. This problem, reviewed by Marchase et al. (1976), has been recently considered by Oppenheimer (1978a) as the possible key to understanding not only developmental processes such as fertilization (Oppenheimer, 1978b) and gastrulation (Phillips and Grayson, 1978), but also pathological processes such as cancer and metastasis (Nicolson, 1978; Oppenheimer, 1978b).

Finally, the general traits of the preconjugant cell interactions of *O. bifaria* clearly demonstrate the precise nature of the time period for the induction of meiosis. These many aspects of the biology of *O. bifaria* suggest that this organism is an excellent cellular model suitable for fruitful investigations on the nature of the mechanisms involved in the process of the induction of meiosis.

ACKNOWLEDGMENTS

I wish to thank my wife and my children who patiently supported and lovingly helped me during the preparation of this manuscript.

I am deeply and truly indebted to Dr. Renzo Nobili, who not only taught me the way to the world of *Oxytricha*, but also suggested how to get along in it properly. It is moreover a pleasure to recall the original contribution to this research of Dr. Fulvio Esposito, a dear friend and esteemed colleague who studied *Oxytricha* with us for many years. I wish to express my thanks to Dr. David L. Nanney and Dr. Akio Miyake for their helpful suggestions and ideas, and to Dr. Danton H. O'Day who kindly rearranged the manuscript. The precious collaboration of Dr. Rosalba Banchetti and Rosanna Cetera is also greatly appreciated.

REFERENCES

Ahkong, Q. F., Fisher, D., Tampion, W., and Lucy, J. A. (1975). Mechanisms of cell fusion. *Nature (London)* 253, 194–195.

Allewell, N. M., Oles, J., and Wolfe, J. (1976). A physicochemical analysis of conjugation in *Tetrahymena pyriformis*. *Exp. Cell Res.* 97, 394–405.

Beisson, J., Lefort-Tran, M., Pouphile, M., Rossignol, M., and Satir, B. (1976). Genetic analysis of membrane differentiation in *Paramecium*. *J. Cell Biol.* 69, 126–143.

Bleyman, L. K. (1964). The inhibition of mating reactivity in *Paramecium aurelia* by inhibitors of protein and RNA synthesis. *Genetics* 50, 236.

Bruns, P. J., and Brussard, T. B. (1974). Pair formation in *Tetrahymena pyriformis*, and inducible developmental system. *J. Exp. Zool.* 188, 337–344.

Bruns, P. J., and Palestine, R. F. (1975). Costimulation in *Tetrahymena pyriformis*: A developmental interaction between specially prepared cells. *Dev. Biol.* 42, 75–83.

Bütschli, O. (1876). Studien über die ersten Entwicklungsvorgänge der Einzelle, die Zellteilung under der Konjugation der Infusorien. *Abhl. Senckenb. Naturforsch. Ges.* 10, 1–150.

Cleffmann, G., Fehrendt, I., and Behrendt, W. (1974). The rate of uptake of ^{3}H-Actinomycin during the cell cycle of *Tetrahymena. Exp. Cell Res.* 87, 139–142.

Cohen, L. W. (1964). Diurnal intracellular differentiation in *Paramecium bursaria. Exp. Cell Res.* 36, 398–406.

Cohen, L. W. (1965). The basis for the circadian rhythm of mating in *Paramecium bursaria. Exp. Cell Res.* 37, 360–367.

Corliss, J. O. (1979). "The Ciliated Protozoa. Characterization, Classification and Guide to the Literature," 2nd ed. Pergamon, Oxford.

Crandall, M. (1977). Mating type interactions in microorganisms. *In* Receptors and Recognition" (P. Cuatrecasas and M. F. Greaves, eds.), Ser. A, Vol. 3, pp. 45–100. Chapman & Hall, London.

Dini, F., and Luporini, P. (1979). Preconjugant cell interaction and cell cycle in the ciliate *Euplotes crassus. Dev. Biol.* 69, 505–516.

Dini, F., and Miyake, A. (1977). Preconjugant cell interaction in *Euplotes crassus. J. Protozool.* 24, 32A.

Dini, F., and Miyake, A. Preconjugant cell interaction in *Euplotes crassus:* Participating factors and their reaction. In preparation.

Dini, F., Bracchi, P., and Luporini, P. (1975). Cellular cell cycle in two ciliates hypotrichs. *Acta Protozool.* 14, 59–66.

Dryl, S. (1958). Photographic registration of movement of Protozoa. *Bull. Acad. Pol. Sci., Cl.* 2 6, 429–432.

Esposito, F., and Ricci, N. (1975). Inhibition of conjugation by Actinomycin D and Cycloheximide in *Oxytricha bifaria. Boll. Zool.* 42, 237–241.

Esposito, F., Ricci, N., and Nobili, R. (1976). Mating-type specific soluble factors (gamones) in cell interaction of conjugation in the ciliate *Oxytricha bifaria. J. Exp. Zool.* 197, 275–282.

Hammersmith, R. L. (1976). Differential cortical degradation in the two members of early conjugant pairs of *Oxytricha fallax. J. Exp. Zool.* 196, 45–69.

Heckmann, K., and Siegel, R. W. (1964). Evidence for the induction of mating-type substances by cell to cell contacts. *Exp. Cell Res.* 36, 688–691.

Heuser, J. E., Reese, T. S., and Landis, D. M. D. (1974). Functional changes in frog neuromuscular functions studied with freeze fracture. *J. Neurocytol.* 3, 109–131.

Honda, H., and Miyake, A. (1975). Taxis to a conjugation inducing substance in the ciliate *Blepharisma. Nature (London)* 257, 678–680.

Jennings, H. S. (1906). "Behavior of the Lower Organisms." Indiana Univ. Press, Bloomington.

Kimball, R. F. (1939). Mating types in *Euplotes. Am. Nat.* 73, 57–71.

Lawson, D., Raff, M. C., Gompert, B., Fewtrell, C., and Gilula, N. B. (1977). Molecular events during membrane fusion. *J. Cell Biol.* 72, 242–259.

Luporini, P., and Dini, F. (1975). Relationships between cell cycle and conjugation in 3 hypotrichs. *J. Protozool.* 22, 541–544.

Marchase, R. B., Vosbeck, K., and Roth, S. (1976). Intercellular adhesive specificity. *Biochem. Biophys. Acta,* 457, 385–416.

Maupas, E. (1889). La rejeunissement karyogamique chez les cilies. *Arch. Zool. Exp. Gen.* 7, 149–517.

Metz, C. B. (1954). Mating substances and the physiology of fertilization in ciliates. *In* "Sex in Microorganisms" (D. H. Wenrich, ed.), pp. 284–334. Am. Assoc. Adv. Sci., Washington, D.C.

Mishima, S. (1978). Feeding and mating in *Paramecium* multimicronucleatum. *J. Protozool.* 25, 75-76.

Miyake, A. (1968). Induction of conjugation by cell-free fluid in the ciliate *Blepharisma*. *Proc. Jpn. Acad.* 44, 837-841.

Miyake, A. (1974). Cell interaction in conjugation of ciliates. *Curr. Top. Microbiol. Immunol.* 64, 49-77.

Miyake, A. (1978). Cell communication, cell union, and initiation of meiosis in ciliate conjugation. *Curr. Top. Dev. Biol.* 12, 37-82.

Miyake, A. (1981). Physiology and biochemistry of conjugation in ciliates. *In* "Biochemistry and Physiology of Protozoa" (M. Levandowsky and S. H. Hutner, eds.), Vol. 4, 2nd ed., pp. 125-198. Academic Press, New York.

Miyake, A., and Beyer, J. (1973). Cell interaction by means of soluble factors (gamones) in conjugation of *Blepharisma intermedium*. *Exp. Cell Res.* 76, 15-24.

Miyake, A., and Honda, H. (1976). Cell union and protein synthesis in conjugation of *Blepharisma*. *Exp. Cell Res.* 100, 31-40.

Miyake, A., and Nobili, R. (1974). Mating reaction and its daily rhythm in *Euplotes crassus*. *J. Protozool.* 21, 584-587.

Nanney, D. L. (1977). Cell-cell interaction in ciliates: Evolutionary and genetic constraints. *In* "Microbial Interactions" (J. Reissig, ed.), pp. 351-397. Chapman & Hall, London.

Nicolson, G. L. (1978). Cell and tissue interactions leading to malignant tumor spread (metastasis). *Am. Zool.* 18, 71-80.

Nobili, R. (1963). Effects of antibiotics base- and amino-acid-analogues on mating reactivity of *Paramecium aurelia*. *J. Protozool.* 10, Suppl., p. 24.

Nobili, R. (1967). Ultrastructure of the fusion region of conjugating *Euplotes*. *Monit. Zool. Ital.* 1, 73-89.

Nobili, R., and Kotopulos De Angelis, F. (1963). Effetti degli antibiotici sulla riproduzione di *Paramecium aurelia*. *Atti Assoc. Genet. Ital.* 8, 45-57.

Ofer, L., Mercari, M., and Loyter, A. (1976). Conjugation in *Tetrahymena pyriformis*. The effect of Polylysine, Concanavaline-A and bivalent metals on the conjugation process. *J. Cell Biol.* 70, 287-293.

Oppenheimer, S. B. (1978a). Introduction to the symposium: Cellular Adhesion. *Am. Zool.* 18, 11-12.

Oppenheimer, S. B. (1978b). Cell surface carbohydrates in adhesion and migration. *Am. Zool.* 18, 13-23.

Phillips, H. M., and Grayson, S. D. (1978). Liquid-tissue mechanics in Amphibian gastrulation: Germ layer assembly in *Rana pipiens*. *Am. Zool.* 18, 81-83.

Plattner, M., Miller, F., and Bachmann, L. (1973). Membrane specialization in the form of the regular membrane to membrane attachment sites in *Paramecium aurelia*: Correlated freeze etching and ultrathin sectioning analysis. *J. Cell Sci.* 13, 687-719.

Pollock, E. G. (1978). Fine structural analysis of animal cell surfaces: Membranes and cell surface topography. *Am. Zool.* 18, 25-69.

Poste, G., and Allison, A. C. (1973). Membrane fusion. *Biochim. Biophys. Acta* 300, 421-467.

Preparata, R. M., and Nanney, D. L. (1977). Cytogenetics of triplet conjugation in *Tetrahymena*: Origin of haploid and triploid clones. *Chromosoma* 60, 49-57.

Reissig, J. L. (1974). Decoding of regulatory signals at the microbial surface. *Curr. Top. Microbiol. Immunol.* 67, 43-96.

Reissig, J. L. (1977). An overview. *In* "Microbial Interactions" (J. Reissig, ed.), pp. 399-415. Chapman & Hall, London.

Ricci, N. (1980). Ethogram of *Oxytricha bifaria* Stokes. In preparation.
Ricci, N., Banchetti, R., Nobili, R., and Esposito, F. (1975a). Conjugation in *Oxytricha* sp. (Hypotrichida, Ciliata): I: Morphocytological aspects. *Acta Protozool.* 13, 335-342.
Ricci, N., Esposito, F., and Nobili, R. (1975b). Conjugation in *Oxytricha bifaria:* Cell interaction. *J. Exp. Zool.* 192, 343-348.
Ricci, N., Cetera, R., and Banchetti, R. (1980a). Cell to cell contacts mediating mating type dependent recognition(s) during the preconjugant cell interactions of *Oxytricha bifaria. J. Exp. Zool.* 171-183.
Ricci, N., Banchetti, R., and Cetera, R. (1980b). Initiation of meiosis and other nuclear changes in two species of *Oxytricha. Protistologica* (in press).
Ricci, N., Banchetti, R., and Cetera, R. (1980c). Messa a punto di una tecnica di coltura per il ciliato ipotrico *Oxytricha bifaria* stokes. *Atti Soc. Tosc. Sc. Nat.* (in press).
Ricci, N., and Banchetti, R. (1981). Nuclear phenomena in vegatative and sexual reproduction of *Oxytricha bifaria* Stokes. *Acta Protozool.*, submitted.
Sapra, G. R., and Ammermann, D. (1973). RNA synthesis and acquisition of Actinomycin D insensitivity, during conjugation in *Stylonychia mytilus. Exp. Cell Res.* 78, 168-174.
Satir, B., Schooley, C., and Satir, P. (1973). Membrane fusion in a model system. Mucocysts secretion in *Tetrahymena. J. Cell Biol.* 56, 153-157.
Siegel, R. W. (1956). Mating types in *Oxytricha* and the significance of mating type systems in Ciliates. *Biol. Bull. (Woods Hole, Mass.)* 110, 352-357.
Sonneborn, T. M. (1937). Sex, sex inheritance and sex determination in *Paramecium aurelia. Proc. Natl. Acad. Sci. U.S.A.* 23, 378-385.
Sonneborn, T. M. (1950). Methods in general biology and genetics of *Paramecium aurelia. J. Exp. Zool.* 113, 87-148.
Sonneborn, T. M. (1957). Breeding, reproductive methods and species problems in Protozoa. *In* "The Species Problem" (E. Mayr, ed.), pp. 155-324. Am. Assoc. Adv. Sci., Washington, D.C.
Sonneborn, T. M. (1977). Genetics of cell-cell interactions in ciliates. *Birth Defects, Orig. Artic. Ser.* 14(2), 417-427.
Stokes. A. C. (1887). Some new hypotrichous infusoria from american fresh waters. *Ann. Mag. Nat. Hist.* 20, 104-114.
Tyler, L., and Wolfe, J. (1972). Control of cell fusion in conjugating *Tetrahymena. J. Protozool.* 19, 42A.
Verni, F., Rosati, and Luporini, P. (1978). Preconjugant cell-cell interaction in the ciliate *Euplotes crassus.* A possible role of the ciliary ampules. *J. Exp. Zool.* 204, 171-180.
Wolfe, J. (1973). Conjugation in *Tetrahymena:* The relationship between the division cycle and cell pairing. *Dev. Biol.* 35, 221-231.

14

Sexual Interactions of the Cell Surface in *Paramecium*

KOICHI HIWATASHI

I.	Introduction	351
II.	Processes of Sexual Interactions	353
III.	Mating Reaction and Mating Substances	358
	A. Mating Type-Specific Components of the Cell	358
	B. Ciliary Membranes and Membrane Vesicles	361
	C. Nature of Mating Substances	363
IV.	Output of the Mating Substance Interactions and Activation-Initiating Mechanisms	364
	A. Inactivation of Ciliary Movement	364
	B. Early Micronuclear Migration	365
	C. Degeneration of Cilia and Formation of the Holdfast Union	367
	D. Theory of Activation-Initiating Mechanism	369
V.	Control of Mating Type and Mating Activity	370
	A. Genetic Control of Mating Type Specificity within a Sibling Species	370
	B. Genetic Control of Mating Type Specificity between Different Sibling Species	371
	C. Regulation of Mating Activity	372
VI.	Perspectives	373
	References	375

I. INTRODUCTION

Paramecium was one of the first eukaryotic microbes to be used experimentally. Its discovery goes back to the seventeenth century, immediately after the development of the compound microscope. The first observation of the sexual process of conjugation in *Paramecium* appears to have been made as early as that of the discovery of the organism itself (see Wichterman, 1953). Morphological descriptions including cytological details of the process of conjugation in *Paramecium* were almost complete by the end of last century. Recent work on conjugation in *Paramecium*, however, began with the discovery of mating types by Sonneborn (1937). As a result of this discovery, sexual interactions in *Paramecium* are now subject to repeatable, orderly manipulations for modern investigations in cell biology.

Paramecium is easily cultivated in bacterized culture media and can also be grown in a fully defined axenic media (Sonneborn, 1970; Van Wagtendonk and Soldo, 1970; Van Wagtendonk, 1974). In spite of being a microorganism, it is large enough (100–200 μm) for single cell isolation under the binocular microscope without the use of a micromanipulator.

Paramecium has a simple, well-defined life cycle. Conjugation (or autogamy) marks the onset of a new life cycle and the resultant exconjugant clones pass through periods of sexual immaturity, maturity, and senescence ending in clonal death (Fig. 1). The gross anatomy and size of the cells do not change significantly through the life span, although some morphological abnormalities appear in the period of late senescence. Strains of complementary mating types are available for nearly all species of *Paramecium* (Sonneborn, 1957, 1970). When moderately starved, sexually mature cells of complementary mating types are brought together, nearly all of them instantaneously adhere together forming large agglutinates. This reaction is called the mating reaction (Sonneborn, 1939). The mating reaction triggers the subsequent events of conjugation. If the mating reaction is very strong and environmental conditions are constant, the conjugation processes occur rather synchronously. This has enabled us to make a detailed analysis of the sexual interactions in this organism.

The ready isolation of mutants and the ease of genetic analysis in this organism (Sonneborn, 1970) make it possible to take genetic approaches to the problems of sexual interaction. However, mutants of sexual processes are scarce so far. Because of its large size, *Paramecium* easily permits electrode penetration and, as a result, has been a favorite material for electrophysiological analyses (Naitoh and Eckert, 1974). Conjugation in *Paramecium* is a process of fertilization and electrophysiological studies can been carried out which are similar to those done during fertilization of metazoan eggs (Jaffe, 1976; Jaffe and Robinson, 1978). The early work on sexual cell interactions in *Paramecium* has been reviewed by Hiwatashi (1969) and more recently

14. Sexual Interactions in *Paramecium*

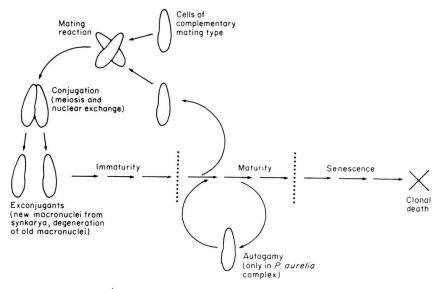

Fig. 1. A schematic representation of the life cycle of *Paramecium*.

by Nanney (1977) and by Miyake (1978). In this chapter, the research that is discussed will focus on the surface interactions during the mating process and the signal transductions that result in *Paramecium*.

II. PROCESSES OF SEXUAL INTERACTIONS

As previously mentioned, when cells of complementary mating types are mixed under appropriate conditions they instantaneously form large agglutinates (Fig. 2). For this initial cell contact, pheromonal interactions are not necessary. Even if cells of complementary mating types are mixed after repeated washing with appropriate buffers, they react in the same manner. In certain other ciliates, such as *Tetrahymena* and *Euplotes,* a period of coexistence of complementary mating types is necessary for the initial cell contacts and this time interval is known as the period of costimulation or, more simply, the waiting period (Bruns and Brussard, 1974; Heckmann, 1963). *Paramecium* has no such waiting period. The initial cell contact of the mating reaction occurs at the ciliary surface, but only those cilia on the ventral surface are reactive (Hiwatashi, 1961; Cohen and Siegel, 1963; Byrne, 1972).

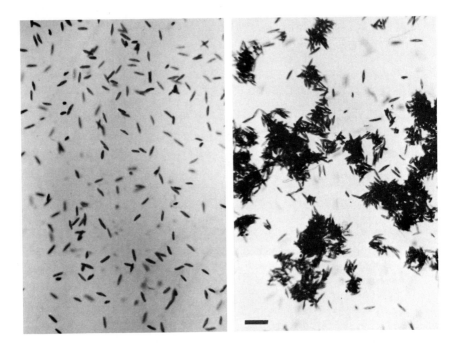

Fig. 2. Mating agglutination in *Paramecium caudatum*. Left, cells of mating type V before mixing with mating type VI; right, 1 min after mixing. Scale, 300 μm.

Thus, when cilia that have been detached from reactive cells are applied to reactive living cells of the complementary mating type, they only adhere to cilia of the ventral surface (Fig. 3).

The first external morphological change after the mating reaction is the degeneration of cilia at the anterior tips of the cells (Hiwatashi, 1955). This occurs about 30 min after the onset of the mating reaction. About 45 to 60 min after the beginning of mating reaction, the second step of sexual cell contact begins. This process called holdfast union (Fig. 4) involves cellular association at the anterior tips where the cilia have begun to degenerate. During this period, the agglutinates begin to disaggregate and the degeneration of the cilia extends posteriorly along the ventral suture of the cells where the patterns of the ciliary rows meet. The third step of sexual cell contact occurs at the regions of the cytostome, and this process is called paroral union (Fig. 4). The paroral union is the last stage in the series which produces firm unions of conjugation, and dissociation of conjugating pairs by mechanical force becomes difficult after this stage.

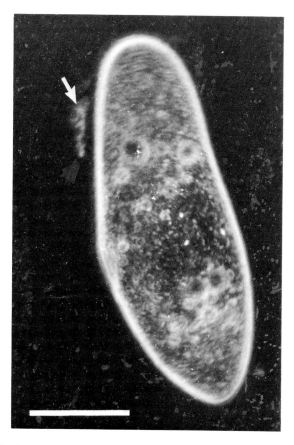

Fig. 3. Mating reaction with detached cilia. Cilia that had been detached from mating reactive cells of mating type VI were applied to mating reactive living cells of mating type V. Scale, 50 μm. The arrow indicates the site of maximum ciliary binding.

In the conjugation of *Paramecium,* the membranes at the contact regions do not disintegrate and thus the two mates do not fuse into single cells as occurs in unicellular algae or yeast (discussed earlier in several chapters). Immediately after the formation of paroral unions, however, cytoplasmic connections occur between the two conjugating cells through small openings (0.2–0.5 μm in diameter) that can be observed with the electron microscope (Schneider, 1963; Vivier and André, 1961; T. Watanabe, unpublished). This cytoplasmic continuity may allow the exchange of cytoplasm during conjugation (Berger, 1976; Hiwatashi *et al.,* 1980).

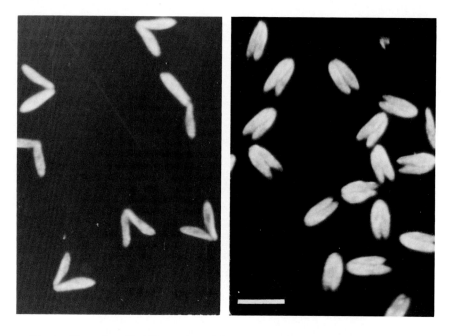

Fig. 4. Examples of holdfast unions (left) and paroral unions (right) of *P. caudatum*. Scale, 200 μm. (Courtesy of A. Kitamura.)

Paramecium, as other ciliates, has two functionally different nuclei: the micronucleus and the macronucleus (Fig. 5). Meiosis occurs in the micronucleus but not in the macronucleus which is the somatic nucleus. During meiosis, the first evident change is micronuclear swelling which occurs at the stage of premeiotic DNA synthesis. In *P. caudatum,* this occurs immediately after the paroral union, i.e., 2 to 2.5 hr after mixing the mating types (M. Fujishima, unpublished). The micronucleus divides a total of three times. The first two divisions are meiotic. The third division which is mitotic occurs in only one of the products of the meiotic divisions while the other three nuclei degenerate. One of the resultant mitotic nuclei from each mate migrates into the partner and fuses with the remaining nucleus to form the synkaryon. Immediately after this exchange of the haploid gametic nuclei, the conjugating pairs separate. Separation of pairs usually begins at the anterior regions and ends at the paroral regions. Immediately after the separation of pairs, the regeneration of the ventral cilia occurs. The synkaryon divides two or three times depending on the species and the

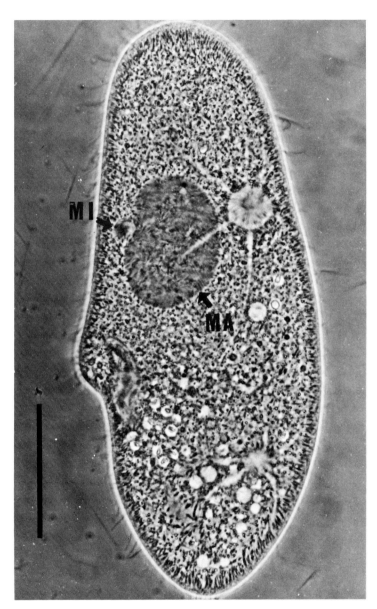

Fig. 5. A negative phase-contrast photomicrograph showing the two functionally different nuclei of P. caudatum. MA, macronucleus; MI, micronucleus. Scale, 50 μm. (From Fujishima and Hiwatashi, 1977, with permission of the Wister Institute Press.)

products differentiate into new micro- and macronuclei. The details of the conjugation process differ in the various species.

In addition to the morphological changes, an important aspect of the process of conjugation is the specificity of the cell contact that occurs. The first step, the mating reaction, is highly mating type-specific and a clump always contains both of the mating types (Hiwatashi, 1951; Jennings, 1938; Larison and Siegel, 1961). The second and the third steps, however, are not necessarily mating type-specific. The occurrence of selfing pairs together with cross pairs in the stages of holdfast and paroral unions can be clearly demonstrated in *P. caudatum* by marking cells with vital stains or by using genetic markers such as erythromycin resistance or a behavioral abnormality (Hiwatashi, 1951; Myohara and Hiwatashi, 1978; Hiwatashi et al., 1980). Since selfing of single mating types can be induced by chemicals or by using detached cilia in the *P. aurelia* complex and in *P. multimicronucleatum* (Miyake, 1964, 1968) this suggests that holdfast and paroral unions are also nonspecific in those species. By the chemical induction of conjugation, in which holdfast unions are directly induced, both selfing (conjugation of the same mating types) and interspecific conjugation (among *P. caudatum, P. multimicronucleatum,* and species of the *P. aurelia* complex) can be induced (Miyake, 1968). This suggests that specificity of holdfast unions and the mechanism for their formation are common among those species.

III. MATING REACTION AND MATING SUBSTANCES

A. Mating Type-Specific Components of the Cell

Since the cell-to-cell contacts that occur after holdfast union are nonspecific, the mating type-specific components of the cell must be involved in the initial mating reaction. An analysis of the mating reaction using cells that had been killed by various physical and chemical agents has already been extensively reviewed (Metz, 1954; Hiwatashi, 1969). The first attempt to isolate mating-reactive components of the cell was made by Cohen and Siegel (1963). These workers succeeded in inducing the mating agglutination of living cells with detached cilia of the complementary mating type in *P. bursaria*. Subsequently, the induction of the mating reaction with detached cilia was also accomplished with *P. multimicronucleatum* (Miyake, 1964), *P. caudatum* (Fukushi and Hiwatashi, 1970), *P. octaurelia* (Cronkite, 1972), and *P. tetraurelia* (Byrne, 1972). In these studies, the cilia not only reacted with the living cells of the opposite mating type but also induced conjugating pairs. Several laboratories have tried to develop a complete *in vitro* system (i.e., without involving living cells) for the mating reaction, but have failed to obtain either a reaction between detached cilia or a reaction between detached cilia and killed cells. Thus, it was thought that the participation of living cells was necessary for ciliary sexual adhesion as it is in flagellar sexual adhesions in *Chlamydomonas* (Goodenough, 1977; see also Chapter 12 by van den Ende). However, Takahashi *et al.* (1974), finally succeeded in obtaining agglutination between cilia which were detached from complementary mating types by a method (Naitoh and Kaneko, 1973) using Triton X-100 and calcium (Fig. 6). Thus, the presence of a soluble cofactor which is secreted from living cells is not essential for the ciliary agglutination. Kitamura (unpublished) has also obtained agglutination of cilia which were removed from complementary mating types by the $MnCl_2$ method of Fukushi and Hiwatashi (1970). Although he tried

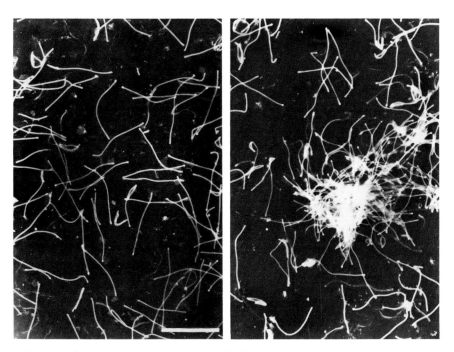

Fig. 6. Electron micrographs of detached cilia which have been fixed with formalin and shadowed with chromium. Left, cilia of mating type V (control); right, agglutination of cilia when those from mating types V and VI were mixed. Scale, 10 μm. (From Takahashi *et al.*, 1974, with permission of Academic Press.)

to assay this mating reaction photometrically, no change of turbidity was detected.

An important problem in these investigations is that only ventral cilia possess mating reactivity. As a result, crude preparations of detached cilia contain a large proportion of nonreactive cilia, and the proportion of the reactive cilia is probably as little as one-tenth of the total cilia. Attempts to isolate mating reactive cilia from crude preparations of detached cilia have been made using living cells of the opposite mating type. In *P. multimicronucleatum,* A. Miyake and V. Klimetzek (personal communication) tried to isolate mating reactive cilia from a $K_2Cr_2O_7$-detached ciliary preparation by first absorbing them to mating-reactive living cells and then by separating them by killing the cells. They succeeded in recovering mating-reactive cilia, but the specific activity of the recovered cilia preparation was not determined. Y. Ito and K. Hiwatashi (unpublished) tried a similar experiment with *P. caudatum* but the recovered cilia were nonreactive. The same line of experiments was performed on a much larger scale by A. Kitamura (unpublished) using $MnCl_2$-detached cilia of *P. caudatum*. When cilia were detached from mating reactive cells of *P. caudatum* by the $MnCl_2$ method and mixed with mating-reactive living cells of the opposite mating type, large mating clumps were formed. These clumps were then separated from remaining free cilia and unclumped living single cells by slow centrifugation or by allowing the clumps to settle. The clumps were then treated with 5 to 7% formalin for a short time at 0°C to dislodge the cilia from the cells. The cilia were recovered by centrifugation and their specific activity was measured. The specific activity was defined as the reciprocal of the lowest protein concentration that could induce more than 10 clumps in about 800 mating reactive tester cells. A. Kitamura (unpublished) succeeded in recovering mating-reactive cilia but the specific activity of the recovered cilia never exceeded that of the original cilia preparation. Thus, inactivation of mating reactivity probably occurred during the isolation process. Possibly the living cells secrete a protease that inactivates the cilia, because mating reactivity of detached cilia is easily inactivated by proteases as will be described later. However, Kitamura used a protease inhibitor, TLCK (N-α-p-tosyl-L-lysine chloromethyl ketone HCl) during the preparation but was unsuccessful in getting cilia preparations with a higher specific activity. Other protease inhibitors might be employed to determine if the secretion of protease is the cause of the inactivation. The mating reaction itself may in some way change the activity of the molecules involved, and once reacted, when cilia are separated, they are weaker in subsequent reactions. If this could be the case, isolation of highly mating-reactive cilia using living cells of the opposite mating type would be impossible, and other methods of separation should be sought.

B. Ciliary Membranes and Membrane Vesicles

Since the mating reaction occurs at the ciliary surface, the ciliary membranes are most likely the site where the mating substances exist. Watanabe (1977) treated mating-reactive, detached cilia with the nonionic detergent, Triton X-100 and followed the relationship between the loss of mating reactivity and the removal of the ciliary membranes. The extent of the loss of mating reactivity always correlated directly with the extent of ciliary membrane removal. Subsequently, he fractionated the ciliary components by dialysis according to a slightly modified method of Gibbons (1965) and assayed the mating reactivity of the fractionated components. His results showed that the mating reactivity was always associated with the ciliary membranes and not with the axonemal or matrix components, and also that purified ciliary membranes have a high mating reactivity.

In *Chlamydomonas,* the culture supernatant of mating-reactive cells contains membrane vesicles which induce an isoagglutination reaction when applied to cells of the opposite mating type (see Goodenough, 1977). The vesicles are as large as 0.5 μm and are thought to be continuously sloughed off from the flagellar membranes. This subject is discussed in detail in Chapter 12 by van den Ende. Under normal conditions such membrane vesicles could not be detected in *Paramecium*. However, when the mating reactive cilia of *P. caudatum* were treated with a solution containing urea and EDTA, membrane vesicles with a high mating activity were obtained (Kitamura and Hiwatashi, 1976). The vesicles are about 100–150 nm in diameter (Fig. 7, left). They not only induce agglutination in the mating-reactive cells of opposite mating type but also induce selfing in the latter. The same kind of membrane vesicles can be obtained with lithium diiodosalicylate (LIS) which has been used to isolate glycoproteins from cell membranes (Marchesi and Andrews, 1971). Kitamura and Hiwatashi (1980) treated mating reactive cilia with 4 mM LIS and isolated membrane vesicles with a diameter of about 50 to 100 nm (Fig. 7, right). The LIS membrane vesicles also induce both mating agglutination and conjugating pairs when added to mating reactive cells of the opposite mating type. However, the LIS membrane vesicles have some peculiar characteristics which differ from the urea–EDTA membrane vesicles. With the LIS membrane vesicles, the specific activity of the pair induc-

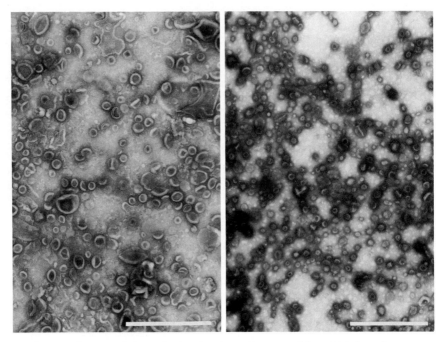

Fig. 7. Electron micrograph of isolated membrane vesicles which have been negatively stained with uranyl acetate. Left, vesicles isolated with urea–EDTA; right, with LIS. Scale, 1 μm. (Left, from Kitamura and Hiwatashi, 1976, with permission of the Rockefeller University Press; right, courtesy of A. Kitamura.)

tion was always higher than that of the induction of agglutination. As a result, with certain dilutions of the membrane vesicles, selfing pairs were induced without the prior occurrence of agglutination. Furthermore, when the vesicles were stored in a refrigerator (4°C) the ability to induce mating agglutination was lost earlier than the ability to induce conjugating pairs, and in some experiments, the LIS membrane vesicles continued to show the conjugation-inducing ability even after 12 days storage. The formation of conjugating pairs without agglutination can be induced chemically in mating reactive cells (e.g., a high concentration of K^+ in Ca^{2+}-poor medium; Miyake, 1968; Hiwatashi, 1969). In the chemical induction of conjugation, however, the chemicals induce conjugation in both mating types. The induction of conjugation by the LIS membrane vesicles is the first case of mating type-specific induction of conjugation in the absence of mating agglutination.

Many laboratories have tried to isolate the mating substances in a soluble form but have not succeeded. It is very probable that the mating substances are intrinsic proteins of the ciliary membrane (Hiwatashi, 1969; Kitamura and Hiwatashi, 1978). Detergent solubilization of the mating reactive cilia or of the urea-EDTA membrane vesicles has been employed in an attempt to isolate the mating substances (A. Kitamura, unpublished). Most ionic detergents inactivate the mating activity. Nonionic detergents have been employed with the intent of solubilizing the ciliary membranes but attempts at recovery of the mating substances in a soluble form were unsuccessful. However, when the urea–EDTA membrane vesicles were treated with 9 mM LIS and the resulting supernatant of a 105,000 g centrifugation was dialyzed to remove the LIS, the membrane vesicles were reconstituted. These reconstituted membrane vesicles had no agglutination-inducing ability but did induce conjugation when added to mating reactive cells of the opposite mating type (Kitamura and Hiwatashi, 1980). We intend to use the reconstituted membrane vesicles to identify the mating substance. If the LIS-soluble components of the ciliary membranes are fractionated and various membrane vesicles are reconstituted by addition or omission of various fractions, we shall be able to identify the molecular components necessary for the induction of conjugation.

C. Nature of Mating Substances

Evidence obtained from experiments with both mating-reactive killed cells and detached cilia strongly suggests that the mating substances are protein in nature (Metz, 1954; Hiwatashi, 1969; Watanabe, 1977; Kitamura and Hiwatashi, 1978). Since the role of surface glycoproteins in sexual cell contact has often been suggested (Brock, 1965; Crandall *et al.*, 1974; Wiese, 1974), Kitamura and Hiwatashi (1978) examined the possible involvement of sugar residues in the mating agglutination using the detached cilia method. Five different glycosidases, α-mannosidase, α-L-fucosidase, β-galactosidase, β-glucosidase, and neuraminidase, all had no demonstrable effect on the mating activity of the cilia, while trypsin destroyed the mating activity even in a concentration as low as 0.00001% (Fig. 8). Treatment of the mating-reactive cilia with sodium periodate (in the dark at 2°C) markedly decreased the mating activity of the cilia. However, the inactivation by $NaIO_4$ was inhibited by sodium aspartate or by pretreatment with 5% formalin. These workers suggested that the inactivation of mating activity by the $NaIO_4$ treatment was caused not by the direct inactivation of the mating substances but by the steric hindrance that

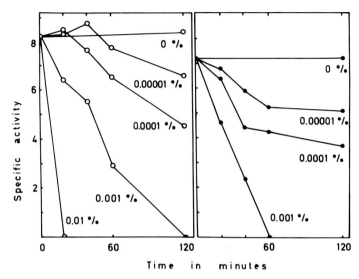

Fig. 8. Decrease of mating reactivity of cilia detached from mating type V (left) and VI (right) after incubation with various concentrations of trypsin. Specific activity is indicated as $\log_2$ of the reciprocal of the lowest concentration (mg/ml) of ciliary proteins that can induce mating agglutination. (From Kitamura and Hiwatashi, 1978, with permission of the Wister Institute Press.)

resulted from cross-linking between sugar and amino residues located near the mating substances. The formation of the cross-link by $NaIO_4$ and inhibition of cross-link formation by pretreatment with formalin were demonstrated by gel electrophoresis (A. Kitamura, unpublished). These results exclude the possible involvement of sugar residues in the active site of mating substances, though they do not exclude that the mating substances contain sugars outside their active sites.

IV. OUTPUT OF THE MATING SUBSTANCE INTERACTIONS AND ACTIVATION-INITIATING MECHANISMS

A. Inactivation of Ciliary Movement

As outlined in Section II, conjugation in *Paramecium* comprises a series of events which occur in a precise sequence. This chain of events is triggered by the mating reaction—an interaction of mating substances. When the mating reaction occurs, the earliest change observed by light microscopy is inactivation of ciliary movement. Cells in the agglutinate move slowly and sometimes stop ciliary movement. Two alternative interpretations are possible: (1) the directions of swimming of different cells in the mating clumps are different and this cancels the net movement of the cells, or (2) interaction of mating substances involves some mechanism that directly inactivates ciliary movement. Kitamura et al. (1979) have observed complete inactivation of ciliary movement in the cells reacted with LIS membrane vesicles of the opposite mating type. In this work, they used LIS membrane vesicles which had conjugation-inducing activity but lacked agglutination-inducing activity. Thus, when the cells were reacted with LIS membrane vesicles they stopped their movement without forming mating clumps. If the reaction between the LIS membrane vesicles and cells of the opposite mating type is the same in its nature as the mating reaction between living cells, the result supports the second interpretation. Ciliary movement is controlled by membrane potential (Naitoh and Eckert, 1974). Hyperpolarization augments ciliary movement and depolarization retards it. Kitamura et al. (1979) measured the membrane potential before and after addition of the LIS membrane vesicles but found no significant difference between treated and untreated cells. When mating-reactive cells are treated with conjugation-inducing chemicals, they show a whirling motion for a few minutes and then gradually become sluggish (Miyake, 1958; Cronkite, 1972; Tsukii and Hiwatashi, 1978). When sexually immature cells were treated with the conjugation-inducing chemicals, no such change of ciliary movement was observed. These results suggest that the inactivation of ciliary movement is intimately correlated with the mating-receptive condition of the cells.

B. Early Micronuclear Migration

In stationary phase cells of *P. caudatum*, the micronucleus sits within a concavity of the macronucleus. Fujishima and Hiwatashi (1977) found that the micronucleus moves out of this macronuclear concavity and migrates into the cytoplasm when the mating reaction begins. This phenomenon is called "early micronuclear migration" (EMM) (Fig. 9). The EMM begins in some cells as early as 10 min after the start of the mating reaction and is evident in 90% of the cells by about 20 min. This is the earliest nuclear change observed after the onset of mating reaction. The EMM is closely related to the sexual responsiveness of cells. When mating reactive cells were treated with conjugation-inducing chemicals (e.g., 6 mM KCl in 2 mM phosphate buffer) they underwent EMM, but when stationary phase cells in the

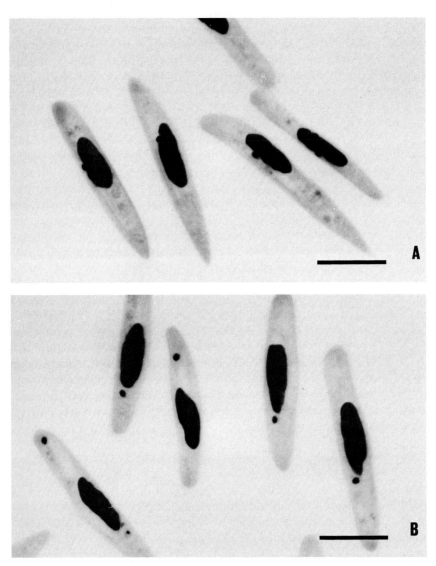

Fig. 9. Early micronuclear migration (EMM) in *P. caudatum*. Photomicrographs of Carnoy fixed and Feulgen-Fast green stained preparations. (A) Cells in stationary phase, (B) cells 30 min after the beginning of the mating reaction. Scale, 50 μm. (From Fujishima and Hiwatashi, 1977, with permission of the Wister Institute Press.)

immature period of clonal life cycle were similarly treated, EMM was not observed.

Cronkite (1977) used this micronuclear reaction in his analysis of the activation-initiating mechanism in *Paramecium*. When mating reactive cells were cooled for 60 min and then rapidly warmed, EMM occurred spontaneously. Treatment of mating reactive cells with the calcium ionophore A23187 also induced EMM. Neither treatment induced EMM in sexually immature cells. Cronkite's (1977) experiments with ^{45}Ca showed that cooled cells accumulate Ca^{2+} and then rapidly lose the ion upon warming. As a result he suggested that changes in free Ca^{2+} concentration play a key role in the occurrence of EMM. These results strongly suggest that conjugation in *Paramecium* has an activation-initiating mechanism which is similar to that of fertilization of animal eggs, since changes in the concentration of free Ca^{2+} play an important role in fertilization of sea urchin (Steinhardt *et al.*, 1977), fish (Medaka) (Ridgway *et al.*, 1977), and mammalian eggs (Fulton and Whittingham, 1978).

C. Degeneration of Cilia and Formation of the Holdfast Union

When the mating reaction proceeds to conjugating pair formation, the cilia and trichocysts at the anterior tip and on the ventral surface disappear and the union of conjugating pairs occurs at these cilia-free surfaces (Hiwatashi, 1955; Miyake, 1966). Watanabe (1978) has made detailed observations on the degeneration of cilia during early stages of conjugation using the scanning electron microscopy. He noted that the ciliary degeneration begins about 30 min after the mixing of complementary mating types, when the cells are still clumped. It begins first at the anterior tips of the cells and then proceeds posteriorly along the ventral surfaces. The ciliary degeneration is likely due to resorption, since cilia of various lengths that are shorter than normal cilia are observed during the degeneration process (Fig. 10). The same kind of ciliary degeneration was observed when autogamy was chemically induced in both *P. caudatum* (Watanabe, 1978) and *P. tetraurelia* (T. M. Sonneborn, personal communication). However, ciliary degeneration was not observed during natural autogamy in *P. tetraurelia* (Watanabe, 1978). This discrepancy between chemically induced and natural autogamy has an important implication. In the chemical induction of autogamy, cells are also induced to conjugate but pair formation is prevented by the addition of protease or by isolating single cells (Tsukii and Hiwatashi, 1979; T. M. Sonneborn, personal communication). When cells of *P. tetraurelia* undergo natural autogamy, they

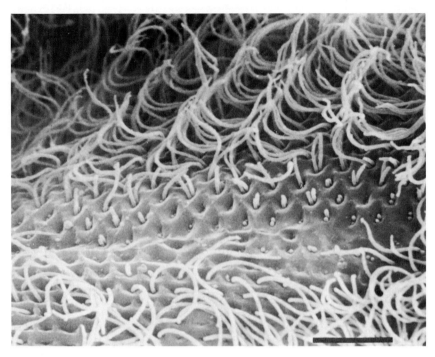

Fig. 10. Scanning electron micrograph of degenerating cilia, 60 min after the beginning of the mating reaction. Scale, 5 μm. (Courtesy of T. Watanabe.)

have no mating activity. These facts suggest that the degeneration of cilia is intimately connected with the induction of conjugation.

A common mechanism regulated by the interaction of mating substances may induce both the degeneration of cilia and the synthesis of cementing substances for the holdfast union (the holdfast substances). Holdfast union is stopped by the addition of protein synthesis inhibitors (Miyake, 1969). Whether protein synthesis is necessary for ciliary degeneration is not known. Though ciliary degeneration and holdfast union formation may be triggered by a common mechanism, the processes are separable. Con A inhibits the holdfast union and the formation of tight pairs (paroral unions) when it is induced by either the mating reaction or the conjugation-inducing chemical. It also inhibits chemical induction of autogamy. However, the Con A does not interfere with the initial agglutinative mating reaction or the inactivation of ciliary movement that is caused by the conjugation-inducing chemicals (Tsukii and Hiwatashi, 1978). Furthermore, Con A does not inhibit degeneration of the ventral cilia (Watanabe, personal communication). Thus, degeneration of cilia must be regulated by a different process than pair formation and nuclear activation.

D. Theory of Activation-Initiating Mechanism

In the fertilization of multicellular animals, the specific contact of the sperm and egg activates the egg inducing a series of sequential changes. In the same way, specific contact of *Paramecium* cells during the mating reaction "activates" the cells causing the sequential changes of conjugation which ultimately lead to the development of a new clone. Just as eggs can be activated without sperm in artificial parthenogenesis, *Paramecium* can be chemically activated without the interaction of cells of the opposite mating types. Thus, *Paramecium* is considered to be a good model for the study of the activation of metazoan eggs (Metz, 1954; Hiwatashi, 1969). An advantage provided by *Paramecium* is the availability of mutants. Selfing among cells of a single mating type in the *P. aurelia* complex can be induced in Ca^{2+} poor conditions by a solution containing K^+ or Mg^{2+} and acriflavin (Miyake, 1968). Cronkite (1974, 1975) has isolated mutants of *P. octaurelia*, which can conjugate normally but cannot be chemically induced to conjugate. Two recessive mutants, *kau*-1 and *kau*-2, cannot be induced to conjugate either by K^+ and acriflavin or by Mg^{2+} and acriflavin. The third mutant gene he found is a dominant suppressor, $Su(kau$-2). $Su(kau$-2) restores the Mg^{2+} inducibility but not the K^+ inducibility of *kau*-2. Since conditions for the chemical induction of conjugation resemble the conditions that stimulate ciliary reversal (i.e., a low concentration of Ca^{2+} and relatively high concentration of some other cation like K^+ or Mg^{2+}), Cronkite (1975) examined the mutants' response to stimuli that normally would induce ciliary reversal. When wild-type cells were stimulated by 1 mg% acriflavin, they showed backward swimming for a short time, but when cells of *kau*-1 or *kau*-2 were stimulated, they showed no such ciliary reversal. In *P. tetraurelia*, mutants lacking the "avoiding reaction" (i.e., ciliary reversal in the face of stimuli) have been isolated and are referred to as "pawns" (Kung, 1971a,b). Cronkite (1976) tested the pawn mutants for their ability to respond to the induction of conjugation. Cells of the unconditional pawn (*pwB*) and of the temperature sensitive pawn (*pwC*) at the stringent temperature (32°C) did not respond, while *pwC* was induced to conjugate at the permissive temperature (23°C). Ciliary reversal is induced by the influx of Ca^{2+} across the membrane upon stimulation. Pawns have altered membranes which prevent this influx of Ca^{2+} (Kung and Eckert, 1972). From this evidence, Cronkite (1976) proposed that the influx of Ca^{2+} across the membrane is necessary for the chemical induction of conjugation. Another group of mutants called CNR mutants of *P. caudatum* are like the pawn mutants of *P. tetraurelia*. The CNR mutants lack "avoiding reaction" in the face of stimuli (Takahashi and Naitoh, 1978). Unlike pawns, however, CNRs are capable of being chemically induced to conjugate (Takahashi, 1979). Mating reactive cells and sexually immature cells show the same membrane potential changes in the presence of conjugation-inducing chemical (Kitamura *et al.*, 1979). If Ca^{2+} influx is associated with the chemical induction of conjugation, then differences in the membrane potential changes between mating reactive and nonreactive (immature) cells would be expected. Using *P. caudatum*, Cronkite (1977) tested the effect of $LaCl_3$ on the avoiding reaction and on the chemical induction of conjugation, both of which are induced by solutions high in K^+ and low in Ca^{2+}. With K^+ the $LaCl_3$ (1 μM) inhibited both events. However, when complementary mating types were mixed, the same concentration of $LaCl_3$ inhibited neither the mating reaction of complementary mating types nor the induction of conjugating pairs. $LaCl_3$ is known to block Ca^{2+} influx across membranes (Narahashi, 1974). These results suggest that the normal induction of conjugation by the mating reaction does not require Ca^{2+} influx. Cronkite (1977) has proposed that the release of internally bound Ca^{2+} is necessary for activation and this may be induced either by influx of Ca^{2+} or by some other mechanisms.

V. CONTROL OF MATING TYPE AND MATING ACTIVITY

A. Genetic Control of Mating Type Specificity within a Sibling Species

A detailed analysis of the genetic control of the mating types is beyond the scope of this chapter and has already been extensively reviewed by Sonneborn (1974), Butzel (1974), Nanney (1977), and others. As a result, only a brief outline will be presented here. The taxonomic species of *Paramecium* are subdivided into a number of sibling species called syngens (Sonneborn, 1957). In some species, clones belonging to each syngen are grouped into two complementary mating types (two-type system), and in other species each syngen contains 4, 8, or more complementary mating types (multiple-type system). In the *P. aurelia* complex, 14 syngens have been characterized and recently assigned species names (Sonneborn, 1975). In species of *Paramecium* with the two-type system, mating types of different species are homologous and are grouped in O (odd-numbered) and E (even-numbered) types according to their numerical designation. A pair of alleles (mt^+ and mt^O or Mt and mt) controls the potentiality of the mating type expression. The recessive allele (mt^O or mt) restricts homozygotes to the category O while the dominant allele (mt^+ or Mt) permits expression of the E mating type. In some species, clones with the dominant allele are determined to become E type but in other species those with the dominant allele can express either E or O with the actual expression depending on cytoplasmic controlling factors, environmental factors, or the action of other loci. An important common principle is that when the action of the dominant allele is absent either by recessive mutation or by repression of the allele, clones are always of the O type. This suggested that mating type specificity of the O type itself is not controlled by the *mt* locus but by some other locus. This prediction was proved by the results of the study on *P. caudatum* which is presented in Section V,B. In the species belonging to the multiple mating type system, *P. bursaria* syngen 1 is the only species in which mating type determination has been extensively studied (Siegel and Larison, 1960; Siegel, 1965). Two pairs of alleles in two different loci control four different mating types in *P. bursaria* syngen 1. Though the action of the dominant allele in each locus seems to have a common feature with that of the two-type system, this remains to be clarified by future analysis.

B. Genetic Control of Mating Type Specificity between Different Sibling Species

In the *P. aurelia* complex, conjugation between certain sibling species has been known to occur, but all these interspecific conjugations were either lethal or sterile (Sonneborn and Dippell, 1946; Haggard, 1974). This indicates that gene flow between sibling species is completely blocked and thus gives a genetic basis for referring to the interbreeding groups (syngens) as species.

However, in *P. caudatum,* which also consists of many syngens, recent studies by Tsukii (1980) have revealed that intersyngenic hybrids are completely fertile. By mixing mating reactive cells of four mating types belonging to two different syngens, he obtained clones of intersyngenic hybrids. Though the initial agglutinative mating reaction is highly syngen specific, pair formation is not as was discussed in reference to the chemical induction of conjugation (see Section II). Thus, if two mating types of one syngen are marked by a mutant gene such as CNR, we can easily separate the intersyngenic pairs from the intrasyngenic pairs. Intersyngenic F_1 hybrids were obtained when cells from syngen 3 were mated with members of syngen 1, 12, or 13. All the resultant hybrid clones were fertile and were able to cross with each other or with various parental clones. Extensive genetic analyses using these various crosses showed that mating type specificity is controlled by at least three loci. Syngen specificity of even mating types (E types) is controlled by multiple codominant alleles Mt^n (n designates the number of syngen, Mt^1, Mt^3, etc.) and that of odd types (O types) by multiple codominant alleles Om^n (Om^1, Om^3, etc.). Mt is epistatic to Om and Om can be expressed only when the former locus is homozygous for the recessive allele mt. Intersyngenic heterozygotes of Mt (e.g., Mt^1/Mt^3) express two E types while those of Om (e.g., Om^1/Om^3) express two O types. Unexpectedly, isolates that expressed neither mating type were obtained when double O types were chemically induced to undergo selfing. With selfing among O type (e.g., Om^1/Om^3, expressing mating types I and V) a segregation of Om^1/Om^1(I), Om^1/Om^3(I-V), and Om^3/Om^3(V) would be expected to occur in a ratio of 1:2:1. However, the results showed a frequency of Om^1/Om^3 significantly lower than the expected ratio, and some clones expressing no mating type appeared. These mating type-less clones did not react with any mating type of syngen 1 and 3 but, nevertheless, became sexually mature since they could be chemically induced to undergo conjugation. The segregation data from the selfing progeny of the double O type suggest that another one or two loci (A^n and B^n) are necessary for the expression of O types, and these loci also act syngen specifically (Table I). When the specificity of those loci is different from the specificity of Om, the O type does not get expressed and the clones become mating typeless. The mating typeless clones should prove to be very important for biochemical analysis since they appear to be deficient only in the synthesis of mating type substances.

C. Regulation of Mating Activity

As mentioned, *Paramecium* cells express their mating activity only when sexually mature and moderately starved. During the period of sexual immaturity which usually occurs after conjugation, cells do not respond sexually to the starvation stimulus. The length of this postconjugation immature period varies in different species and is measured in terms of the number of cell divisions rather than in absolute time

TABLE I

Segregation of Mating Type in Selfing Progeny of Double Odd Type (Mating Type I-V)[a]

Mating type	I	V	I-V	Mating typeless
Two-loci hypothesis				
Genotype	$Om^1/Om^1, A^1/A^1$	$Om^3/Om^3, A^3/A^3$	$Om^1/Om^3, A^1/A^3$	$Om^1/Om^1, A^3/A^3$
	$Om^1/Om^1, A^1/A^3$	$Om^3/Om^3, A^3/A^1$	$Om^1/Om^3, A^3/A^1$	$Om^3/Om^3, A^1/A^1$
	$Om^1/Om^1, A^3/A^1$	$Om^3/Om^3, A^1/A^3$	$Om^3/Om^1, A^1/A^3$	
	$Om^1/Om^3, A^1/A^1$	$Om^3/Om^1, A^3/A^3$	$Om^3/Om^1, A^3/A^1$	
	$Om^3/Om^1, A^1/A^1$	$Om^1/Om^3, A^3/A^3$		
Expected ratio (a)	5	5	4	2
Three-loci hypothesis				
Expected ratio (b)	19	19	8	18
Experimental result (c)	40	35	15	20

[a] P value between (a) and (c), $0.02 - 0.03$ ($\chi^2 = 9.27$); between (b) and (c), $0.1 - 0.2$ ($\chi^2 = 5.84$).

(Sonneborn, 1957; Miwa and Hiwatashi, 1970). To know what controls the difference between the immature and the mature period, Miwa *et al.* (1975) injected cytoplasm from immature cells into mature cells. The injected cells lost their mating activity and, sometimes, more than ten fissions were required before they again became sexually mature. Haga and Hiwatashi (1980) succeeded in isolating the factor from the soluble fraction of immature cells. The factor was identified as a protein of about 10,000 molecular weight and was named "immaturin." Miwa (1979a,b) subsequently found that the immaturin of *P. caudatum* is effective on cells of the *P. aurelia* and *P. multimicronucleatum* complexes but it is not effective on *P. bursaria*. He also discovered that immature cells of *P. bursaria* have their own immaturin. Immaturin is possibly a repressor-like protein but its mechanism of action has not yet been explored.

Almost nothing is known about how starvation induces mating activity in *Paramecium*. Probably some metabolic shift from the logarithmic growth phase to the stationary phase induces synthesis of mating substances. M. Takahashi (unpublished) tried to change the timing of the expression of mating activity by adding cAMP or theophylline to log-phase cells which had been washed free of culture medium but was unsuccessful.

IV. PERSPECTIVES

The sexual interaction in *Paramecium* begins with agglutination which occurs immediately upon mixing cells of complementary mating types. In the mating reaction, cells of complementary mating types stick together at their ventral cilia. The substances involved in this specific cell adhesion are called mating substances. Indirect evidence including the results of enzymatic digestion show that the mating substances or at least their active sites are protein and that sugar components, if any exist, are not involved in the active site. The mating substances probably exist as proteins intrinsic to the ciliary membranes. The isolation and characterization of the substances have not been done but solubilization of the ciliary membranes with LIS and reconstitution of mating reactive membrane vesicles from the LIS soluble fraction seem like a promising way to identify the mating substances.

The mating reaction triggers the following sequential changes in the process of conjugation. The first observable change induced by the mating reaction is the inactivation of ciliary movement. No remarkable change in the membrane potential was observed in association with the inactivation of ciliary movement. Since this inactivation is also observed during the chemical induction of conjugation, it is probably the earliest conjugation specific response of the cells. The in-depth analysis of the mechanism of ciliary inactivation may lead to the discovery of the initial step of the signal transduction in the mating reaction. Moreover, when we fractionate ciliary membrane proteins, the mating type-specific inactivation of ciliary movement should provide a more quantitative and simpler assay method for the detection of mating substance activity than agglutination or pair formation.

Recent studies on the genetic control of mating types using intersyngenic crosses in *P. caudatum* have revealed that many genes are involved in the determination of mating type specificity. The sexually mature but mating type-less strains described in this chapter should prove especially useful for the analysis of mating substances using the method of membrane vesicle reconstitution.

As described in Section IV,D, the activation-initiating mechanisms during conjugation of *Paramecium* are closely related to the ion physiology of the cell membranes. *Paramecium* has been favored material for the study of membrane physiology (Naitoh and Eckert, 1974) and membrane mutants have been proved to be powerful tools for the study of membrane excitation (Nelson and Kung, 1978; Byrne and Byrne, 1979; Cronkite, 1979). These mutants should also be useful for the study of the sexual interaction in *Paramecium* because membrane excitation and sexual interaction are both the phenomena occurring in the ciliary membranes.

ACKNOWLEDGMENTS

The author wishes to thank Drs. D. L. Cronkite, D. H. O'Day, and A. Kitamura for suggestions and help in preparation of the manuscript. He also thanks Drs. A. Kitamura, M. Fujishima, T. Watanabe, and M. Takahashi for invaluable sharing of unpublished work, and Drs. T. M. Sonneborn and A. Miyake for generously providing personal communications. Work was supported by a grant-in-aid for the special project research, Mechanisms of Animal Behavior, from the Ministry of Education, Science and Culture.

REFERENCES

Berger, J. D. (1976). Gene expression and phenotypic change in *Paramecium tetraurelia* exconjugants. *Genet. Res.* 27, 123–134.

Brock, T. D. (1965). The purification and characterization of an intercellular sex-specific mannan protein from yeast. *Proc. Natl. Acad. Sci. U.S.A.* 54, 1104–1112.

Bruns, P. J., and Brussard, T. B. (1974). Pair formation in *Tetrahymena pyriformis*, an inducible developmental system. *J. Exptl. Zool.* 188, 337–344.

Butzel, H. M. (1974). Mating type determination and development in *Paramecium aurelia*. In "Paramecium, A Current Survey" (W. J. van Wagtendonk, ed.), pp. 91–130. Elsevier, Amsterdam.

Byrne, B. C. (1972). Mutagenic analysis of mating type and isolation of reactive cilia of both mating types in the ciliated protozoan, *Paramecium aurelia* syngen 4. Ph.D. Thesis, Indiana Univ., Bloomington.

Byrne, B. J., and Byrne, B. C. (1978). Behavior and the excitable membrane in *Paramecium*. *Crit. Rev. Microbiol.* 6, 53–108.

Cohen, L. W., and Siegel, R. W. (1963). The mating-type substances of *Paramecium bursaria*. *Genet. Res.* 4, 143–150.

Crandall, M., Lawrence, L. M., and Saunders, R. M. (1974). Molecular complementarity of yeast glycoprotein mating factors. *Proc. Natl. Acad. Sci. U.S.A.* 71, 26–29.

Cronkite, D. L. (1972). Genetics of chemical induction of conjugation in *Paramecium aurelia*. Ph.D. Thesis, Indiana Univ., Bloomington.

Cronkite, D. L. (1974). Genetics of chemical induction of conjugation in *Paramecium aurelia*. *Genetics* 76, 706–714.

Cronkite, D. L. (1975). A suppressor gene involved in chemical induction of conjugation in *Paramecium aurelia*. *Genetics* 80, 13–21.

Cronkite, D. L. (1976). A role of calcium ions in chemical induction of mating in *Paramecium tetraurelia*. *J. Protozool.* 23, 431–433.

Cronkite, D. L. (1977). An analysis of the mechanism of activation of rapid micronuclear migration, a very early event in conjugation of *Paramecium caudatum*. *Proc. Int. Congr. Protozool., 5th, New York*, 284.

Cronkite, D. L. (1979). The genetics of swimming and mating behavior in *Paramecium*. *In* "Biochemistry and Physiology of Protozoa" (M. Levandowsky and S. H. Hutner, eds.), Vol. 3, 2nd ed., pp. 221–273. Academic Press, New York.

Fujishima, M., and Hiwatashi, K. (1977). An early step in initiation of fertilization in *Paramecium:* Early micronuclear migration. *J. Exp. Zool.* 201, 127–134.

Fukushi, T., and Hiwatashi, K. (1970). Preparation of mating reactive cilia from *Paramecium caudatum* by $MnCl_2$. *J. Protozool.* 17, Suppl., p. 21.

Fulton, B. P., and Whittingham, D. G. (1978). Activation of mammalian oocytes by intracellular injection of calcium. *Nature (London)* 273, 149–151.
Gibbons, I. R. (1965). Chemical dissection of cilia. *Arch. Biol.* 76, 317–352.
Goodenough, U. W. (1977). Mating interactions in *Chlamydomonas*. In "Microbial Interactions" (J. L. Reissig, ed.), pp. 323–350. Chapman & Hall, London.
Haga, N., and Hiwatashi, K. (1980). Immaturin: A protein controlling sexual immaturity in *Paramecium*. *Nature (London)* (in press).
Haggard, B. W. (1974). Interspecies crosses in *Paramecium aurelia* (syngen 4 by syngen 8). *J. Protozool.* 21, 152–159.
Heckmann, K. (1963). Paarungssystem und genabhängige Paarungstypdifferenzierung bei dem hypotrichen Ciliaten *Euplotes vannus* O. F. Müller. *Arch. Protistenkd.* 106, 393–421.
Hiwatashi, K. (1951). Studies on the conjugation of *Paramecium caudatum*. IV. Conjugating behavior of individuals of two mating types marked by a vital staining method. *Sci. Rep. Tohoku Univ., Ser. 4* 19, 95–99.
Hiwatashi, K. (1955). Studies on the conjugation of *Paramecium caudatum*. VI. On the nature of the union of conjugation. *Sci. Rep. Tohoku Univ., Ser. 4* 21, 207–218.
Hiwatashi, K. (1961). Locality of mating reactivity on the surface of *Paramecium caudatum*. *Sci. Rep. Tohoku Univ., Ser. 4* 27, 93–99.
Hiwatashi, K. (1969). *Paramecium*. In "Fertilization II" (C. B. Metz and A. Monroy, eds.), pp. 255–293. Academic Press, New York.
Hiwatashi, K., Haga, N., and Takahashi, M. (1980). Restoration of membrane excitability in a behavioral mutant of *Paramecium caudatum* during conjugation and by microinjection of wild-type cytoplasm. *J. Cell Biol.* 84, 476–480.
Jaffe, L. A. (1976). Fast block to polyspermy in sea urchin eggs is electrically mediated. *Nature (London)* 261, 68–71.
Jaffe, L. A., and Robinson, K. R. (1978). Membrane potential of the unfertilized sea urchin egg. *Dev. Biol.* 62, 215–228.
Jennings, H. S. (1938). Sex reaction types and their interrelation in *Paramecium bursaria*. I and II. *Proc. Natl. Acad. Sci. U.S.A.* 24, 112–120.
Kitamura, A., and Hiwatashi, K. (1976). Mating-reactive membrane vesicles from cilia of *Paramecium caudatum*. *J. Cell Biol.* 69, 736–740.
Kitamura, A., and Hiwatashi, K. (1978). Are sugar residues involved in the specific cell recognition of mating in *Paramecium*? *J. Exp. Zool.* 203, 99–108.
Kitamura, A., and Hiwatashi, K. (1980). Reconstitution of mating active membrane vesicles in *Paramecium*. *Exp. Cell Res.* 125, 486–489.
Kitamura, A., Onimaru, H., Naitoh, Y., and Hiwatashi, K. (1979). Relation between swimming and sexual behaviors in *Paramecium caudatum*. *Dobutsugaku Zasshi* 88, 528 (Abstr., in Jpn.)
Kung, C. (1971a). Genic mutants with altered system of excitation in *Paramecium aurelia*. I. Phenotypes of the behavioral mutants. *Z. Vgl. Physiol.* 71, 142–164.
Kung, C. (1971b). Genic mutants with altered system of excitation in *Paramecium aurelia*. II. Mutagenesis, screening and genetic analysis of the mutants. *Genetics* 69, 29–45.
Kung, C., and Eckert, R. (1972). Genetic modification of electric properties in an excitable membrane. *Proc. Natl. Acad. Sci. U.S.A.* 69, 93–97.
Larison, L. L., and Siegel, R. W. (1961). Illegitimate mating in *Paramecium bursaria* and the basis for cell union. *J. Gen. Microbiol.* 26, 499–508.
Marchesi, V. T., and Andrews, E. P. (1971). Glycoproteins: Isolation of cell membranes with lithium diiodosalicylate. *Science* 174, 1247–1248.

Metz, C. B. (1954). Mating substances and the physiology of fertilization in ciliates. *In* "Sex in Microorganisms" (D. H. Wenrich, ed.), pp. 284–334. Am. Assoc. Adv. Sci., Washington, D.C.

Miwa, I. (1979a). Specificity of the immaturity substances in *Paramecium*. *J. Cell Sci.* 36, 253–260.

Miwa, I. (1979b). Immaturity substances in *Paramecium primaurelia* and their specificity. *J. Cell Sci.* 38, 193–199.

Miwa, I., and Hiwatashi, K. (1970). Effect of mitomycin C on the expression of mating ability in *Paramecium caudatum*. *Jpn. J. Genet.* 45, 269–275.

Miwa, I., Haga, N., and Hiwatashi, K. (1975). Immaturity substances: Material basis for immaturity in *Paramecium*. *J. Cell Sci.* 19, 369–378.

Miyake, A. (1958). Induction of conjugation by chemical agents in *Paramecium caudatum*. *J. Inst. Polytech., Osaka City Univ., Ser. D.* 9, 251–296.

Miyake, A. (1964). Induction of conjugation by cell-free preparations in *Paramecium multimicronucleatum*. *Science* 146, 1583–1585.

Miyake, A. (1966). Local disappearance of cilia before the formation of holdfast union in conjugation of *Paramecium multimicronucleatum*. *J. Protozool.* 13, Suppl., p. 28.

Miyake, A. (1968). Induction of conjugation by chemical agents in *Paramecium*. *J. Exp. Zool.* 167, 359–380.

Miyake, A. (1969). Mechanism of initiation of sexual reproduction in *Paramecium multimicronucleatum*. *Jpn. J. Genet.* 44, Suppl. 1, 388–395.

Miyake, A. (1978). Cell communication, cell union, and initiation of meiosis in ciliate conjugation. *Curr. Top. Dev. Biol.* 12, 37–82.

Myohara, K., and Hiwatashi, K. (1978). Mutants of sexual maturity in *Paramecium caudatum* selected by erythromycin resistance. *Genetics* 90, 227–241.

Naitoh, Y., and Eckert, R. (1974). The control of ciliary activity in Protozoa. *In* "Cilia and Flagella" (M. A. Sleigh, ed.), pp. 305–352. Academic Press, New York.

Naitoh, Y., and Kaneko, H. (1973). Control of ciliary activities by adenosine triphosphate and divalent cations in Triton-extracted models of *Paramecium caudatum*. *J. Exp. Biol.* 58, 657–676.

Nanney, D. L. (1977). Cell-cell interaction in ciliates: Evolutionary and genetic constraints. *In* "Microbial Interactions" (J. L. Reissig, ed.), pp. 351–397. Chapman & Hall, London.

Narahashi, T. (1974). Chemicals as tools in the study of excitable membranes. *Physiol. Rev.* 54, 813–889.

Nelson, D. L., and Kung, C. (1978). Behavior of *Paramecium:* chemical, physiological and genetic studies. *In* "Taxis and Behavior" (G. L. Hazelbauer, ed.), pp. 75–100. Chapman & Hall, London.

Ridgway, E. B., Gilkey, J. C., and Jaffe, L. F. (1977). Free calcium increases explosively in activating medaka eggs. *Proc. Natl. Acad. Sci. U.S.A.* 74, 623–627.

Schneider, L. (1963). Elektronenmikroskopische Untersuchungen der Konjugation von *Paramecium*. I. Die Auflösung und Neubildung der Zellmembran bei der Konjugation. *Protoplasma* 56, 109–140.

Siegel, R. W. (1965). Hereditary factors controlling development in *Paramecium*. *Brookhaven Symp. Biol.* 18, 55–65.

Siegel, R. W., and Larison, L. L. (1960). The genetic control of mating type in *Paramecium bursaria*. *Proc. Natl. Acad. Sci. U.S.A.* 46, 344–349.

Sonneborn, T. M. (1937). Sex, sex inheritance and sex determination in *Paramecium aurelia*. *Proc. Natl. Acad. Sci. U.S.A.* 23, 378–385.

Sonneborn, T. M. (1939). *Paramecium aurelia:* Mating types and groups; lethal interactions; determination and inheritance. *Am. Nat.* 73, 390-413.
Sonneborn, T. M. (1957). Breeding systems, reproductive methods, and species problems in Protozoa. *In* "The Species Problem" (E. Mayr, ed.), pp. 155-324. Am. Assoc. Adv. Sci., Washington, D.C.
Sonneborn, T. M. (1970). Methods in *Paramecium* research. *Methods Cell Physiol.* 4, 241-339.
Sonneborn, T. M. (1974). *Paramecium aurelia. In* "Handbook of Genetics" (R. C. King, ed.), Vol. 2, pp. 469-594. Plenum, New York.
Sonneborn, T. M. (1975). The *Paramecium aurelia* complex of fourteen sibling species. *Trans. Am. Microsc. Soc.* 94, 155-178.
Sonneborn, T. M., and Dippell, R. V. (1946). Mating reactions and conjugation between varieties of *Paramecium aurelia* in relation to conceptions of mating type and variety. *Physiol. Zool.* 19, 1-18.
Steinhardt, R., Zucker, R., and Schatten, G. (1977). Intracellular calcium release at fertilization in the sea urchin egg. *Dev. Biol.* 58, 185-196.
Takahashi, M. (1979). Behavioral mutants in *Paramecium caudatum. Genetics* 91, 393-408.
Takahashi, M., and Naitoh, Y. (1978). Behavioral mutants of *Paramecium caudatum* with defective membrane electrogenesis. *Nature (London)* 271, 656-659.
Takahashi, M., Takeuchi, N., and Hiwatashi, K. (1974). Mating agglutination of cilia detached from complementary mating types of *Paramecium. Exp. Cell Res.* 87, 415-417.
Tsukii, Y. (1980). Genetics of mating type in *Paramecium caudatum.* Ph.D. thesis, Tohoku Univ., Sendai, Japan.
Tsukii, Y., and Hiwatashi, K. (1978). Inhibition of early events of sexual processes in Paramecium by Concanavalin A. *J. Exp. Zool.* 205, 439-446.
Tsukii, Y., and Hiwatashi, K. (1979). Artificial induction of autogamy in *Paramecium caudatum. Genet. Res.* 34, 163-172.
Van Wagtendonk, W. J. (1974). Nutrition of *Paramecium. In "Paramecium,* A Current Survey" (W. J. Van Wagtendonk, ed.), pp. 339-376. Elsevier, Amsterdam.
Van Wagtendonk, W. J., and Soldo, A. T. (1970). Methods used in the axenic cultivation of *Paramecium aurelia. Methods Cell Physiol.* 4, 117-130.
Vivier, E., and André, J. (1961). Données structurales et ultrastructurales nouvelles sur la conjugaison de *Paramecium caudatum. J. Protozool.* 8, 416-426.
Watanabe, T. (1977). Ciliary membranes and mating substances in *Paramecium caudatum. J. Protozool.* 24, 426-429.
Watanabe, T. (1978). A scanning electron-microscopic study of the local degeneration of cilia during sexual reproduction in *Paramecium. J. Cell Sci.* 32, 55-66.
Wichterman, R. (1953). "The Biology of *Paramecium.*" Blakiston, New York.
Wiese, L. (1974). Nature of sex specific glycoprotein agglutinins in *Chlamydomonas. Ann N.Y. Acad. Sci.* 234, 383-395.

15

The Genetics and Cellular Biology of Sexual Development in *Ustilago Violacea*

ALAN W. DAY AND JOSEPH E. CUMMINS

I. Introduction 379
   A. Life Cycle 380
   B. Culture Methods 382
II. Summary of the Literature 382
   A. Historical Background 382
   B. Mutants and Mapping by Sexual and Parasexual Techniques 383
   C. Cytology and Ultrastructure 384
   D. Conjugation (Plasmogamy) and Cell-to-Cell Communication ... 384
   E. Nuclear Fusion (Karyogamy) 387
III. Current Research 388
   A. Fimbriae and Intercellular Communication 388
   B. The Genetic Program Leading to Mating and the Assembly of the Conjugation Tube 392
   C. The Mating Type Locus as a Developmental Master Switch ... 396
   D. The Induction of the Parasitic Infection Hyphae by Host Products Acting on the Mating Type Locus 397
IV. Perspectives 398
   References 400

I. INTRODUCTION

Smut fungi are economically important and are beginning to be recognized as valuable tools for genetic research. They combine the dual advantages of a yeastlike saprophytic growth phase with a pathogenic phase on many plants. The techniques that have been developed for yeast genetics may be applied to the smut fungi and in addition complex problems of morphogenesis and host parasite interaction may be examined.

Much has been learned about the genetics of the corn smut, *Ustilago maydis,* by Robin Holliday, P. R. Day, and their colleagues (Holliday, 1974; Day, 1974) and of the anther smut, *Ustilago violacea* (Pers.) Roussel, by ourselves and more recently by E. D. Garber and his colleagues (see below). In *U. violacea* the earlier work on methods of sexual and parasexual genetic analysis (Day and Jones, 1969) has provided the basis for the following description of the mating type locus and its effect on morphogenesis.

A. Life Cycle

Ustilago violacea attacks over 70 species of host plant in the *Caryophyllaceae* (Carnation family) (Zillig, 1921; Liro, 1924). Physiological races specific for many of the host species were described by Goldschmidt (1928). We employ a physiological race obtained from the dioecious species *Silene alba* (White Campion). Infection is systemic but sporulation occurs only in the anthers of the host flower, where the fungal diploid sexual spores (brandspores or teliospores) replace the pollen. When the fungus attacks a female host plant (XX chromosome type) it causes a "sex change" and the plant develops like an XY male, producing anthers and reabsorbing female tissues. The physiological basis of this fascinating and potentially important "sex change" has not been determined and remains an inviting project.

A diagrammatic representation of the life cycle of *U. violacea* is shown in Fig. 1. The diploid brandspores remain viable for many decades (Garber *et al.,* 1978) and germinate well on water agar or nutrient media. A short germ tube (promycelium) is produced which divides into three cells when meiosis is completed. Yeastlike uninucleate haploid cells (sporidia) are budded from each of the three cells and from the brandspore itself, forming a meiotic tetrad. The sporidia continue to bud indefinitely on nutritive media. Segregation for mating types occurs during brandspore germination, two of the haploid sporidia being a_1 in mating type and two being a_2. Sporidia and/or promycelial cells of opposite mating type conjugate on nonnutritive media by means of a conjugation tube. The fate of a conjugated cell pair depends on the environment. On nutritive media each haploid conjugant buds off haploid cells and resumes vegetative growth. When, however, auxotrophic conjugants are plated on minimal medium, diploid cells produced by

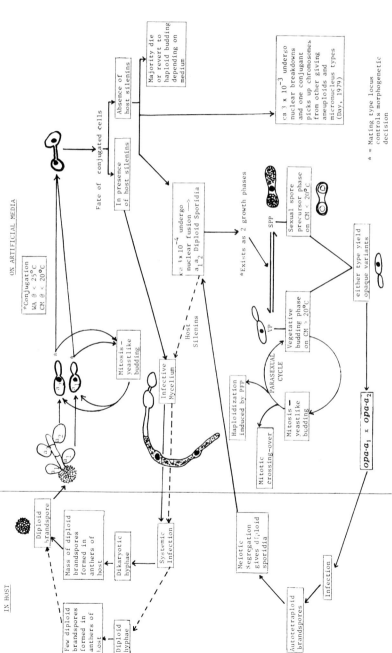

Fig. 1. The life cycle of *Ustilago violacea* showing the growth phases that occur both in natural conditions and on artificial media.

nuclear fusion (karyogamy) may be selected at frequencies varying from 0.01 to 1% of the conjugation pairs (Day and Day, 1974).

Along with the stable euploid forms a variety of unusual types grow up from auxotrophic conjugants on minimal medium or selective media. These forms include aneuploids incorporating individual chromosomes containing prototrophic alleles from part of the genome of one conjugant along with the complete genome of the other conjugant. In addition there are forms in which acquired chromosomes are apparently maintained as a separate group (a micronucleus) which may be expressed or lost preferentially (Day, 1978).

When sporidia mate on the leaves of a host plant the conjugated cells produce infection hyphae which penetrate leaf axils, flower tissue, etc., and invade the plant. While the hyphae are said to be dikaryotic this has been questioned by some workers (see Fischer and Holton, 1957). Sporulation occurs only in the tissue of the anthers, the hyphae first round off and produce binucleate dikaryotic cells. The cells become surrounded in a gelatinous matrix and the brandspore develops as a spherical body within the cell. Karyogamy does not take place until the brandspore is almost mature.

Vegetative diploid cells, heterozygous for mating type, will infect the host plant producing few brandspores but normal segregation of haploid progeny. Homozygotes will not infect unless they are mated with cells of the opposite mating type. In the case of an infection by two strains which were probably $a_1 a_1$ and $a_2 a_2$ the resulting brandspores were large, and yielded diploid sporidia with typical autotetraploid segregation ratios.

B. Culture Methods

The culture methods and methods for genetic manipulation of *U. violacea* are summarized by Cummins and Day (1977). These methods are similar to the methods of yeast genetics and molecular biology with a main complication being the involvement of a host plant to complete the parasitic phases of the life cycle which include meiosis.

II. SUMMARY OF THE LITERATURE

A. Historical Background

Ustilago violacea was a popular organism among mycologists at the turn of this century. In particular the works of Brefeld, Harper, Kniep,

Zillig, Goldschmidt, Liro, and Bauch provided the basis of our knowledge of its life cycle, mating type, specialization into physiological races, cytology, etc. Many other workers described the parasitic phases and their effect on the host plant. An excellent and comprehensive review of this early work is available by judicious use of the index in Fischer and Holton (1957), while the main details are summarized in Day (1968). We review here the "modern phase" of work with this organism starting with its use, on the recommendation of Robin Holliday, as an undergraduate project by A. W. Day in 1964.

B. Mutants and Mapping by Sexual and Parasexual Techniques

Auxotropic and color mutants were isolated in the 1960's and Day and Jones (1968, 1969) used some 33 different isolates of a total of over 50 isolated, to carry out genetic mapping by means of sexual and parasexual methods. A large number of carotenoid mutants have also been isolated and studied by Garber *et al.* (1975, 1978). Replica plating by means of velvet made possible the speedy and efficient analysis of genotypes. Sexual analysis was carried out through random spore analysis (Day and Jones, 1969) although recently (Cattrall *et al.*, 1978; Castle and Day, 1981) techniques for unordered tetrad analysis have been devised. As centromere-linked markers have been found, it is now possible to locate the centromere in maps of each linkage group. Garber's group is continuing to develop these mapping techniques and to use them to study the mechanisms of crossing-over and nondisjunction and the importance of sexual recombination in nature (Garber *et al.*, 1975, 1978; Cattrall *et al.*, 1978; Baird and Garber, 1979a,b).

Mapping techniques that utilize parasexual or mitotic recombination depend on the isolation of somatic diploid strains. A technique for the efficient production of somatic diploid strains from mated pairs or complementary auxotrophs was devised by Day and Jones (1968). These somatic diploids could be induced to undergo mitotic crossing-over by uv light at 253 nm, and to haploidize by random loss of chromosomes in the presence of *p*-fluorophenylalanine (PFP) (Day and Jones, 1969, 1971). The haploidization technique is particularly useful as many hundreds of haploid genotypes may be recovered from a single plate and these allow unambiguous identification of linkage groups (Day and Jones, 1969, 1971). Two particular chromosomes, however, tended to remain disomic after PFP treatment, although monosomics could be selected for quite easily (Day and Jones, 1971). Later work indicates that each chromosome has its own characteristic frequency of residual disomy after PFP treatment, i.e., a constant proportion of between 0 and 45% of the "haploid" colonies may in fact be disomic for a particular chromosome.

C. Cytology and Ultrastructure

The mapping data indicate that there are at least 12 genetically marked chromosomes and a probable haploid chromosome number of around 20, a result that conflicted with the observation of earlier cytologists that $n = 2$ in this species. Accordingly, the cytology of somatic cells was reexamined. Using acetic orcein staining of synchronous cultures, Day and Jones (1972) described the nuclear division cycle in which the nucleus moves into the bud to divide, one daughter nucleus later returning to the mother cell. While at one time the chromatin is condensed into two bars, giving an appearance of two chromosomes, examination of earlier stages showed unequivocally that there were many (around 20) chromosomes. We now use acridine–orange fluorescence microscopy which gives vital staining and spectacular results, although it is still impossible to count accurately the many tiny chromosomes (Poon and Day, 1974b). A detailed study of the ultrastructure of the sporidium of *U. violacea* was completed by Poon in 1975 (Poon and Day, 1976a,b; Poon, 1978). In the 1976b study the structure and development of the spindle pole body (SPB) was described and its possible multifunctional roles were discussed. This light and electron microscopic work led to consideration of the genetic implications of current models of somatic nuclear division in fungi (Day, 1972).

D. Conjugation (Plasmogamy) and Cell-to-Cell Communication

In *U. violacea* the mating process consists of the cooperative construction of a cylindrical fusion tube between a pair of uninucleate cells of opposite mating type. As indicated earlier the mating type locus forms a simple bipolar system and conjugation and nuclear fusion are temporally separated. The stage of the cell cycle of a cell bearing a particular mating allele (a_1 or a_2) influences expression of the allele and the two alleles appear to be influenced differently by the cell cycle. We used the term temporal allelic interaction to describe the resultant interaction in diploid cells and suggested that there were *cis* acting control loci that serve to regulate gene action in relation to other cell cycle events (see reviews in Cummins and Day, 1974b; Day and Cummins, 1975).

Development of the conjugation tube between compatible cells involves several stages (Poon et al., 1974). These range from a period of courtship where mating partners show no visible morphogenetic change to the initiation, fusion, and final elongation of the organelle. The courtship period has stringent requirements including low temperature, low salt levels, and the persistent undisturbed presence of the two mating types. During courtship transcription and translation of a genetic program (the sex message) is completed prior to assembly of the conjugation tube (Cummins and Day, 1974a). Experiments on the influence of ultraviolet light on assembly using uv sensitive and resistant strains (Day and Cummins, 1974) and the influence of an inhibitor of protein synthesis, cycloheximide, in sensitive and resistant strains (Cummins and Day, 1976a) both showed that mating cells mutually exchange information that regulates "readout" of the genes governing morphogenetic induction. This information exchange occurs prior to cell fusion or assembly of the conjugation tube.

During courtship the mating cells may contact each other or be separated by a distance of up to 20 μm (Poon and Day, 1974a, 1975; Day and Poon, 1975). At the end of courtship the conjugation tube is initiated as a peg in the cell bearing mating type a_2 which seems to grow toward and induce the later development of a similar peg in the a_1 cell (Day, 1976). Long hairs (fimbriae) on the cell surface have been observed in *Ustilago* (Poon and Day, 1974a, 1975; Day and Poon, 1975). Day and Poon (1975) showed that fimbriation and conjugation have the same maximum temperature, similar response to temperature shifts, similar response to enzyme treatment, and both are blocked before translation by high temperatures. Cell growth is not prevented in conditions that limit both conjugation and fimbriation. Strong circumstantial evidence was obtained that the fimbriae of smut establish the first connection between the mating cells and that the conjugation tube grows along this fimbrial path. It was hypothesized that the fimbrial connection is used to transfer sex-specific molecules between the two conjugants (Day, 1976). It was also noted that many yeasts and yeastlike fungi including *Saccharomyces, Schizosaccharomyces, Hansenula, Candida* spp. also form very short fimbriae and an interesting correlation between fimbriation and flocculation was reported for brewing strains of *Saccharomyces cerevisiae* (Day et al., 1975). In summary, information governing readout of the genetic program for conjugation is reciprocally exchanged between mating cells during courtship. Exchange of mating information appears to be governed wholly or partly by fine hairs called fimbriae that join the cells. The appearance of these hairs and their association with pair formation is shown in Fig. 2.

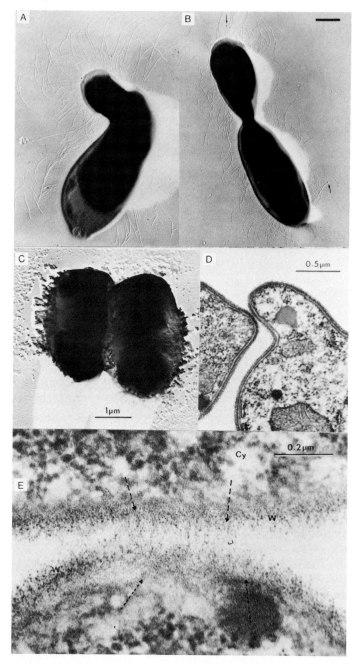

Fig. 2. (A and B) Scanning and transmission electron micrographs of tungsten oxide shadowed cells of *Ustilago violacea* showing numerous fimbriae. Note the apparent branching due to intertwining of individual fimbriae and the occasional knobbed fimbriae

E. Nuclear Fusion (Karyogamy)

As indicated earlier, karyogamy does not automatically follow plasmogamy in the smut fungi. Indeed, fusion of nuclei is a relatively rare event until late in the life cycle just prior to brandspore maturation. This situation contrasts with budding yeasts where karyogamy normally occurs immediately after plasmogamy. In yeast a gene (kar^+) regulates obligate karyogamy and its allele (kar^-) establishes a state similar to *in vitro* spordial fusion in smut (Fink and Condé, 1977). Clements *et al.* (1969) began to study the regulation of karyogamy in smut first noting that uv irradiation (253 nm) induced nuclear fusion in conjugated pairs of cells and that photoreactivation or conditions favoring excision repair reduced the level of induction. Later studies using several uv-sensitive mutants, some with defects in excision repair (Day and Day, 1970), suggested that karyogamy is repressed in conjugated cells. The transcription of repressor may be disrupted by uv irradiation allowing greater frequencies of karyogamy (Day and Day, 1974). Presumably, the karyogamy repressor is turned off in premeiotic cells.

Turning to a comparison of yeast and smut in their regulation of karyogamy it is interesting to note that the yeast gene kar^- that prevents karyogamy is dominant to kar^+ when the alleles are in separate nuclei suggesting that a product is produced that prevents nuclear fusion. However, the gene kar^- is not expressed when kar^+ and kar^- reside within the same nucleus suggesting that regulation of karyogamy may be a complex affair in yeast (Fink and Condé, 1977). Constitutive high frequency karyogamy would be predicted if mutations occur in the karyogamy repressor gene of *U. violacea* but such mutants have not yet been observed.

(arrowed). Bar represents 1 μm. [From Poon and Day (1975). Reproduced by permission of the National Research Council of Canada from the *Canadian Journal of Microbiology* 21, 537–546.] (C) Cell pairing. This shadow-cast preparation taken 3 hr after mating in a culture defimbriated by sonication shows the large area of cell pairing and globular masses present between the cells. Some of the loose materials on the walls are undoubtedly artifacts produced by the sonication treatment, but nonsonicated cells show similar masses between the cells at the time of pairing and the photograph is reasonably representative of cells at the time of pairing. (D) Thin-section electron micrograph of pair of conjugating cells about 20 min after cell pairing showing development of conjugation pegs. Note that the wall layers and plasma membranes of each peg are still intact and that the walls appear to be in close contact over a small area. (E) High-power detail of an area of close contact similar to that shown in Fig. 4. Note the fibrils appear to penetrate the wall and plasma membrane into the cytoplasm (Cy). The fibrils (dotted lines) are about 120 Å diameter. The fibrils may represent fimbrial connections between the mating cells. [From Day and Poon (1975). Reproduced by permission of the National Research Council of Canada from the *Canadian Journal of Microbiology* 21, 547–557.]

III. CURRENT RESEARCH

A. Fimbriae and Intercellular Communication

As indicated earlier there is evidence that fimbriae are necessary for intercellular communication during mating. They appear to be significant in the final posttranslational assembly of the conjugation tube but their role in the earlier mutual and reciprocal exchange of information regulating reading of the sex message genes has not been clearly established. Certainly, distant mating experiments in which mating cells are separated by as much as 20 µm on an agar surface suggest that fimbriae may serve as communication channels for macromolecules. It may very well be true that fimbriae have sensory functions beyond their role in mating. An understanding of the function of fimbriae depends on an elucidation of their structure in wild-type cells and the isolation and study of mutants that have modified fimbriae.

The structure of fimbriae has recently been studied both at the level of their ultrastructure and the level of their molecular composition. The ultrastructure of the fiber was studied by high resolution electron microscopy following negative staining. A diffraction analysis of high resolution micrographs shows that each fiber is constructed as a tightly coiled helix quite similar to F-actin (R. Gardiner, unpublished). Fimbriae can be isolated by a simple procedure which involves stripping smut cells in a homogenizer, followed by centrifugation of the cells to recover pure fimbriae in the supernatant (Poon and Day, 1975). The fibers behave in the centrifuge as a heterogeneous array of sizes achieving s values greater than 100 in a sucrose gradient made without salts. In a salt containing gradient (i.e., 0.8 M NaCl) the fibers are dissociated to form a single peak about 2 S.

After intact fibers are dissociated in 8 M urea or 1.5 M guanidine chloride then dialyzed against water at room temperature, they reassociate to form fibers longer than the original. Finally, the dissociated and reassembled fibers will rejoin to stripped cells and their appearance is normal. Thus, it is possible to dissociate and reassemble the fiber. The biological consequences of reassembly and interaction of the reassembled fiber with mating type are currently being studied.

SDS gel electrophoresis of the isolated fimbriae shows that they are made up of proteins of a single molecular weight of 74,000 (Fig. 3). Peptides (less than 15,000 MW) are found in the stripped fraction but do not centrifuge with the long fibers (experiments in progress). In spite of the apparent homogeneity in monomer protein in SDS at least two major bands appear on isoelectric focusing of the fiber protein.

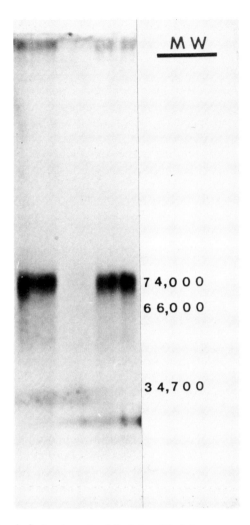

Fig. 3. SDS gel electrophoresis of fimbriae. Fimbriae were prepared by stripping cells then concentrating isolated fibers by centrifugation at 45,000 rpm for 3 hr using the 50 rotor of the Beckman Ultracentrifuge. Fimbriae were dissociated in 0.8 M NaCl or 8 M urea and heated to 60°C. Following chromatography on Sephadex G-50 the fibers are reconstituted in water and recentrifuged as above. The SDS gel electrophoresis employed a procedure slightly modified from Laemmli (1970) using a 8.4% gel. The gel was stained with silver as described by Merril et al. (1979). Molecular weight standards included bovine serum albumin and pepsin (Sigma).

Current experiments show that one band contains phosphorylated protein while the other does not, but we have not exhaustively studied other modifications such as methylation or acetylation. The role of protein modification in assembly of the fibers and in their communication and sensory functions is currently being studied. Protein modification appears to play an important biological role in a number of sensory phenomena (Springer et al., 1979).

The amino acid composition of various filamentous proteins is detailed in Table I. It is of interest to note that the amino acid composition of fimbriae is more similar to the composition of actin and tubulin than it is to the composition of bacterial pilin, the prokaryotic counterpart of fungal fimbriae. Fungal fimbriae may therefore have a functional similarity to the cytoplasmic fibers of eukaryotes and resemble only superficially the fimbriae (pili) of bacteria.

The synthesis of fungal fimbriae is regulated at the level of transcription. This conclusion is based on inhibitor studies (Poon and Day, 1975) and on recent work showing that synthesis and assembly of the fiber is sensitive to uv light (256 nm) and can be photoreactivated by visible light. Pyrimidine dimers, the major uv photoproduct and site for photoreactivation, behave like a natural transcription termination signal in a structural gene. It is therefore possible to measure the size of a structural gene and to determine its position relative to a promoter by relating the inactivating dose of ultraviolet light to a gene target of known size (e.g., the ribosomal RNA genes) (reviewed in Sauerbier and Hercules, 1978). We have found it possible to relate the inactivating dose of uv light for the ribosomal genes of *Ustilago violacea* to the inactivating dose for fimbrial synthesis and assembly using both radioactive labeling and electron microscopy in both excision deficient and proficient strains of the smut. The fimbrial monomer contains 658 amino acids specified by a structural gene at least 1944 base pairs in length. From the inactivating UV dose we find that the fimbrial structural gene and its promoter span 2200 base pairs maximally (Fig. 4). Thus the fimbrial gene is a relatively simple one and it is unlikely that it is a part of a complex transcription array.

The synthesis and assembly of fimbriae is crucial in intercellular communication during mating. Fimbriae can be dissociated to monomer proteins and reassembled and rejoined to cells stripped of fimbriae. Fimbriae appear to be made up of a single protein species that can be modified as the protein functions. The genetic organization of the fimbrial gene is not complex; it entails a short promoter and a structural gene as detected by an ultraviolet mapping technique.

TABLE I
The Amino Acid Composition of Some Filamentous Proteins[a]

Amino acid	Fimbriae (MW 74,000)	Pilin (MW 17,000)	Actin (MW 42,000)	Tubulin A (MW 55,000)	Tubulin B (MW 55,000)
Glycine	12.3	10.4	7.7	7.3	8.3
Glutamic acid	12.0	8.0	10.9	13.5	13.6
Aspartic acid	11.5	12.3	9.4	11.2	10.7
Serine	9.5	6.1	6.7	4.9	4.9
Threonine	9.4	12.3	7.9	5.6	5.7
Alanine	9.3	21	8.2	7.7	7.8
Valine	7.0	8.0	5.1	6.5	7.0
Leucine	6.3	6.1	7.0	7.8	7.4
Proline	5.3	1.2	5.1	5.1	4.8
Arginine	4.1	1.8	4.9	5.6	5.0
Lysine	4.0	1.8	5.0	6.0	5.0
Isoleucine	3.5	2.5	7.5	4.9	4.9
Phenylalanine	2.6	5.0	3.2	4.1	4.2
Half-cystine	1.3	1.2	1.3	1.5	1.7
Tyrosine	1.2	1.2	3.9	3.3	3.3
Methionine	0.4	0	3.4	2.4	2.5

[a] The data in this table are molar percentage of each amino acid. The composition of fimbriae is from current studies (R. Gardiner, unpublished), composition of pilin is from Brinton (1964), composition of actin from Oosawa and Kasai (1971), and tubulins from Stephens (1971).

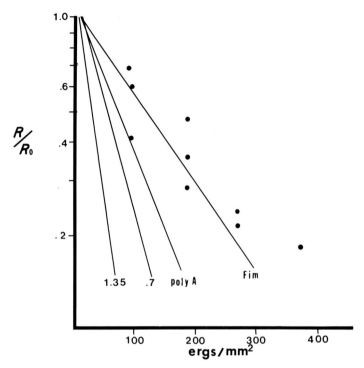

Fig. 4. Reduced synthesis of fimbriae following ultraviolet irradiation. Cells of a uv-sensitive smut strain (2.716u4) which is deficient in pyrimidine dimer excision were irradiated with varying doses of uv (254 nm), then immediately stripped of fimbriae, and placed in a tritiated amino acid mixture in water for 4 hr. The new fimbriae were stripped from the cell then recovered and counted. The data points are the ratio of counts in fimbriae (R) at a given UV dose relative to counts in an unirradiated control (R_0). Some of the lines are from other experiments showing similar ratios for synthesis of 1.35×10^6 daltons and 0.7×10^6 daltons ribosomal RNA along with uv sensitivity of the total polyadenylated RNA of the smut cell.

B. The Genetic Program Leading to Mating and the Assembly of the Conjugation Tube

As indicated earlier, there is a period of courtship prior to visible morphological modification of the mating cells and their assembly of the mating tube. During courtship genes governing sexual morphogenesis are transcribed and translated, this apparently depending on a mutual and reciprocal exchange of information. Thus, physical interruption of mating inhibits complete reading of the sex message (program) and prevents final assembly of the organelle. Cells interrupted during courtship require at least one cycle of cell growth before the mating process will reinitate. Assembly of the conjugation tube therefore requires orderly reading of a genetic program and reciprocal information exchange prior to cell fusion. Currently we have been attempting to identify and enumerate the gene products (mRNA and protein) required for mating, by means of nucleic acid hybridization, acrylamide gel electrophoresis, selection of mutants temperature sensitive in mating, and examination of the uv light dosage required to inhibit mating.

As mating and growth are mutually exclusive under normal circumstances, it is necessary to consider the metabolic changes that accompany "shift down" from rich to poor growth media and to extract from that complex pattern those parts that are exclusively attributable to mating when cells of the two mating types are present together.

First, in considering changes in transcription as the sex program unfolds we have identified the major stable and unstable RNA molecules in the smut cell. The stable RNA molecules include cytoplasmic ribosomal RNA molecules [1.35×10^6 daltons, 0.7×10^6 daltons, 58,000 daltons (5.8 S), 40,000 daltons (5 S), and 25,000 daltons (4 S)] transfer RNA, and mitochondrial ribosomal RNA molecules (1.1×10^6 daltons and 0.55×10^6 daltons) (Dunne and Cummins, 1980). There is an unusual stable 0.23×10^6 minor RNA component that appears in all of the smut strains we have studied. Polyadenylated RNA makes up about 2.75% of the total during log phase and is polydisperse from 0.1×10^6 to 1×10^6 daltons (Dunne and Cummins, 1980). During the shift from rich growth to starvation accompanying mating there is a stringent cessation of ribosomal RNA synthesis in auxotrophic strains along with a concurrent increase in transfer RNA synthesis and the synthesis of a polydisperse fraction with maximum levels near 0.2×10^6 daltons. Polyadenylated RNA remains at less than 10% of the total labeled RNA after a 3-hr exposure to radioactive adenine in both log phase and starving–mating cells. Thus, the heterogeneous peak of low molecular weight RNA that appears during starvation mating is not exclusively polyadenylated RNA. Earlier work showed that RNA synthesized during starving–mating was less sensitive to uv irradiation treatment than was ribosomal RNA synthesis, and doses of uv irradiation that inhibited mating did not interfere with total RNA synthesis in mating cells (Day and Cummins, 1975; Dunne and Cummins, 1980). These observations suggest that much of the RNA made in starving–mating cells is not messenger RNA but is a low molecular weight RNA species made in response to a stringent restriction of ribosomal RNA synthesis. However, fuller study will be needed to clarify the function of the heterogeneous RNA synthesized during starvation–mating.

A preliminary analysis of the DNA-DNA hybridization characteristics of *U. violacea* DNA showed that nuclear DNA contained mainly (90%) unique sequences and 10% highly redundant and foldback sequences (Cummins and Day, 1976b). Highly labeled unique smut DNA was hybridized with whole cell RNA in a R_0t analysis (Firtel, 1972). The results of such analyses suggested that mating-specific RNA was barely detectable as it was transcribed from less than 1% of the single copy DNA while starving cells transcribe about 12% of their DNA (Fig. 5). Since the haploid genome of *U. violacea* contains 3×10^{10} daltons of DNA (Cummins and Day, 1977) it is possible to set the upper limit of mating-specific genes at less than 500 genes of 10^3 base pairs. A more precise estimate of the number of mating specific gene products is being attempted using more sophisticated techniques.

We have undertaken a comprehensive analysis of the final products of gene expression, the mating specific proteins, using one- (Laemlli, 1970) and two-dimensional (O'Farrell, 1975) acrylamide gel electrophoresis. This study has not yet been completed but the results presently available indicate that both qualitative and quantitative changes are readily detected in the soluble proteins of mating cells, while starving and log phase cultures show far fewer differences between each other. In particular, strong bands of 15,000 and 60,000 daltons readily distinguish mating (courtship) cultures from starving

DNA–RNA Hybridization at or near $R_0t \infty$

				Total
22° log phase	8%			8%
22° water agar 1.5 hr	7%	4%		11%
22° mating 1.5 hr	7%	4%	1%	12%

Fig. 5. R_0t analysis of whole cell RNA. Highly labeled DNA was prepared after uniformly labeling smut cells with [H³]adenine using hydroxyapatite (HA) chromatography as described in Cummins and Day (1977). The DNA was sonicated then denatured with alkali and multiple copy molecules were eliminated by hybridization to C_0t 3 prior to rechromatography on HA. Whole cell RNA was prepared from log phase, starving or mating cells as described in Cummins and Day (1977). DNA was removed from the RNA preparation using a brief DNase treatment or passage through an HA column. R_0t hybridizations were as described by Firtel (1972) using 0.48 M phosphate buffer, 0.1% sodium dodecyl sulfate at 65° for up to 20 hr. 2 to 3 µg DNA (3×10^3 to 5×10^3 cpm) was employed in each hybridization and a correction was made for self-hybridization (normally about 2%). DNA-RNA hybrids were analyzed using HA chromatography. Hybridization was with RNA from log phase, starving or mating cells and from mixtures of the RNA samples. Hybridizations included R_0t values from 1×10^3 to 20×10^3. Saturation was approached at R_0t values above 10×10^3.

or log phase cultures after Coomassie blue staining of the SDS gels. Presently, it can be safely concluded that there are major changes in a few proteins during mating; however, the completion of the more comprehensive studies should fully elucidate many of the other more subtle changes that accompany mating.

Another approach to studying the orderly transcription of genes during mating is to study changes in uv sensitivity as the sex message is expressed. Earlier we discussed the uv mapping technique that relates the inactivating dose of uv light to a gene target of known size. When an excisionless strain is mated with a proficient strain, the dose of UV irradiation required to reduce mating by 50% increases from 8 ergs/mm^2 to 16 ergs/mm^2 during the first 2 hr of courtship and finally to 20 ergs/mm^2 shortly before assembly of the conjugation tube (A. J. Castle, A. W. Day, and J. E. Cummins, unpublished data). These values suggest that the gene targets are 3.75, 2.14, and 1.50 times larger than the 7 kilobase (kb) ribosomal gene coding for the two large ribosomal RNA molecules. They suggest that the DNA target is large enough to code for 26 proteins, then 19 proteins, and finally 11 average-sized proteins (about 300 amino acids) as mating progresses. However, it is unlikely that gene targets regulating mating are physically linked because doubling gene dosage by mating an excisionless diploid with a proficient haploid increases the initial dose of uv light causing a 50% decrease in mating from 8 ergs/mm^2 in the haploid to 24 ergs/mm^2 in the diploid. We reason that a pyrimidine dimer in any one of 26 genes is sufficient to prevent mating, but a different protein may be inhibited in each of 26 cells. When gene dosage is doubled as in the diploid, the dose of uv required to inhibit both copies of any of the 26 genes should be squared but now will approach a limit set by the size of the largest gene target in the sex message. In this system an excisionless triploid or tetraploid could be used to safely measure the size of the largest genes in the sex message while the haploid can be used to measure the number of genes essential for sexual morphogenesis. The excisionless triploid strain is being constructed and will be used to verify the theory described above. At any rate, the decrease in UV sensitivity as courtship progresses is good evidence for orderly and sequential reading of a sex message.

Finally, we have selected haploid mutants incapable of mating, mutants temperature-sensitive in mating, and mutants temperature-sensitive for cell cycle and metabolic events. Mutants of the types described above will serve as useful tools for studying orderly transcription during courtship. Another mutant (*op-C*) selected by A. Castle has greatly influenced our concepts about mating type and regulation of mating. Several such mutants arose as chance opaque (see Section II,C) isolates following uv treatment of an a_1a_2 diploid (Castle and Day, 1980). These mutants were probably still diploid and were normal in appearance and growth at 22°C but at 15°C on CM they mated vigorously and synchronously with fellow clone members, disregarding both the normal metabolic signals not to mate in rich medium and the mating type barrier. Furthermore, such strains mated with haploids of either mating type on normal mating medium.

An analysis of log phase and mating cells of the mutant strain showed that the protein bands on SDS gels that were typical of mating cells were present in both mating and log phase cells of the mutant. This evidence thus suggests that the mutant is constitutive for the mating tube and that assembly of this organelle is triggered by the shift to low temperature. Since the mutant mates with haploids of either mating type the mutant must either circumvent the normally stringent mating type control or that barrier does not operate when one cell of the mating pair has begun to assemble the conjugation tube.

The previous discussion on the orderly genetic program can be summarized as follows: First, mating is normally associated with reduced nutrition (shift down) and during courtship ribosome synthesis is reduced while the synthesis of transfer RNA and heterogeneous low molecular weight RNA continues. Mating-specific RNA is barely detectable by DNA-RNA hybridization. A few mating-specific proteins are strongly enhanced during courtship but more subtle changes are yet to be resolved. The uv light sensitivity of mating decreases during courtship as the reading of the sex message nears completion. Finally, a strain derived from a diploid appears to constitutively produce mating tubes at permissive temperatures.

C. The Mating Type Locus as a Developmental Master Switch

We have established that the mating type locus responds to several environmental factors and in turn directs the appropriate development, either vegetative budding, conjugation tube formation, dikaryon formation, or sexual spore formation (Day, 1979). On artifical media, alteration of temperature and/or nutritional factors appears to control activity of the mating type locus. On CM at temperatures above 20°C the locus appears to be inactive and haploids or a_1a_2 diploids remain in the vegetative phase (VP) and cannot conjugate or develop into the sexual spore precursor phase (SPP). As the nutritive (probably cation) content of the media is lowered, or the temperature of incubation on CM decreased below 20°C the mating-type alleles become activated and permit haploid cells to respond to a cell of opposite mating type and form conjugation tubes. In a_1a_2 diploid cells, activity of the two mating type alleles does not allow conjugation or vegetative growth but instead they begin to develop into sexual spores. The cells cease budding, swell to 5–6 times their volume, and develop a characteristic dumbbell shape, full of refractile granules. Later they form spherical bodies within these cells which resemble the precursor stages of brandspore formation in the host plant (Day, 1979). It has not been possible so far to induce this sexual precursor phase (SPP) to complete brandspore formation on artificial media, but this is being attempted.

Diploid (a_1a_2) cells unable to develop into SPP cells even under the inducing conditions, arise spontaneously at very high frequencies (about 5×10^{-3}) (Castle and Day, 1980). These strains are termed *opaques* because of their opaque colonies compared to the translucent colonies of normal SPP diploids. Some opaques are neutral in mating type (*op-N*), others mate as either a_1 or a_2 (*op-a$_1$*; *op-a$_2$*), and some are constitutive maters (*op-C*, see Section III,B). Ultraviolet light induces opaque formation at the same rate as it induces mitotic crossing-over of marker loci (Castle and Day, 1980). This together with the classical autotetraploid segregation data obtained from an *op-a$_1$* × *op-a$_2$* cross suggest that opaques form as a result of a hot spot of mitotic crossing-over near the mating type locus yielding a_1a_1 (*op-a$_1$*) and a_2a_2 (*op-a$_2$*) types (Castle and Day, 1981). The origin of *op-N* and *op-C* types is not yet clear, but could be explained by the same mechanism assuming that regulatory loci are also involved in the recombination. Further work on these derivative strains is in progress.

D. The Induction of the Parasitic Infection Hyphae by Host Products Acting on the Mating Type Locus

In many smut fungi the dikaryotic mycelial stage is obligately parasitic and has not been induced on artifical media. Recently, we have discovered the mechanism by which it is induced (Day *et al.*, 1981). Simple aqueous extracts of host (*Silene alba*) leaves contain one or more compounds (which we term silenins) which stimulate cells carrying both mating types (i.e., conjugated cells *or* diploids) to produce infection hyphae. Continued development of these hyphae depends on continued feeding with silenin. If feeding is stopped the cells revert to the saprophytic morphogenetic phases discussed above (VP, conjugation or SPP). The response of a_1a_2 cells to silenin overrides the response to low temperature regimes and SPP cells are not formed. Preliminary work with a variety of physiologic races of *U.*

violacea tested against many host species suggest that: (1) Silenins are common in the Caryophyllaceae and rare to absent in other groups. (2) All of the physiological races respond to any of the host extracts. Thus physiological race specialization does not appear to depend on the ability to respond to different host silenins. (3) Silenin receptors in the fungus act on the two mating type alleles to trigger the parasitic stages. We have coined the term "mycoboethin" to describe agents such as the silenins that are produced by a plant and utilized by the parasite to induce highly specific growth responses. The compounds have great potential interest for (a) the study of fungal phytopathogenicity and the formation of physiological races and (b) the control of fungal diseases of host plants (Day et al., 1981).

The mating type locus appears to be the major element controlling morphogenesis in *U. violacea*. It can sense and respond to (a) cell cycle signals; (b) temperature changes; (c) nutritional factors; and (d) host silenins. Its exact response depends not only on the nature of the signal, but on whether or not both mating type alleles are present in the same cell. It may direct development accordingly to be (a) vegetative (sporidial budding); (b) conjugative; (c) sexual, forming precursors of brandspores; (d) parasitic, forming infective hyphae. A likely hypothesis at present is that the mating type locus may be associated with a number of sensor elements (Britten and Davidson, 1969) responding to environmental stimuli. There is some evidence that the response to cell cycle changes is governed through such a sensor region (*cc* locus) that is adjacent to the mating type gene and acts in a *cis*-dominant manner (Day and Cummins, 1975). It is possible that the frequent *opaque* derivatives including the *op-N* and *op-C* types (Castle and Day, 1980) may arise through mitotic crossing-over leading to new combinations of control elements.

IV. PERSPECTIVES

The smut fungus, *Ustilago violacea*, is proving to be a powerful tool for understanding the genetics and molecular biology of growth and morphogenesis. The organism is but one representative of a large group of plant pathogens, the smuts and rusts; nevertheless, the current results should shed considerable light on the entire group. In recent years our laboratory has concentrated on fimbriae as organelles of intercellular communication; the genetic and molecular program for sexual morphogenesis; the mating type locus as a developmental master switch; and finally, host products that interact with the mating type alleles to regulate parasitic infection. The perspectives for further investigations in these areas are discussed below.

Fimbriae, fibers projecting from the cell surface, modulate intercellular communication during mating. The fibers are made up of simple protein subunits, some of which are modified by phosphorylation. The structure can be dissociated and reassociated into a tightly coiled helix and the reassembled structure may be rejoined to a cell. Synthesis of fimbriae is regulated at the transcription level and mapping by ultraviolet sensitivity of the gene indicates that its structural gene has a relatively short promoter sequence. Ongoing research is directed toward fully elucidating the structure and determining the possible role of protein modification in intercellular communication. An antifimbrial antibody has been prepared and is being used to study the distribution and relatedness of fimbriae in the Ustilaginales (Gardiner *et al.*, 1981) and to select afimbriate mutants.

The genetic program regulating synthesis and assembly of a conjugation tube has been studied. Previously, we observed that the genes governing mating morphogenesis were transcribed and translated during a "courtship" period preceeding assembly of the mating organelle. Currently, we are directing our research toward enumerating the genes regulating morphogenesis and determining the temporal sequence of their action. We have established that sensitivity to uv light decreases as courtship proceeds and that mating-specific RNA and proteins are synthesized during courtship. The goal of this study is to enumerate the genes and gene products essential for mating and to study their regulation using the temperature-sensitive mutants for growth and mating currently available and to be selected in the future. This biological system is highly suitable for studying the genetic regulation of morphogenesis.

In *Ustilago violacea* the mating type locus acts as a developmental master switch. In response to several environmental signals it directs development along the paths of vegetative budding, mating tube formation, dikaryon formation or sexual spore formation, and meiosis. The genetic organization of the mating type locus has been clarified and the evidence indicates that mating type variants arise from diploid strains by mitotic crossing-over. Further work will entail a fuller elucidation of the genetic fine structure of the locus, including its sensor elements. Achievement of that goal will be greatly facilitated by the identification of genetic markers flanking mating type. The long-term benefit to be gained from studying mating type is in the understanding of developmental master switches.

Finally, the dikaryotic mycelial stage of *U. violacea* which is obligately parasitic has been induced to form on artificial media using extracts of the host plant containing a fungal growth regulator, silenin. Evidence suggests that the growth regulator is a phenolic compound and a fuller characterization of it is currently under way. The use of the growth regulator allowed us to determine that initiation of the dikaryon is regulated by the mating type locus and it will allow us to study the molecular changes accompanying this initiation of infection hyphae. Previously, such studies were restricted by the obligately parasitic nature of the dikaryon. The identification of plant products that regulate fungal development (mycoboethins) is of potential importance in plant breeding programs aimed at improving disease resistance. It may be possible to synthesize antagonistic regulatory chemicals to control certain fungal diseases.

ACKNOWLEDGMENTS

We are grateful for the enthusiastic help of Marc Canton III, Alan Castle, Ken Dunne, and Richard Gardiner, and also for research support from the Natural Sciences and Engineering Research Council (Grant No. A5062).

REFERENCES

Baird, M. L., and Garber, E. D. (1979a). Genetics of *Ustilago violacea*. IV. An electrophoretic survey for urease variants in wild strains. *Bot. Gaz. (Chicago)* 140, 84–88.

Baird, M. L., and Garber, E. D. (1979b). Genetics of *Ustilago violacea*. V. Outcrossing and selfing in teliospore inocula. *Bot. Gaz. (Chicago)* 140, 89–93.

Brinton, C. C. (1964). The structure, function, synthesis and genetic control of bacterial pili and a molecular model for DNA and RNA transport in gram negative bacteria. *Trans. N.Y. Acad. Sci.* 27, 1003–1054.

Britten, R. J., and Davidson, E. H. (1969). Gene regulation for higher cells: A theory. *Science* 165, 349–357.

Castle, A. J., and Day, A. W. (1980). Diploid derivatives of *Ustilago violacea* with altered mating-type activity. I. Isolation and properties. *Bot. Gaz. (Chicago)* 141, 85–93.

Castle, A. J., and Day, A. W. (1981). Diploid derivatives of *Ustilago violacea* with altered mating-type activity. II. Polyploid segregations and mechanism of origin. *Bot. Gaz. (Chicago)* (in press).

Cattrall, M. E., Baird, M. L., and Garber, E. D. (1978). Genetics of *Ustilago violacea*. III. Crossing-over and nondisjunction. *Bot. Gaz. (Chicago)* 139, 266–270.

Clements, L. L., Day, A. W., and Jones, J. K. (1969). Effect of ultraviolet light on nuclear fusion in *Ustilago violacea*. *Nature (London)* 223, 961–963.

Cummins, J. E., and Day, A. W. (1974a). Transcription and translation of the sex message of the smut fungus *Ustilago violacea*. II. The effects of inhibitors. *J. Cell Sci.* 16, 49–62.

Cummins, J. E., and Day, A. W. (1974b). The cell cycle regulation of sexual morphogenesis in a basidiomycete, *Ustilago violacea*. *In* "Cell Cycle Controls (G. M. Padilla, I. L. Cameron, and A. Zimmerman, eds.), pp. 181–200. Academic Press, New York.

15. Sexual Development in *Ustilago violacea*

Cummins, J. E., and Day, A. W. (1976a). Exchange of information between mating types prior to cell fusion in the anther smut *Ustilago violacea. Can. J. Genet. Cytol.* 18, 555. (Abstr.)
Cummins, J. E., and Day, A. W. (1976b). DNA from the anther smut *Ustilago violacea. Can. J. Genet. Cytol.* 18, 555-556. (Abstr.)
Cummins, J. E., and Day, A. W. (1977). Genetic and cell cycle analysis of a smut fungus, *Ustilago violacea. Methods Cell Biol.* 15, 445-469.
Day, A. W. (1968). The genetics of *Ustilago violacea*. Ph.D. Thesis, Univ. of Reading, Reading, England.
Day, A. W. (1972). Genetic implications of current models of somatic nuclear division in fungi. *Can. J. Bot.* 50, 1337-1347.
Day, A. W. (1976). Communication through fimbriae during conjugation in a fungus. *Nature (London)* 262, 583-584.
Day, A. W. (1978). Chromosome transfer in dikaryons of a smut fungus. *Nature (London)* 273, 753-755.
Day, A. W. (1979). Mating type and morphogenesis in *Ustilago violacea. Bot. Gaz. (Chicago)* 140, 94-101.
Day, A. W., and Cummins, J. E. (1974). Transcription and translation of the sex message in the smut fungus, *Ustilago violacea* I. The effect of ultraviolet light. *J. Cell Sci.* 15, 619-632.
Day, A. W., and Cummins, J. E. (1975). Evidence for a new kind of regulatory gene controlling expression of genes for morphogenesis during the cell cycle in *Ustilago violacea. Genet. Res.* 25, 253-266.
Day, A. W., and Day, L. L. (1970). Ultraviolet light sensitive mutants of *Ustilago violacea. Can. J. Genet. Cytol.* 12, 891-904.
Day, A. W., and Day, L. L. (1974). The control of karyogamy in somatic cells of *Ustilago violacea. J. Cell Sci.* 15, 619-632.
Day, A. W., and Jones, J. K. (1968). The production and characterization of diploids in *Ustilago violacea. Genet. Res.* 11, 63-81.
Day, A. W., and Jones, J. K. (1969). Sexual and parasexual analysis of *Ustilago violacea. Genet. Res.* 14, 195-221.
Day, A. W., and Jones, J. K. (1971). P-fluorophenylalanine induced mitotic haploidization in *Ustilago violacea. Genet. Res.* 18, 299-309.
Day, A. W., and Jones, J. K. (1972). Somatic nuclear division in the sporidia of *Ustilago violacea* I. Acetic orcein staining. *Can. J. Microbiol.* 18, 663-670.
Day, A. W., and Poon, N. H. (1975). Fungal Fimbriae II. Their role in conjugation in *Ustilago violacea. Can. J. Microbiol.* 21, 547-557.
Day, A. W., Poon, N. H., and Stewart, G. G. (1975). Fungal fimbriae III. The effect on flocculation in Saccharomyces. *Can. J. Microbiol.* 21, 558-564.
Day, A. W., Castle, A. J., and Cummins, J. E. (1981). Regulation of parasitic development of the smut fungus, *Ustilago violacea* by extracts from host plants. *Bot. Gaz. (Chicago)* (in press).
Day, P. R. (1974). "Genetics of Host-Parasite Interaction." Freeman, San Francisco, California.
Dunne, K. D., and Cummins, J. E. (1980). Interruption of gene transcription by UV light in excision proficient and deficient strains of *Ustilago violacea. Can. J. Genet. & Cytol.* (in press).
Fink, G. R., and Condé, J. (1977). Studies on Kar_1, a gene required for nuclear fusion in yeast. *In* "International Cell Biology" (B. Brinkley and K. R. Porter, eds.), pp. 414-419. Rockefeller Univ. Press, New York.
Firtel, R. A. (1972). Changes in the expression of single-copy DNA during development of the cellular slime mold *Dictyostelium discoideum. J. Mol. Biol.* 66, 363-377.

Fischer, G. W., and Holton, C. S. (1957). "Biology and Control of the Smut Fungi." Ronald Press, New York.

Garber, E. D., Baird, M. L., and Chapman, D. J. (1975). Genetics of *Ustilago violacea* I. Carotenoid mutants and carotenogenesis. *Bot. Gaz. (Chicago)* 136, 341–346.

Garber, E. D., Baird, M. L., and Weiss, L. M. (1978). Genetics of *Ustilago violacea* II. Polymorphism of colour and nutritional requirements of sporidia from natural populations. *Bot. Gaz. (Chicago)* 139, 261–265.

Gardiner, R. B., Canton, M., and Day, A. W. (1981). Fimbrial variation in smuts and heterobasidiomycete fungi. *Bot. Gaz. (Chicago)* (in press).

Goldschmidt, V. (1928). Vererbungsversuche mit den biologischen Arten des Antherenbrandes (*Ustilago violacea* Pers.) *Z. Bot.* 21, 1–90.

Holliday, R. (1974). *Ustilago maydis*. In "Handbook of Genetics" (R. C. King, ed.), pp. 575–595. Plenum, New York.

Laemlli, U. K. (1970). Cleavage of structural proteins during the assembly of the head of bacteriophage T4. *Nature (London)* 227, 680–682.

Liro, J. I. (1924). Die Ustilagineen Finnlands. *Ann. Acad. Sci. Fenn., Ser. A* 17, 30–49, 258–343.

Merril, C., Switzer, R., and van Keuren, M. (1979). Trace polypeptides in cellular extracts and human body fluids detected by two-dimensional electrophoresis and a highly sensitive silver stain. *Proc. Natl. Acad. Sci. U.S.A.* 76, 4335–4339.

O'Farrell, P. H. (1975). High resolution two-dimensional electrophoresis of proteins. *J. Biol. Chem.* 250, 4007–4029.

Oosawa, F., and Kasai, M. (1971). Actin. In "Subunits in Biological Systems," Part A (S. N. Timashef and G. D. Fasman, eds.), pp. 261–322. Dekker, New York.

Poon, N. H. (1978). The structure of the sporidium of *Ustilago violacea* during growth and conjugation. Ph.D. Thesis, Univ. of Western Ontario, London, Ontario.

Poon, N. H., and Day, A. W. (1974a). Fimbriae in the fungus *Ustilago violacea*. *Nature (London)* 250, 648–649.

Poon, N. H., and Day, A. W. (1974b). Somatic nuclear division in the sporidia of *Ustilago violacea* II. Observations on living cells with phase contrast and fluorescence microscopy. *Can. J. Microbiol.* 20, 739–746.

Poon, N. H., and Day, A. W. (1975). Fungal fimbriae I. Structure, origin and synthesis. *Can. J. Microbiol.* 21, 537–546.

Poon, N. H., and Day, A. W. (1976a). Somatic nuclear division in the sporidia of *Ustilago violacea* III. Ultrastructural observations. *Can. J. Microbiol.* 22, 495–506.

Poon, N. H., and Day, A. W. (1976b). Somatic nuclear division in the sporidia of *Ustilago violacea* IV. Microtubules and the spindle pole body. *Can. J. Microbiol.* 22, 507–522.

Poon, N. H., Martin, J., and Day, A. W. (1974). Conjugation in *Ustilago violacea* I. Morphology. *Can. J. Microbiol.* 20, 187–191.

Sauerbier, W., and Hercules, K. (1978). Gene and transcription unit mapping by radiation effects. *Annu. Rev. Genet.* 12, 329–363.

Springer, M. S., Goy, M. F., and Adler, J. (1979). Protein methylation in behavioural control mechanisms and in signal transduction. *Nature (London)* 280, 279–384.

Stephens, R. E. (1971). Microtubules. In "Subunits in Biological Systems," Part A (S. N. Timashef and G. D. Fasman, eds.), pp. 355–391. Dekker, New York.

Zillig, H. (1921). Uber spezialisierte Formen beim Antherenbrand, *Ustilago violacea* (Pers.) Fuckel. *Zentralbl. Bakteriol., Parasitenkd. Infektionskr.* 53, 33–73.

Index

A

a factor
  functions, 8, 33–47
  purification, 32, 33
Achlya, 7, 8, 155–178
Actinomycin D, 61, 331
Adenyl-cyclase, 85
Adhesion sites
  Chlamydomonas, 305–315
  yeast, 272–276
Agglutination, 23, 24, 37, 38, 234–235, 263–291, 299–301, 310–315
Aggregation, slime molds, 211–212
Albino mutant, 95
Allomyces, 53–72
  A. macrogynus life cycle, 53
  macromolecular synthesis, 66
  pheromone gradient, 64
Alpha factor
  agglutination, 275–287
  amino acid sequence, 31
  functions, 9, 33–47
  purification, 29–30
Amino acid composition, 391
Amino acid sequence, 30–32
Anchoring step, 333, 334
Androgonia, 75
Antheridiol, 5, 7, 8, 159–175
Ascus, 132
Asexual reproduction
  Allomyces, 54–55
  Blepharisma, 97–99
  Dictyostelium, 199–201
  Euplotes, 322–323

Mucor, 180–181
Neurospora, 132–134
Paramecium, 353
Saccharomyces, 23
Ustilago, 380–381
Volvox, 74–78
Attenuation, 270, 271

B

Beta carotene, 184, 194
Binding experiments, *Volvox*, 86–88
Bioassay
  cyclic AMP, 209–211
  yeast mating factor, 28–29
Blakeslea trispora, 184
Blepharisma, 95–129
  B. japonicum, 98
  gamones, 103–109
Blepharismin, 95
  deficient mutants, 95–96
Blepharismone, 103, 104
Blepharmone, 104–106
Budding, yeast, 39–41

C

Calcium ion, 64, 214
Carboxypeptidase Y, 266
Cell adhesion, 11–12, *see also* Agglutination
  in *Euplotes*, 109–114

403

Cell-cell interactions
  Chlamydomonas, 297-318
  Oxytricha, 319-350
  Paramecium, 351-378
  Saccharomyces, 261-295
  Schizosaccharomyces, 225-259
Cell cycle, 322
  G1 arrest, 12, 33-37, 239, 284
Cell fusion, 10, 213-217, 227, 319-350
Cell lysis, 239-240
Cell surface interactions, 224-402
Cell volume, 43-47
Cell wall, 12, 201-203, 225-259, 281-285
Cellular continuities, 7
Cellular slime molds, 199-221
Chemistry of mating pheromones, 8-9
  Achlya, 156-159
  Allomyces, 56-59
  Blepharisma, 103-109
  Mucor, 183-195
  Neurospora, 147-150
  Saccharomyces, 30-33
  Volvox, 81-84
Chemosensory activities, 56
Chemoattractant, 208-211
Chemotaxis
  Achlya, 157
  Allomyces, 53-72
  bacterial, 53
  Blepharisma, 100-111
  Dictyostelium, 54, 200, 208-211
Chlamydomonas, 297-318
Chromatography, 105, 148, 168, 274, 288
Chromatin, 165-168
Cilia
  adhesion, 359
  agglutination, 359
  degeneration, 354, 367-369
  membranes, 361
  union, 111-113
Ciliary movement, inactivation, 364, 365
Colchicine, 67
Competence, 235-38
Conidia, 133
Conjugation
  Escherichia coli, 226
  Paramecium, 355-358
  Schizosaccharomyces, 225-255
  Ustilago, 384-387

Conjugation tube formation, 238
Contact inhibition, 4
Culture methods
  Achlya, 158-159
  Allomyces, 54-56
  Blepharisma, 99-100
  Chlamydomonas, 302
  Dictyostelium, 202
  Euphotes, 324-325
  Neurospora, 134
  Oxytricha, 324-325
  Paramecium, 352
  Saccharomyces, 26-27, 263
  Ustilago, 382
Cyclic AMP, 6, 85, 209-211
Cycloheximide, 61, 122, 331
Cytochalasin B, 68
Cytophagic giant cell, 202, 203
Cytosol, 162-164

D

Deflagellation, 314
Deflocculation, 246
2-Deoxyglucose, 242-247
Dictyostelium, 199-220
Diffusable molecules, 5, see also Part II, Phermonal Interactions
DNA, 156, 161, 167
  synthesis, 33-34
DNA-RNA hybridization, 394
Dose-effect, *Neurospora*, 142

E

EDTA-EGTA, 213, 214
Electrophoresis, 81-84, 173-174, 305-306, 389
Electrophysiological studies, *Paramecium*, 325
Encystment-excystment cycle
  Blepharisma, 97
  Dictyostelium, 200
Enzyme splitting of cell couples, 336-341
Escherchia coli, 226
Extracellular matrix, 12-14, see also *Volvox*

Index

F

Fimbriae, 386-396
  composition, 391
  synthesis, 392
Filamentous proteins, amino acid composition, 391
Filopodial contact, 11
Fission yeast, 225-229
Flagella
  morphology, 302
  as sensory organelle, 297-301
  surface components, 302, 306-310
Flocculation, 232-248
5-fluorouracil, 184
Fuscannels, 249

G

Gamma carotene, 55
Gamma ray mutations, 147
Gametogenesis, 66, 67, 302-311
Gamone
  biosynthesis, 106-108
  Oxytricha, 329-333
  receptors, 118-123
Gap junctions, 7, 10
Gas chromatography, 143, 148-150
Gene expression
  Achlya, 160-178
  Ustilago, 392-396
Geotropism, 100
Giant cell, 200-212
Glucan synthetase, 240
Glucanases, 239-240
Glusulase, 39, 265
Glycoproteins, 80-84, 104-106, 120, 263, 305, 361
Glycosyltransferases, 12
Gonadotropic hormones, 85
Gonidia, 74

H

Hansenula wingei, 263, 269
Histones, 159, 160

HMG-CoA
  reductase, 192-193
  synthetase, 193
Hoechst stain, 215
Holdfast union, 367
Homotypic union, 114
Hormones, 5
  receptor complexes, 7, 61-70, 118, 162-164
Hypotrichs, 321-350

I

Immunofluorescence, 306-307
Immunoelectrophoresis, 306, 308
Immunological assays, 306-310
Inactivation, cilary movement, 364-365
Inhibitor, giant cell formation, 218
Induction of agglutination, 282-284
Intercellular bridges, 10
Intercellular communication, *see* Part II, Pheromonal Interactions
Intracellular reception, 6
  Achlya, 162-164
  Allomyces, 61-76
  Blepharisma, 118-120
Ions, 67, 214
Isoagglutinins, 31, 303, 313

K

Karyogamy
  Dictyostelium, 215
  Paramecium, 353
  Ustilago, 387
Keerosin, 59

L

Lectins, 189
Life cycle
  Allomyces, 52-56
  Blepharisma, 97-99
  Dictyostelium, 199-201
  Mucorales, 179-182
  Neurospora, 132-134
  Oxytricha, 322-324

Life cycle (*cont.*)
  Paramecium, 352-353
  Saccharomyces, 21-26, 261-263
  Schizosaccharomyces, 228-230
  Ustilago, 380-382
  Volvox, 73-77
Lysis, 239, 240

M

Macrocyst, 201-226
Macromolecular changes
  Achlya, 154-160
  Allomyces, 60, 61
  Schizosaccharomyces, 239
  Ustilago, 388-398
Macronucleus, 97, 356, 357
Mass mating, yeast, 262-266
Mating type
  Blepharisma, 101
  Chlamydomonas, 182
  Dictyostelium, 205-207
  Mucor, 166
  Neurospora, 142
  Paramecium, 358
  Saccharomyces, 23, 262
  Schizosaccharomyces, 247
  Ustilago, 396
Meiosis
  Blepharisma, 121
  Oxytricha, 342
  Schizosaccharomyces, 230, 236
Membrane
  glycoconjugates, 305
  vesicles, 361-363
Membranelles, 95
Ménage à trois, 251-253
Messenger RNA, 160, 161, 170
Microcyst, 200
Micronucleus, 97, 323, 356
  migration, 365-367
Modes of cellular communication, 5-14
Molecular cloning, 22
Mucor, 179-198
  M. mucedo life cycle, 181
Mutants
  nonagglutinable, 287
  Paramecium, 352
  pigment deficient, 95-96
  Ustilago, 383

N

Neurospora, 131-154
  life cycle, 133
Nonhistone protein, 171-174
Nuclear events, 232, 323-324, 365-367
Nucleosomes, 167

O

Oögoniol, 157
Oxytricha, 319-350
  life cycle, 322
  O. bifaria, 319

P

P and d cells, 242, 255
Paramecium, 351-379
  life cycle, 353
Parasexual cycle, 383
Parasitic infection hyphae, 397
Paroral union, 354, 355
Particle data celloscope, 44
Peristome, 111
Phagocytosis, 202, 203, 218
Phenols, 7
Pheromones, 7, 20-222
  Achlya, 155-178
  Allomyces, 58-82
  Blepharisma, 95-129
  Dictyostelium, 199-221
  gradients of, 64-67
  Mucor, 179-198
  Neurospora, 131-153
  Volvox, 73-92
  yeast, 21-50, 280-284
Phototropism, 186
Plasma papilla, 300
Propheromones, 184
Preconjugant interactions
  Blepharisma, 101-111
  Oxytricha, 325-350
Protein
  acetylation, 160
  synthesis, 114-116, 159-160
  phosphorylation, 85
Protoperithecia, 141-151

… Index

R

Responder, 3
Receptors
  Achlya, 162–164
  Allomyces, 61–70
  Blepharisma, 118–120
  intracellular, 162–164
  surface, 85, 118–120
Red bread mold, 121
Retinal, 190
Rhizopus, 180
Rifampicin challenge assay, 166–168
Rotation step, 334
RNA polymerases, 164–170

S

Saccharomyces, 21–51, 261–295
  cell volume changes, 42
  differentiation, 35
  pheromone purification, 27–33
Schizosaccharomyces, 227–259
  S. pombe, 229, 264
Scanning electron microscopy, 25, 298, 368, 386
Schmoos, 37–40
Secretor-responder system, 205
Sesquiterpene, 56
Serotonin, 124, 125
Sex hairs, 237
Sexual agglutination, 226, 234–235, 263–291, 299–301, 310–315
Signal transduction, 6, 62
Signaler, 3
Silenins, 398
Sirenin
  production, 60
  reception, 61
  structure, 57
Sorocarp, 206
Spheroids, 74
Sporopollenin, 181
Starvation, 302, 326, 352
Sterility mutants, 145–147
Steroids, 6–7, 156–178
Subunits, RNA polymerase II, 169
Surface interactions, 13
Surface receptors, 6, 85–88, 120–121

T

Taxis, 108
Temperature and sexual agglutination, 278–286
Template activity, 165
Terpene biosynthesis, 191–195
Thermosensitivity, 240
T_m of deflocculation, 231
Translation products, 177
Transcription, 162–174, 393–396
Trisporic acid, 184–195
  biosynthesis, 185–191
  physiological effects, 188–195
  structure, 184–187
Turbidity, 232
Twitching flagella, 310

U

Ustilago
  genetics, 392
  life cycle, 381
  U. violacea, 379
U-tubes, 135–146
Ultrastructure, 17, 25, 116–117, 229, 237–238, 362, 386
UV irradiation, 395
UV spectrometry, 149–150

V

Visible reaction, *Oxytricha*, 282
Vesicles, *Chlamydomonas*, 211
Volatile pheromones, 206–211, 218
Volvox, 73–92
  life cycle, 76

W

Waiting period, 323–331

Z

Zygote formation, 23, 37, 39–40, 211, 226

CELL BIOLOGY: A Series of Monographs

EDITORS

D. E. BUETOW

Department of Physiology
and Biophysics
University of Illinois
Urbana, Illinois

I. L. CAMERON

Department of Anatomy
University of Texas
Health Science Center at San Antonio
San Antonio, Texas

G. M. PADILLA

Department of Physiology
Duke University Medical Center
Durham, North Carolina

A. M. ZIMMERMAN

Department of Zoology
University of Toronto
Toronto, Ontario, Canada

G. M. Padilla, G. L. Whitson, and I. L. Cameron (editors). THE CELL CYCLE: Gene-Enzyme Interactions, 1969

A. M. Zimmerman (editor). HIGH PRESSURE EFFECTS ON CELLULAR PROCESSES, 1970

I. L. Cameron and J. D. Thrasher (editors). CELLULAR AND MOLECULAR RENEWAL IN THE MAMMALIAN BODY, 1971

I. L. Cameron, G. M. Padilla, and A. M. Zimmerman (editors). DEVELOPMENTAL ASPECTS OF THE CELL CYCLE, 1971

P. F. Smith. The BIOLOGY OF MYCOPLASMAS, 1971

Gary L. Whitson (editor). CONCEPTS IN RADIATION CELL BIOLOGY, 1972

Donald L. Hill. THE BIOCHEMISTRY AND PHYSIOLOGY OF *TETRAHYMENA*, 1972

Kwang W. Jeon (editor). THE BIOLOGY OF AMOEBA, 1973

Dean F. Martin and George M. Padilla (editors). MARINE PHARMACOGNOSY: Action of Marine Biotoxins at the Cellular Level, 1973

Joseph A. Erwin (editor). LIPIDS AND BIOMEMBRANES OF EUKARYOTIC MICROORGANISMS, 1973

A. M. Zimmerman, G. M. Padilla, and I. L. Cameron (editors). DRUGS AND THE CELL CYCLE, 1973

Stuart Coward (editor). DEVELOPMENTAL REGULATION: Aspects of Cell Differentiation, 1973

I. L. Cameron and J. R. Jeter, Jr. (editors). ACIDIC PROTEINS OF THE NDCLEUS, 1974

Govindjee (editor). BIOENERGETICS OF PHOTOSYNTHESIS, 1975

James R. Jeter, Jr., Ivan L. Cameron, George M. Padilla, and Arthur M. Zimmerman (editors). CELL CYCLE REGULATION, 1978

Gary L. Whitson (editor). NUCLEAR-CYTOPLASMIC INTERACTIONS IN THE CELL CYCLE, 1980

Danton H. O'Day and Paul A. Horgen (editors). SEXUAL INTERACTIONS IN EUKARYOTIC MICROBES, 1981

In preparation

Ivan L. Cameron and Thomas B. Pool (editors). THE TRANSFORMED CELL, 1981